Springer Optimization and Its Applications

Volume 201

Aims and Scope

Optimization has continued to expand in all directions at an astonishing rate. New algorithmic and theoretical techniques are continually developing and the diffusion into other disciplines is proceeding at a rapid pace, with a spot light on machine learning, artificial intelligence, and quantum computing. Our knowledge of all aspects of the field has grown even more profound. At the same time, one of the most striking trends in optimization is the constantly increasing emphasis on the interdisciplinary nature of the field. Optimization has been a basic tool in areas not limited to applied mathematics, engineering, medicine, economics, computer science, operations research, and other sciences.

The series **Springer Optimization and Its Applications (SOIA)** aims to publish state-of-the-art expository works (monographs, contributed volumes, textbooks, handbooks) that focus on theory, methods, and applications of optimization. Topics covered include, but are not limited to, nonlinear optimization, combinatorial optimization, continuous optimization, stochastic optimization, Bayesian optimization, optimal control, discrete optimization, multi-objective optimization, and more. New to the series portfolio include Works at the intersection of optimization and machine learning, artificial intelligence, and quantum computing.

Volumes from this series are indexed by Web of Science, zbMATH, Mathematical Reviews, and SCOPUS.

Alexander J. Zaslavski

Turnpike Phenomenon in Metric Spaces

Springer

Alexander J. Zaslavski
Haifa, Israel

ISSN 1931-6828 ISSN 1931-6836 (electronic)
Springer Optimization and Its Applications
ISBN 978-3-031-27210-3 ISBN 978-3-031-27208-0 (eBook)
https://doi.org/10.1007/978-3-031-27208-0

Mathematics Subject Classification: 49J20, 49J99, 49K20, 49K27, 49K40

This Springer imprint is published by the registered company Springer Nature Switzerland AG
The registered company address is: Gewerbestrasse 11, 6330 Cham, Switzerland

Preface

In this book we study the turnpike phenomenon arising in the optimal control theory. The term was first coined by P. Samuelson in 1948 when he showed that an efficient expanding economy would spend most of the time in the vicinity of a balanced equilibrium path (also called a von Neumann path). To have the turnpike property means, roughly speaking, that the approximate solutions of the problems are determined mainly by the objective function and are essentially independent of the choice of interval and endpoint conditions, except in regions close to the endpoints. The turnpike property discovered by P. Samuelson is well known in the economic literature, where it was studied for various models of economic growth. Usually for these models, a turnpike is a singleton.

In our previous research, it was shown that the turnpike property is a general phenomenon which holds for large classes of finite-dimensional variational and optimal control problems. In that research, using the Baire category (generic) approach, we showed that the turnpike property holds for a generic (typical) variational problem [117] and for a generic optimal control problem [130]. According to the generic approach, we say that a property holds for a generic (typical) element of a complete metric space (or the property holds generically) if the set of all elements of the metric space possessing this property contains a G_δ everywhere dense subset of the metric space which is a countable intersection of open everywhere dense sets. This means that the property holds for most elements of the metric space.

In this book, we are interested in turnpike results, in sufficient and necessary conditions for the turnpike phenomenon and in its stability under small perturbations of objective functions. The most important feature of this book is that we study a large, general class of optimal control problems in metric space. Probably, at the first time in the literature to our knowledge. All the main results obtained in the book are new.

The monograph contains nine chapters. Chapter 1 is an introduction. In Chap. 2 we discuss Banach space valued functions, set-valued mappings in infinite dimensional spaces, and related continuous-time dynamical systems. Some convergence results are obtained. In Chap. 3, we study a discrete-time dynamical system with a Lyapunov function in a metric space induced by a set-valued mapping. For this discrete-time dynamical system, we establish a turnpike result. Chapter 4 is devoted to the study of a class of continuous-time dynamical systems which is an analog of the class of discrete-time dynamical systems considered in Chap. 3. This class also contains the class of dynamical systems studied in Chap. 2. In Chap. 5, we develop a turnpike theory for a class of general dynamical systems in a metric space with a Lyapunov function. This class of dynamical systems contains as particular cases the classes of dynamical systems studied in Chaps. 3 and 4. We obtain several turnpike results and show that the turnpike phenomenon is stable under small perturbations of objective functions. Chapter 6 contains our study of the turnpike phenomenon for discrete-time nonautonomous problems on subintervals of half-axis in metric spaces, which are not necessarily compact. For these optimal control problems, the turnpike is not a singleton. We show that the turnpike phenomenon is stable under small perturbations of an objective function. Using the Baire category approach, we show that for some classes of problems a typical (generic) problem has a turnpike property. Chapter 7 contains preliminaries which we need in order to study turnpike properties of infinite dimensional optimal control problems. We discuss unbounded operators, semigroups of linear operators, evolution equations, and admissible control operators. In Chap. 8, we establish sufficient and necessary conditions for the turnpike phenomenon for continuous-time optimal control problems on subintervals of half-axis in metric spaces. For these optimal control problems, the turnpike is not a singleton. The results of this chapter will be obtained for three large classes of problems which will be treated simultaneously. In Chap. 9, we continue to study the turnpike phenomenon for the continuous-time optimal control problems on subintervals of half-axis in metric spaces which are discussed in Chap. 8. We show that the turnpike phenomenon is stable under small perturbations of an objective function. Using the Baire category approach, we show that for some classes of problems a typical (generic) problem has a turnpike property.

Rishon LeZion, Israel
October 19, 2022

Alexander J. Zaslavski

Contents

Chapter 1
Introduction

In this chapter, we discuss turnpike properties in the calculus of variations. To have the turnpike property means, roughly speaking, that the approximate solutions of the problems are determined mainly by the integrand and are essentially independent of the choice of interval and endpoint conditions, except in regions close to the endpoints. We also consider some classes of optimal control problems studied in the book.

1.1 Turnpike Property for Variational Problems

The study of the existence and structure of solutions of optimal control problems and dynamic games defined on infinite intervals and sufficiently large intervals has been a rapidly growing area of research [9, 10, 24, 31, 39, 43, 44, 46, 60, 63, 64, 76, 88, 117, 119, 121, 124, 129, 133, 138, 139, 142, 144, 147] that has various applications in engineering [6, 31, 70, 117], models of economic growth [7, 11, 29–32, 42, 47, 58, 59, 62, 69, 75, 81, 90, 97, 99, 105, 117, 127, 134, 143, 144], infinite discrete models of solid-state physics related to dislocations in one dimensional crystals [14, 94, 107], model predictive control [38, 50], and the theory of thermodynamical equilibrium for materials [36, 71, 78–80]. Discrete-time optimal control problems were considered in [8, 15, 16, 27, 45, 55, 108, 109, 114, 116, 120, 123, 125, 127, 131, 132, 135, 140, 141, 143], finite dimensional continuous-time problems were analyzed in [23, 25, 26, 28, 33, 69, 72, 74, 77, 91, 112, 115, 118, 130, 145, 146], infinite dimensional optimal control was studied in [31, 32, 51–53, 84, 86, 87, 92, 101, 102, 111, 113, 122], while solutions of dynamic games were discussed in [22, 48, 49, 54, 56, 66, 93, 126, 128, 136, 137]. Sufficient and necessary conditions for the turnpike phenomenon were obtained in our previous research [115, 116, 118, 131].

A. J. Zaslavski, *Turnpike Phenomenon in Metric Spaces*, Springer Optimization and Its Applications 201, https://doi.org/10.1007/978-3-031-27208-0_1

In this section, which is based on [115], we discuss the structure of approximate solutions of variational problems with continuous integrands $f : [0, \infty) \times R^n \times R^n \to R^1$ that belong to a complete metric space of functions. We do not impose any convexity assumption. The main result of this section, obtained in [115], deals with the turnpike property of variational problems.

We consider the variational problems

$$\int_{T_1}^{T_2} f(t, z(t), z'(t))dt \to \min, \; z(T_1) = x, \; z(T_2) = y, \tag{P}$$

$$z : [T_1, T_2] \to R^n \text{ is an absolutely continuous function,}$$

where $T_1 \geq 0$, $T_2 > T_1$, $x, y \in R^n$, and $f : [0, \infty) \times R^n \times R^n \to R^1$ belongs to a space of integrands described below.

It is well known that the solutions of the problems (P) exist for integrands f that satisfy two fundamental hypotheses concerning the behavior of the integrand as a function of the last argument (derivative): one that the integrand should grow superlinearly at infinity and the other that it should be convex [100]. Moreover, certain convexity assumptions are also necessary for properties of lower semicontinuity of integral functionals that are crucial in most of the existence proofs, although there are some interesting theorems without convexity [34, 35, 83, 85]. For integrands f that do not satisfy the convexity assumption, the existence of solutions of the problems (P) is not guaranteed, and in this situation, we consider *δ-approximate solutions*.

Let $T_1 \geq 0$, $T_2 > T_1$, $x, y \in R^n$, $f : [0, \infty) \times R^n \times R^n \to R^1$ be an integrand, and let δ be a positive number. We say that an absolutely continuous (a.c.) function $u : [T_1, T_2] \to R^n$ satisfying $u(T_1) = x$, $u(T_2) = y$ is a δ-approximate solution of the problem (P) if

$$\int_{T_1}^{T_2} f(t, u(t), u'(t))dt \leq \int_{T_1}^{T_2} f(t, z(t), z'(t))dt + \delta$$

for each a.c. function $z : [T_1, T_2] \to R^n$ satisfying $z(T_1) = x, \; z(T_2) = y$.

The main result of [115] deals with the turnpike property of the variational problems (P). As usual, to have this property means, roughly speaking, that the approximate solutions of the problems (P) are determined mainly by the integrand and are essentially independent of the choice of interval and endpoint conditions, except in regions close to the endpoints.

In the classical turnpike theory, it was assumed that a cost function (integrand) is convex. The convexity of the cost function played a crucial role there. In [115], we get rid of convexity of integrands and establish necessary and sufficient conditions for the turnpike property for a space of nonconvex integrands $\mathcal{M}$ described below.

Let us now define the space of integrands. Denote by $|\cdot|$ the Euclidean norm in R^n. Let a be a positive constant, and let $\psi : [0, \infty) \to [0, \infty)$ be an increasing function such that $\psi(t) \to +\infty$ as $t \to \infty$. Denote by $\mathcal{M}$ the set of all continuous functions $f : [0, \infty) \times R^n \times R^n \to R^1$ that satisfy the following assumptions:

A(i) The function f is bounded on $[0, \infty) \times E$ for any bounded set $E \subset R^n \times R^n$.

A(ii) $f(t, x, u) \geq \max\{\psi(|x|), \psi(|u|)|u|\} - a$ for each $(t, x, u) \in [0, \infty) \times R^n \times R^n$.

A(iii) For each $M, \epsilon > 0$, there exist $\Gamma, \delta > 0$ such that

$$|f(t, x_1, u) - f(t, x_2, u)| \leq \epsilon \max\{f(t, x_1, u), f(t, x_2, u)\}$$

for each $t \in [0, \infty)$ and each $u, x_1, x_2 \in R^n$ that satisfy

$$|x_i| \leq M, \ i = 1, 2, \ |u| \geq \Gamma, \quad |x_1 - x_2| \leq \delta.$$

A (iv) For each $M, \epsilon > 0$, there exists $\delta > 0$ such that $|f(t, x_1, u_1) - f(t, x_2, u_2)| \leq \epsilon$ for each $t \in [0, \infty)$ and each $u_1, u_2, x_1, x_2 \in R^n$ that satisfy

$$|x_i|, |u_i| \leq M, \ i = 1, 2, \quad \max\{|x_1 - x_2|, |u_1 - u_2|\} \leq \delta.$$

It is easy to show that an integrand $f = f(t, x, u) \in C^1([0, \infty) \times R^n \times R^n)$ belongs to $\mathcal{M}$ if f satisfies assumption A(ii), and if $\sup\{|f(t, 0, 0)| : t \in [0, \infty)\} < \infty$ and also there exists an increasing function $\psi_0 : [0, \infty) \to [0, \infty)$ such that

$$\sup\{|\partial f/\partial x(t, x, u)|, |\partial f/\partial u(t, x, u)|\} \leq \psi_0(|x|)(1 + \psi(|u|)|u|)$$

for each $t \in [0, \infty)$ and each $x, u \in R^n$.

For the set $\mathcal{M}$, we consider the uniformity that is determined by the following base:

$$E(N, \epsilon, \lambda) = \{(f, g) \in \mathcal{M} \times \mathcal{M} : |f(t, x, u) - g(t, x, u)| \leq \epsilon$$

$$\text{for each } t \in [0, \infty) \text{ and each } x, u \in R^n \text{ satisfying } |x|, |u| \leq N$$

$$\text{and } (|f(t, x, u)| + 1)(|g(t, x, u)| + 1)^{-1} \in [\lambda^{-1}, \lambda]$$

$$\text{for each } t \in [0, \infty) \text{ and each } x, u \in R^n \text{ satisfying } |x| \leq N\},$$

where $N > 0$, $\epsilon > 0$, $\lambda > 1$.

It is not difficult to show that the space $\mathcal{M}$ with this uniformity is metrizable (by a metric ρ_w). It is known (see [115]) that the metric space $(\mathcal{M}, \rho_w)$ is complete. The metric ρ_w induces in $\mathcal{M}$ a topology.

We consider functionals of the form

$$I^f(T_1, T_2, x) = \int_{T_1}^{T_2} f(t, x(t), x'(t))dt,$$

where $f \in \mathcal{M}$, $0 \leq T_1 < T_2 < \infty$, and $x : [T_1, T_2] \to R^n$ is an a.c. function.

For $f \in \mathcal{M}$, $y, z \in R^n$, and numbers T_1, T_2 satisfying $0 \leq T_1 < T_2$, we set

$$U^f(T_1, T_2, y, z) = \inf\{I^f(T_1, T_2, x) : x : [T_1, T_2] \to R^n$$

$$\text{is an a.c. function satisfying } x(T_1) = y,\ x(T_2) = z\}.$$

It is easy to see that $-\infty < U^f(T_1, T_2, y, z) < \infty$ for each $f \in \mathcal{M}$, each $y, z \in R^n$, and all numbers T_1, T_2 satisfying $0 \leq T_1 < T_2$.

Let $f \in \mathcal{M}$. A locally absolutely continuous (a.c.) function $x : [0, \infty) \to R^n$ is called an *(f)-good function* [117] if for any a.c. function $y : [0, \infty) \to R^n$ there is a number M_y such that

$$I^f(0, T, y) \geq M_y + I^f(0, T, x) \text{ for each } T \in (0, \infty).$$

The following result was proved in [115].

Proposition 1.1 *Let $f \in \mathcal{M}$, and let $x : [0, \infty) \to R^n$ be a bounded a.c. function. Then the function x is (f)-good if and only if there is $M > 0$ such that*

$$I^f(0, T, x) \leq U^f(0, T, x(0), x(T)) + M \text{ for any } T > 0.$$

Let us now give the precise definition of the turnpike property.

Assume that $f \in \mathcal{M}$. We say that f has the turnpike property, or briefly TP, if there exists a bounded continuous function $X_f : [0, \infty) \to R^n$ that satisfies the following condition:

For each $K, \epsilon > 0$, there exist constants $\delta, L > 0$ such that for each $x, y \in R^n$ satisfying $|x|, |y| \leq K$, each $T_1 \geq 0$, $T_2 \geq T_1 + 2L$, and each a.c. function $v : [T_1, T_2] \to R^n$ that satisfy

$$v(T_1) = x,\ v(T_2) = y,\ I^f(T_1, T_2, v) \leq U^f(T_1, T_2, x, y) + \delta,$$

the inequality $|v(t) - X_f(t)| \leq \epsilon$ holds for all $t \in [T_1 + L, T_2 - L]$.

The function X_f is called the turnpike of f.

Assume that $f \in \mathcal{M}$ and $X : [0, \infty) \to R^n$ is a bounded continuous function. How to verify if the integrand f has TP and X is its turnpike? In [115], we introduced two properties (P1) and (P2) and show that f has TP if and only if f possesses the properties (P1) and (P2). The property (P2) means that all (f)-good functions have the same asymptotic behavior, while the property (P1) means that if an a.c. function $v : [0, T] \to R^n$ is an approximate solution and T is large enough, then there is $\tau \in [0, T]$ such that $v(\tau)$ is close to $X(\tau)$.

The next theorem is the main result [115].

Theorem 1.2 *Let $f \in \mathcal{M}$ and $X_f : [0, \infty) \to R^n$ be a bounded absolutely continuous function. Then f has the turnpike property with X_f being the turnpike if and only if the following two properties hold:*

(P1) For each $K, \epsilon > 0$, there exist $\gamma, l > 0$ such that for each $T \geq 0$ and each a.c. function $w : [T, T + l] \to R^n$ that satisfy

$$|w(T)|, |w(T+l)| \leq K,\ I^f(T, T+l, w) \leq U^f(T, T+l, w(T), w(T+l))+\gamma,$$

there is $\tau \in [T, T + l]$ for which $|X_f(\tau) - v(\tau)| \leq \epsilon$.

(P2) For each (f)-good function $v : [0, \infty) \to R^n$,

$$|v(t) - X_f(t)| \to 0 \text{ as } t \to \infty.$$

In [115], we proved the following theorem that is an extension of Theorem 1.2.

Theorem 1.3 *Let $f \in \mathcal{M}$, and $X_f : [0, \infty) \to R^n$ be an (f)-good function. Assume that the properties (P1), (P2) hold. Then for each $K, \epsilon > 0$, there exist $\delta, L > 0$, and a neighborhood $\mathcal{U}$ of f in $\mathcal{M}$ such that for each $g \in \mathcal{U}$, each $T_1 \geq 0$, $T_2 \geq T_1 + 2L$, and each a.c. function $v : [T_1, T_2] \to R^n$ that satisfy*

$$|v(T_1)|, |v(T_2)| \leq K,\ I^g(T_1, T_2, v) \leq U^g(T_1, T_2, v(T_1), v(T_2)) + \delta,$$

the inequality $|v(t) - X_f(t)| \leq \epsilon$ holds for all $t \in [T_1 + L, T_2 - L]$.

In [148], we studied sufficient and necessary conditions for the turnpike phenomenon, using the approach developed in [115, 116, 118, 131], for discrete-time optimal control problems in metric spaces, which are not necessarily compact, and for continuous-time infinite dimensional optimal control problems. The main results of [148] have Theorem 1.2 as their prototype.

Since the discovery of the turnpike phenomenon by Paul Samuelson in 1948, different versions of the turnpike property were considered in the literature. In this book as well as in [115, 116, 118, 131, 148], we study the turnpike property introduced and used in our previous research [112, 117, 130, 131, 142]. This turnpike property differs from other versions and has important features. Our turnpike property is a property of approximate solutions, and the turnpike is a nonstationary trajectory. As it was shown in [112, 130], our turnpike property holds for most problems belonging to large classes of variational and optimal control problems.

The main features of this book are that we study optimal control problems in metric spaces, obtain sufficient and necessary conditions for the turnpike phenomenon, and show that the turnpike phenomenon is stable under small perturbations of objective function.

1.2 Notation

In this section, we collect the notation that will be used in the book.

Let (X, ρ_X) be a metric space equipped with the metric ρ_X that induces the topology in X. For each $x \in X$ and each $r > 0$, set

$$B_X(x, r) = \{y \in X : \rho(x, y) \le r\}.$$

Let (X_i, ρ_{X_i}), $i = 1, 2$, be metric spaces. The set $X_1 \times X_2$ is equipped with the metric

$$\rho_{X_1 \times X_2}((x_1, x_2), (y_1, y_2)) = \rho_{X_1}(x_1, y_1) + \rho_{X_2}(x_2, y_2),$$

$$(x_1, x_2), \ (y_1, y_2) \in X_1 \times X_2.$$

We denote by $\mathrm{mes}(\Omega)$ the Lebesgue measure of a Lebesgue measurable set $\Omega \subset R^1$ and define

$$\chi_\Omega(x) = 1 \text{ for all } x \in \Omega, \ \chi_\Omega(x) = 0 \text{ for all } x \in R^1 \setminus \Omega.$$

For each function $f : X \to [-\infty, \infty]$, where X is nonempty, we set

$$\inf(f) = \inf\{f(x) : x \in X\}.$$

For each $s \in R^1$, set $s_+ = \max\{s, 0\}$, $s_- = \min\{s, 0\}$, $\lfloor s \rfloor = \max\{z : z$ is an integer and $z \le s\}$ and assume that $s \ge -\infty$, $s > -\infty$.

If $(X, \|\cdot\|)$ is a normed space, then X is equipped with the metric $\rho_X(x, y) = \|x - y\|$, $x, y \in X$.

Let (X, ρ_X) be a metric space and T be a Lebesgue measurable subset of R^1. A function $f : T \to X$ is called Lebesgue measurable if for any Borel set $D \subset X$ the set $f^{-1}(D)$ is a Lebesgue measurable set.

In the sequel, we denote by $\mathrm{Card}(D)$ the cardinality of a set D and suppose that the sum over an empty set is zero, the infimum over an empty set is ∞, and the supremum over an empty set is $-\infty$.

Let $(X, \langle \cdot, \cdot \rangle_X)$ be a Hilbert space equipped with an inner product $\langle \cdot, \cdot \rangle_X$ that induces the norm $\|\cdot\|_X$. If the space X is understood, we use the notation $\langle \cdot, \cdot \rangle = \langle \cdot, \cdot \rangle_X$ and $\|\cdot\| = \|\cdot\|_X$. Let X be a Banach space equipped with the norm $\|\cdot\|_X$ and X^* be its dual space with the norm $\|\cdot\|_{X^*}$. If the space X is understood, we use the notation $\|\cdot\| = \|\cdot\|_X$. For $x \in X$ and $l \in X^*$, we set $l(x) = \langle l, x \rangle_{X^*, X}$. The symbol $\langle \cdot, \cdot \rangle_{X^*, X}$ is referred to as the duality pairing between X^* and X. When the pair X, X^* is understood, we use the notation $\langle \cdot, \cdot \rangle = \langle \cdot, \cdot \rangle_{X^*, X}$.

1.3 General Dynamical Systems with a Lyapunov Function

In Chap. 5, we develop a turnpike theory for a class of general dynamical systems in a metric space with a Lyapunov function. This class of dynamical systems contains as particular cases the classes of dynamical systems studied in Chaps. 3 and 4. Our goal is to obtain several turnpike results and show that the turnpike phenomenon is stable under small perturbations of objective functions.

Now we describe this class of problems.

Assume that (X, ρ) is a metric space. For each $x \in X$ and each $r > 0$, set

$$B(x, r) = \{y \in X : \ \rho(x, y) \le r\} \text{ and } B^0(x, r) = \{y \in X : \ \rho(x, y) < r\}.$$

For each $x \in X$ and each pair of nonempty sets $A, B \subset X$, define

$$\rho(x, B) = \inf\{\rho(x, y) : \ y \in B\}$$

and

$$\mathrm{dist}(A, B) = \sup\{\rho(x, B) : \ x \in A\}.$$

Let Γ be either R^1 or the set of all integers $\mathcal{Z} = \{0, \pm 1, \pm 2, \dots\}$.

Assume that for each pair $T_1, T_2 \in \Gamma$ satisfying $T_1 < T_2$, $Y(T_1, T_2)$ is a nonempty set of functions $x : [T_1, T_2] \cap \Gamma \to X$ such that the following properties hold:

(A1) For each pair $T_1, T_2 \in \Gamma$, each pair $S_1, S_2 \in \Gamma$ satisfying $T_1 \le S_1 < S_2 \le T_2$, and each $x \in Y(T_1, T_2)$, the restriction of x to $[S_1, S_2]$ belongs to $Y(S_1, S_2)$.

(A2) For each pair $T_1, T_2 \in \Gamma$ satisfying $T_1 < T_2$, each $S \in \Gamma$, and each $x \in Y(T_1, T_2)$, the function $x(t+S)$, $t \in [T_1, T_2] \cap \Gamma$, belongs to $Y(T_1+S, T_2+S)$.

(A3) For each triplet $T_1, T_2, T_3 \in \Gamma$ satisfying $T_1 < T_2 < T_3$, each $x_1 \in Y(T_1, T_2)$, and each $x_2 \in Y(T_2, T_3)$ satisfying $x_1(T_2) = x_2(T_2)$, the function $x_3 : [T_1, T_3] \cap \Gamma \to X$ defined by

$$x_3(t) = x_1(t), \ t \in [T_1, T_2] \cap \Gamma, \ x_3(t) = x_2(t), \ t \in (T_2, T_3] \cap \Gamma$$

belongs to $Y(T_1, T_3)$.

If $\Gamma = \mathcal{Z}$, then we set

$$\mathcal{A} = \{(y, z) \in X \times X : \ \text{there is } x \in Y(0, 1) \text{ such that } x(0) = y, \ x(1) = z\},$$

define

$$a(x) = \{y \in X : \ (x, y) \in \mathcal{A}\}, \ x \in X,$$

and assume that $\mathcal{A}$ is closed in the metric space $X \times X$.

For each pair of numbers $T_1, T_2 \in \Gamma$ satisfying $T_1 < T_2$, elements of the set $Y(T_1, T_2)$ are called trajectories. Let $T_1 \in \Gamma$. A function $x : [T_1, \infty) \cap \Gamma \to X$ is a trajectory if for every $T_2 \in (T_1, \infty) \cap \Gamma$, the restriction of x to $[T_1, T_2] \cap \Gamma$ belongs to $Y(T_1, T_2)$. The set of all trajectories $x : [T_1, \infty) \cap \Gamma \to X$ is denoted by $Y(T_1, \infty)$.

Denote by $Y(\Gamma)$ the set of all functions $x : \Gamma \to X$ such that for each pair of numbers $T_1, T_2 \in \Gamma$ satisfying $T_1 < T_2$ the restriction of x to $[T_1, T_2] \cap \Gamma$ belongs to $Y(T_1, T_2)$. The elements of $Y(\Gamma)$ are called trajectories too.

Fix $\widehat{\theta} \in X$.

Assume that $\phi : X \to R^1$ is a continuous function such that for each pair of numbers $T_1 \in \Gamma$, $T_2 \in (T_1, \infty) \cap \Gamma$, each $x \in Y(T_1, T_2)$, and each $S_1, S_2 \in [T_1, T_2] \cap \Gamma$ satisfying $S_1 < S_2$,

$$\phi(x(S_2)) \leq \phi(x(S_1)). \tag{1.1}$$

Assume that X_0 is nonempty, closed subset of X such that for each $r > 0$ the set

$$X_0 \cap B(\widehat{\theta}, r) \text{ is compact} \tag{1.2}$$

and that the following assumptions hold:

(A4) For each pair of numbers $T_1, T_2 \in \Gamma$ satisfying $T_2 > T_1$, each $M > 0$, and each sequence $\{x_n\}_{n=1}^{\infty} \subset Y(T_1, T_2)$ that satisfy

$$\lim_{n \to \infty} \sup\{\rho(x_n(t), X_0 \cap B(\widehat{\theta}, M)) : t \in [T_1, T_2] \cap \Gamma\} = 0,$$

there exists a subsequence $\{x_{n_k}\}_{k=1}^{\infty}$ that converges uniformly on $[T_1, T_2] \cap \Gamma$ to a trajectory $x \in Y(T_1, T_2)$ as $k \to \infty$.

(A5) For each bounded, nonempty set B, there exists $M > 0$, and for each $\epsilon > 0$, there exists a positive number $T(B, \epsilon) \in \Gamma$ such that for each $T \in \Gamma \cap [T(B, \epsilon), \infty)$ and each $x \in Y(0, T)$ satisfying $x(0) \in B$,

$$\rho(x(t), X_0 \cap B(\widehat{\theta}, M)) \leq \epsilon, \ t \in [T(B, \epsilon), T] \cap \Gamma.$$

(Note that in the case $\Gamma = \mathcal{Z}$, (A4) follows from the compactness of the set $X_0 \cap B(\widehat{\theta}, M)$ and the closedness of $\mathcal{A}$.)

We suppose that $Y(0, \infty) \neq \emptyset$. Here we consider a general and abstract dynamical system and treat discrete-time and continuous-time case simultaneously. For examples of such systems, see Sects. 2.8 and 3.1. There are also infinite dimensional examples considered in [18, 82].

Define

$$\Omega = \{z \in X : \ \text{for very } \epsilon > 0 \text{ there exists } x \in Y(0, \infty)$$

$$\text{for which } \liminf_{t\to\infty} \rho(z, x(t)) \le \epsilon\}.$$

Clearly, Ω is a closed subset of the metric space (X, ρ). We will show that

$$\Omega \neq \emptyset, \ \Omega \subset X_0,$$

and for each $x \in Y(0, \infty)$,

$$\lim_{t\to\infty} \rho(x(t), \Omega) = 0.$$

We also assume that the following assumption holds:

(A6) If $x \in Y(\Gamma)$ is bounded and for all $t \in \Gamma$,

$$x(t) \in X_0 \text{ and } \phi(x(t)) = \phi(x(0)),$$

then $x(t) \in \Omega$ for every $t \in \Gamma$.

If $\Gamma = \mathcal{Z}$ and $E \subset \Gamma$, then we set

$$\text{mes}(E) = \text{Card}(E).$$

For each $x \in X$, set

$$\pi(x) = \lim_{T\to\infty} \sup\{\phi(x(T)) : \ x \in Y(0, T) \text{ and } x(0) = x\}.$$

(We assume that the supremum over an empty set is $-\infty$.)

It is clear that

$$\pi(x) \le \phi(x), \ x \in X.$$

The function π plays an important role in our study.

Evidently, the following result is true.

Proposition 1.4 *Let $T > 0$ and $x \in Y(0, T)$. Then $\pi(x(s)) \le \pi(x(0))$ for all $s \in [0, T] \cap \Gamma$.*

The following result is proved in Sect. 5.3.

Proposition 1.5 *Let $x \in X$. The value $\pi(x)$ is finite if and only if for every $T \in \Gamma \cap (0, \infty)$, there exists $y \in Y(0, T)$ satisfying $y(0) = x$.*

In Sect. 5.4, we prove the following result.

Proposition 1.6 *Let $M_0 > 0$. Then there exists $M > 0$ such that for each $x \in B(\widehat{\theta}, M_0)$ satisfying $\pi(x) > -\infty$,*

$$B(\widehat{\theta}, M) \cap \Omega \cap \phi^{-1}(\pi(x)) \neq \emptyset.$$

The following turnpike result is proved in Sect. 5.5.

Theorem 1.7 *Let $\epsilon > 0$ and $\widehat{x} \in X$ satisfy $\pi(\widehat{x}) > -\infty$. Then there exist $\delta > 0$ and $L > 0$ such that for every $T \in \Gamma \cap (2L, \infty)$ and every $x \in Y(0, T)$ that satisfy*

$$x(0) = \widehat{x} \text{ and } \phi(x(T)) \geq \pi(\widehat{x}) - \delta,$$

the inequality

$$\rho(x(t), \Omega) \leq \epsilon$$

holds for all $t \in [L, T - L] \cap \Gamma$.

Clearly, the theorem above establishes the turnpike property for approximate optimal trajectories that have a fixed starting point with respect to the objective function ϕ. The turnpike is the set Ω, and δ, L depend on ϵ and the initial point $\widehat{x}$.

For each $x \in X$ satisfying $\pi(x) > -\infty$, set

$$F(x) = \Omega \cap \phi^{-1}(\pi(x)),$$

which is nonempty.

The following result, which is a generalization of the previous theorem, is proved in Sect. 5.7. According to the previous theorem, for all $t \in [L, T - L] \cap \Gamma$, the distance between $x(t)$ and the turnpike Ω does not exceed ϵ. According to our next result, for all $t \in [L, T - L] \cap \Gamma$, the distance between $x(t)$ and the set $F(\widehat{x}) \cap B(\widehat{\theta}, M)$ does not exceed ϵ, where a positive constant M depends on ϵ and the initial point $\widehat{x}$ and $F(\widehat{x})$ is the set of all points of the turnpike Ω where the value of the Lyapunov function ϕ is $\pi(\widehat{x})$.

Theorem 1.8 *Let $\epsilon > 0$ and $\widehat{x} \in X$ satisfy $\pi(\widehat{x}) > -\infty$. Then there exist $M, L, \delta > 0$ such that for every $T \in \Gamma \cap (2L, \infty)$ and every $x \in Y(0, T)$ that satisfy*

$$x(0) = \widehat{x} \text{ and } \phi(x(T)) \geq \pi(\widehat{x}) - \delta,$$

the inequality

$$\rho(x(t), F(\widehat{x}) \cap B(\widehat{\theta}, M)) \leq \epsilon$$

holds for all $t \in [L, T - L] \cap \Gamma$.

In Chap. 5, we obtain several extensions and generalizations of these two turnpike results. In particular, we show that if the function π is continuous on the turnpike set Ω, then the constants L, δ do not depend on the starting point $\widehat{x}$. We also show that the turnpike phenomenon is stable under small perturbations of objective functions.

1.4 Continuous-Time Nonautonomous Problems on Half-Axis

In Chapter 8, we establish sufficient and necessary conditions for the turnpike phenomenon for continuous-time optimal control problems on subintervals of half-axis in metric spaces. For these optimal control problems, the turnpike is not a singleton. The results of this chapter are obtained for three large classes of problems that will be treated simultaneously. In Chap. 9, for the same classes of problems, we show that the turnpike phenomenon is stable under small perturbations of objective functions.

Let us describe these classes of problems.

Let (E, ρ_E) be a complete metric space and (F, ρ_F) be a metric space. We suppose that $\mathcal{A}$ is a nonempty subset of $[0, \infty) \times E$, and $\mathcal{U} : \mathcal{A} \to 2^F$ is a point to set mapping with a graph

$$\mathcal{M} = \{(t, x, u) : \ (t, x) \in \mathcal{A}, \ u \in \mathcal{U}(t, x)\}.$$

We suppose that $\mathcal{M}$ is a Borel measurable subset of $[0, \infty) \times E \times F$. Assume that for each pair of numbers T_1, T_2 satisfying $0 \leq T_1 < T_2 < \infty$, we are given a set $X(T_1, T_2)$ of pairs (x, u) (called as trajectory–control pairs) such that $x : [T_1, T_2] \to E$ is a continuous function, and $u : [T_1, T_2] \to F$ is a Lebesgue function satisfying

$$(t, x(t)) \in \mathcal{A}, \ t \in [T_1, T_2],$$

$$u(t) \in \mathcal{U}(t, x(t)), \ t \in [T_1, T_2] \text{ almost everywhere (a.e.).}$$

We suppose that the following property holds:

if $0 \leq T_1 \leq S_1 < S_2 \leq T_2 < \infty$, $(x, u) \in X(T_1, T_2)$, $\tilde{x}$ is the restriction of x to $[S_1, S_2]$, and $\tilde{u}$ is the restriction of u to $[S_1, S_2]$, then $(\tilde{x}, \tilde{u}) \in X(S_1, S_2)$;

for each triplet of nonnegative numbers $T_1 < T_2 < T_3$, each $(x_1, u_1) \in X(T_1, T_2)$, and each $(x_2, u_2) \in X(T_2, T_3)$ satisfying

$$x_1(T_2) = x_2(T_2)$$

a pair $(x_3, u_3) \in X(T_1, T_3)$, where

$$x_3(t) = x_1(t), \ t \in [T_1, T_2], \ x_3(t) = x_2(t), \ t \in (T_2, T_3],$$

$$u_3(t) = x_1(t), \ t \in [T_1, T_2], \ u_3(t) = u_2(t), \ t \in (T_2, T_3].$$

Let $a_0 > 0$, $\theta_0 \in E$, $\theta_1 \in F$, and let $\psi : [0, \infty) \to [0, \infty)$ be an increasing function such that

$$\psi(t) \to \infty \text{ as } t \to \infty,$$

and let $\psi_1 : \mathcal{M} \to [0, \infty)$ satisfy

$$\psi_1(t, x, u) \geq \psi(\rho_E(x, \theta_0)), \ (t, x, u) \in \mathcal{M}.$$

Denote by $\mathfrak{M}_\psi$ the set of all Borelian functions $g : \mathcal{M} \to R^1$ such that for each $(t, x, u) \in \mathcal{M}$,

$$g(t, x, u) \geq \psi_1(t, x, u) - a_0, \ (t, x, u) \in \mathcal{M}.$$

For each pair of numbers $T_2 > T_1 \geq 0$, each $(x, u) \in X(T_1, T_2)$, and each $g \in \mathfrak{M}_\psi$, define

$$I^g(T_1, T_2, x, u) = \int_{T_1}^{T_2} g(t, x(t), u(t))dt \in (-\infty, \infty].$$

We consider functionals of the form $I^g(T_1, T_2, x, u)$, where $0 \leq T_1 < T_2$, $(x, u) \in X(T_1, T_2)$, and $g \in \mathfrak{M}_\psi$. Let $g \in \mathfrak{M}_\psi$. For each pair of numbers $T_2 > T_1 \geq 0$ and each pair of points $(T_1, y), (T_2, z) \in \mathcal{A}$, we define

$$U^g(T_1, T_2, y, z) = \inf\{I^g(T_1, T_2, x, u) :$$

$$(x, u) \in X(T_1, T_2), \ x(T_1) = y, \ x(T_2) = z\},$$

$$U^g(T_1, T_2, y) = \inf\{I^g(T_1, T_2, x, u) : (x, u) \in X(T_1, T_2), \ x(T_1) = y\},$$

$$\widehat{U}^g(T_1, T_2, z) = \inf\{I^g(T_1, T_2, x, u) : (x, u) \in X(T_1, T_2), \ x(T_2) = z\},$$

$$U^g(T_1, T_2) = \inf\{I^g(T_1, T_2, x, u) : (x, u) \in X(T_1, T_2)\}.$$

We equip the space $\mathfrak{M}_\psi$ with the uniformity determined by the base

$$E(N, \epsilon, \lambda) = \{(f, g) \in \mathfrak{M}_\psi \times \mathfrak{M}_\psi : |f(t, x, u) - g(t, x.u)| \leq \epsilon$$

$$\text{for each } (t, x, u) \in \mathcal{M} \text{ satisfying } \rho_E(x, \theta_0), \ \rho(u, \theta_1) \leq N\}$$

$$\cap \{(f, g) \in \mathfrak{M}_\psi \times \mathfrak{M}_\psi : (|f(t, x, u)| + 1)(|g(t, x, u)| + 1)^{-1} \in [\lambda^{-1}, \lambda]$$

$$\text{for each } (t, x, u) \in \mathcal{M} \text{ satisfying } \rho_E(x, \theta_0) \leq N\},$$

where $N, \epsilon > 0, \lambda > 1$. Clearly, the uniform space $\mathfrak{M}_\psi$ is Hausdorff and has a countable base. Therefore, $\mathfrak{M}_\psi$ is metrizable (by a metric d_ψ). It is not difficult to see that the uniform space $\mathfrak{M}_\psi$ is complete. It is equipped with a topology induced by the uniformity.

In this chapter, we assume that the following assumption holds:

(A0) For each $L_0 > 0$, there exists $L_1 > 0$ such that for each $T_1 \geq 0$, each $T_2 \in (T_1, T_1 + L_0]$, each $(x, u) \in X(T_1, T_2)$, and each $g \in \mathfrak{M}_\psi$ that satisfy

$$\rho_E(x(T_1), \theta_0) \leq L_0, \; I^g(T_1, T_2, x, u) \leq L_0,$$

the inequality

$$\rho_E(\theta_0, x(t)) \leq L_1$$

is true for all $t \in [T_1, T_2]$.

Let $f \in \mathfrak{M}_\psi$. In some cases, we will use the following assumption:

(A0)' For each $L_0 > 0$, there exists $L_1 > 0$ such that for each $T_1 \geq 0$, each $T_2 \in (T_1, T_1 + L_0]$, and each $(x, u) \in X(T_1, T_2)$ that satisfy

$$I^f(T_1, T_2, x, u) \leq L_0$$

and

$$\min\{\rho_E(\theta_0, x(t)) : \; t \in [T_1, T_2]\} \leq L_0,$$

the inequality

$$\rho_E(\theta_0, x(t)) \leq L_1$$

is true for all $t \in [T_1, T_2]$.

Assumption (A0) easily implies the following result.

Proposition 1.9 *Assume that $M_1 > 0$, $0 < \tau < \tau_0 < \tau_1$. Then there exists $M_2 > 0$ such that for each $T_1 \geq 0$, each $T_2 \in [T_1 + \tau_0, T_1 + \tau_1]$, each $g \in \mathfrak{M}_\psi$, and each $(x, u) \in X(T_1, T_2)$ that satisfy*

$$I^g(T_1, T_2, x, u) \leq M_1,$$

the inequality

$$\rho_E(\theta_0, x(t)) \leq M_2$$

is true for all $t \in [T_1 + \tau, T_2]$.

Usually, a control system is governed by a differential equation, and its trajectories are solutions of this equation. Therefore, properties of trajectories depend on the equation. In this book, we use a different framework. We assume that sets of trajectory–control pairs are given and that they satisfy some natural assumptions that hold in the control theory. No equation is involved in the definition of our control system. Therefore, our results can be implied to different classes of control systems.

Now we describe three classes of optimal control problems that are particular cases of the class of problems introduced above.

The First Class of Problems
We begin with the description of the first class of problems. Let $(E, \|\cdot\|)$ be a Banach space and E^* be its dual. Let $\{A(t) : t \in [0, \infty)\}$ be the family of closed densely defined linear operators with the domain and range in the Banach space E. Let (F, ρ_F) be a metric space,

$$\theta_0 = 0, \ \theta_1 \in F,$$

$\rho_E(x, y) = \|x - y\|, \ x, y \in E$. We suppose that $\mathcal{A}$ is a nonempty subset of $[0, \infty) \times E$, and $\mathcal{U} : \mathcal{A} \to 2^F$ is a point to set mapping with a graph

$$\mathcal{M} = \{(t, x, u) : (t, x) \in \mathcal{A}, \ u \in \mathcal{U}(t, x)\}.$$

We suppose that $\mathcal{M}$ is a Borel measurable subset of $[0, \infty) \times E \times F$ and $G : \mathcal{M} \to E$ is a Borelian function. We consider the homogeneous Cauchy problem

$$x'(t) = A(t)x(t), \ t \in [0, \infty).$$

We assume that there exists a function $U : \{(t, s) \in R^2 : 0 \leq s \leq t < \infty\} \to L(E)$ that has the following properties [21]:

(i) For each $x_0 \in E$, the function $(t, s) \to U(t, s)x_0$ is continuous on the set $\{(t, s) \in R^2 : 0 \leq s \leq t < \infty\}$.
(ii) $U(s, s) = Id$ for all $s \in [0, \infty)$, where Id is the identity operator.
(iii) $U(t, s)U(s, \tau) = U(t, \tau)$ for all numbers $t \geq s \geq \tau \geq 0$.
(iv) For each $s \geq 0$, there exists a densely linear subspace E_s of E such that for each $x_0 \in E_s$ the function $t \to U(t, s)x_0$ is continuously differentiable on $[s, \infty)$ and

$$(\partial/\partial t)U(t, s)x_0 = A(t)U(t, s)x_0, \ t \in [s, \infty).$$

(v) There exists an increasing function $\tau \to \Delta_\tau > 0, \tau > 0$ such that for each $\tau > 0$, each $s \geq 0$, and each $t \in [s, s + \tau]$,

$$\|U(t, s)\| \leq \Delta_\tau.$$

In this case problem, homogeneous Cauchy problem is called well-posed [21].

Let $0 \leq T_1 < T_2$, and consider the following equation:

$$x'(t) = A(t)x(t) + f(t), \ t \in [T_1, T_2], \ x(0) = x_{T_1},$$

where $x_{T_1} \in E$ and $f \in L^1(T_1, T_2; E)$.

A continuous function $x : [T_1, T_2] \to E$ is a solution of the equation above if

$$x(t) = U(t, T_1)x(T_1) + \int_{T_1}^{t} U(t, s)f(s)ds, \ t \in [T_1, T_2].$$

Assume that the equation above holds and $\tau \in [T_1, T_2]$. It is not difficult to see (p. 386, [148]) that $x : [\tau, T_2] \to E$ is a solution of the equation

$$y'(t) = A(t)y(t) + f(t), \ t \in [\tau, T_2], \ y(\tau) = x(\tau).$$

Let $0 \le T_1 < T_2 < T_3$, $z_0 \in E$, $f \in L^1(T_1, T_3; E)$, a continuous function $x_1 : [T_1, T_2] \to E$ is a solution of the equation

$$x'(t) = A(t)x(t) + f(t), \ t \in [T_1, T_2], \ x(T_1) = z_0,$$

and a continuous function $x_2 : [T_2, T_3] \to E$ is a solution of the equation

$$x'(t) = A(t)x(t) + f(t), \ t \in [T_2, T_3], \ x(T_2) = x_1(T_2).$$

Set

$$x(t) = x_1(t), \ t \in [T_1, T_2], \ x(t) = x_2(t), \ t \in [T_2, T_3].$$

Clearly, the function $x : [T_1, T_3] \to E$ is continuous. It is not difficult to see (p. 387, [148]) that $x(\cdot)$ is a solution of the equation

$$x'(t) = A(t)x(t) + f(t), \ t \in [T_1, T_3], \ x(T_1) = z_0.$$

Let $0 \le T_1 < T_2$. We consider the following equation:

$$x'(t) = A(t)x(t) + G(t, x(t), u(t)), \ t \in [T_1, T_2].$$

A pair of functions $x : [T_1, T_2] \to E$, $u : [T_1, T_2] \to F$ is called a (mild) solution of the equation above if $x : [T_1, T_2] \to E$ is a continuous function, $u : [T_1, T_2] \to F$ is a Lebesgue measurable function,

$$(t, x(t)) \in \mathcal{A}, \ t \in [T_1, T_2],$$

$$u(t) \in \mathcal{U}(t, x(t)), \ t \in [T_1, T_2] \text{ almost everywhere (a.e.)},$$

$G(s, x(s), u(s))$, $s \in [T_1, T_2]$ is Bochner integrable, and for every $t \in [T_1, T_2]$,

$$x(t) = U(t, T_1)x(T_1) + \int_{T_1}^{t} U(t, s)G(s, x(s), u(s))ds.$$

The set of all solutions (x, u) is denoted by $X(T_1, T_2)$.

Let $T_1 \geq 0$. A pair of functions $x : [T_1, \infty) \to E$, $u : [T_1, \infty) \to F$ is called a (mild) solution of the system

$$x'(t) = A(t)x(t) + G(t, x(t), u(t)), \ t \in [T_1, \infty)$$

if for every $T_2 > T_1$, the pair $x : [T_1, T_2] \to E$, $u : [T_1, T_2] \to F$ belongs to $X(T_1, T_2)$. The set of all such pairs (x, u), which are solutions of the equation above, is denoted by $X(T_1, \infty)$.

A function $x : I \to E$, where I is either $[T_1, T_2]$ or $[T_1, \infty)$ $(0 \leq T_1 < T_2)$, is called a trajectory if there exists a Lebesgue measurable function $u : I \to F$ (referred to as a control) such that $(x, u) \in X(T_1, T_2)$ or $(x, u) \in X(T_1, \infty)$, respectively.

For this class of problem, for each $(t, x, u) \in \mathcal{M}$, set

$$\psi_1(t, x, u) = \max\{\psi(\|x\|), \ \psi(\|u\|),$$

$$\psi((\|G(t, x, u)\| - a_0\|x\|)_+)(\|G(t, x, u)\| - a_0\|x\|)_+\}.$$

In Sect. 8.4, we show that assumption (A0) holds for this class of problems.

The Second Class of Problems

Let $(E, \langle\cdot,\cdot\rangle)_E$ be a Hilbert space equipped with an inner product $\langle\cdot,\cdot\rangle_E$ that induces the norm $\|\cdot\|_E$, and let $(F, \langle\cdot,\cdot\rangle_F)$ be a Hilbert space equipped with an inner product $\langle\cdot,\cdot\rangle_F$ that induces the norm $\|\cdot\|_F$. For simplicity, we set $\langle\cdot,\cdot\rangle_E = \langle\cdot,\cdot\rangle$, $\|\cdot\|_E = \|\cdot\|$, $\langle\cdot,\cdot\rangle_F = \langle\cdot,\cdot\rangle$, $\|\cdot\|_F = \|\cdot\|$, if E, F are understood. Set

$$\theta_0 = 0, \ \theta_1 = 0,$$

$\rho_E(x, y) = \|x - y\|, \ x, y \in E, \rho_F(x, y) = \|x - y\|, \ x, y \in F.$

We suppose that $\mathcal{A}_0$ is a nonempty subset of E, and $\mathcal{U}_0 : \mathcal{A}_0 \to 2^F$ is a point to set mapping with a graph

$$\mathcal{M}_0 = \{(x, u) : \ x \in \mathcal{A}_0, \ u \in \mathcal{U}_0(x)\}.$$

We suppose that $\mathcal{M}_0$ is a Borel measurable subset of $E \times F$. Define

$$\mathcal{A} = [0, \infty) \times \mathcal{A}_0,$$

$\mathcal{U} : \mathcal{A} \to 2^F$ by

$$\mathcal{U}(t, x) = \mathcal{U}_0(x), \ (t, x) \in \mathcal{A},$$

$$\mathcal{M} = [0, \infty) \times \mathcal{M}_0.$$

Let a linear operator $A : \mathcal{D}(A) \to E$ generate a C_0 semigroup $S(t) = e^{At}$, $t \in [0, \infty)$ on E. As usual, we denote by $S(t)^*$ the adjoint of $S(t)$. Then $S(t)^*$, $t \in [0, \infty)$, is C_0 semigroup, and its generator is the adjoint A^* of A. The domain $\mathcal{D}(A^*)$ is a Hilbert space equipped with the graph norm $\|\cdot\|_{\mathcal{D}(A^*)}$:

$$\|z\|^2_{\mathcal{D}(A^*)} = \|z\|^2_E + \|A^* z\|^2_E, \ z \in \mathcal{D}(A^*).$$

Let $\mathcal{D}(A^*)'$ be the dual of $\mathcal{D}(A^*)$ with the pivot space E. In particular,

$$E_1^d := \mathcal{D}(A^*) \subset E \subset \mathcal{D}(A^*)' = E_{-1}.$$

(Here we use the notation of Chap. 7.)

Let $G : \mathcal{M} \to \mathcal{D}(A^*)' = E_{-1}$, $B \in \mathcal{L}(F, E_{-1})$ is an admissible control operator for e^{At}, $t \geq 0$, and for all $(t, x, u) \in \mathcal{M}$,

$$G(t, x, u) = Bu.$$

Let $0 \leq T_1 < T_2$. We consider the following equation:

$$x'(t) = Ax(t) + Bu(t), \ t \in [T_1, T_2] \text{ a.e.}$$

A pair of functions $x : [T_1, T_2] \to E$, $u : [T_1, T_2] \to F$ is called a (mild) solution of the equation above if $x : [T_1, T_2] \to E$ is a continuous function, $u : [T_1, T_2] \to F$ is a Lebesgue measurable function, $u \in L^2(T_1, T_2; F)$,

$$(t, x(t)) \in \mathcal{A}, \ t \in [T_1, T_2],$$

$$u(t) \in \mathcal{U}(t, x(t)), \ t \in [T_1, T_2] \text{ a.e.},$$

and for each $t \in [T_1, T_2]$,

$$x(t) = e^{A(t-T_1)} x(T_1) + \int_{T_1}^t e^{A(t-s)} Bu(s) ds$$

in E_{-1}. The set of all solutions (x, u) is denoted by $X(T_1, T_2, A, G)$. In the sequel for simplicity, we use the notation $X(T_1, T_2) = X(T_1, T_2, A, G)$ if the pair (A, G) is understood.

Let $T_1 \geq 0$. A pair of functions $x : [T_1, \infty) \to E$, $u : [T_1, \infty) \to F$ is called a (mild) solution of the system

$$x'(t) = Ax(t) + Bu(t), \ t \in [T_1, \infty))$$

if for every $T_2 > T_1$, $x : [T_1, T_2] \to E$, $u : [T_1, T_2] \to F$ belongs to $X(T_1, T_2)$. The set of all such pairs (x, u), which are solutions of the equation above, is denoted by $X(T_1, \infty)$.

A function $x : I \to E$, where I is either $[T_1, T_2]$ or $[T_1, \infty)$ $(0 \le T_1 < T_2)$, is called a trajectory if there exists a Lebesgue measurable function $u : I \to F$ (referred to as a control) such that $(x, u) \in X(T_1, T_2)$ or $(x, u) \in X(T_1, \infty)$, respectively.

For this class of problem for each $(t, x, u) \in \mathcal{M}$, set

$$\psi_1(t, x, u) = \max\{\psi(\|x\|),\ K_0\|u\|^2\} - a_0,$$

where K, a_0 are positive constants.

In Sect. 8.4, we show that assumption (A0) holds for this class of problems.

The Third Class of Problems

Let (E, ρ_E) be a complete metric space, (F, ρ_F) be a metric space,

$$\theta_0 \in E,\ \theta_1 \in F.$$

We suppose that $\mathcal{A}$ is a nonempty subset of $[0, \infty) \times E$, and $\mathcal{U} : \mathcal{A} \to 2^F$ is a point to set mapping with a graph

$$\mathcal{M} = \{(t, x, u) :\ (t, x) \in \mathcal{A},\ u \in \mathcal{U}(t, x)\}.$$

We suppose that $\mathcal{M}$ is a Borel measurable subset of $[0, \infty) \times E \times F$ and $G : \mathcal{M} \to [0, \infty)$ is a Borelian function.

Let $0 \le T_1 < T_2 < \infty$. Denote by $X(T_1, T_2)$ the set of all pairs of functions $x : [T_1, T_2] \to E$ and $u : [T_1, T_2] \to F$ such that x is continuous on $[T_1, T_2]$, u is Lebesgue measurable on function $[T_1, T_2]$,

$$(t, x(t)) \in \mathcal{A},\ t \in [T_1, T_2],$$

$$u(t) \in \mathcal{U}(t, x(t)),\ t \in [T_1, T_2] \text{ almost everywhere (a.e.)},$$

$G(s, x(s), u(s)),\ s \in [T_1, T_2]$ is Bochner integrable, and for every pair $S_1, S_2 \in [T_1, T_2]$ satisfying $S_1 < S_2$,

$$\rho_E(x(S_1), x(S_2)) \le \int_{S_1}^{S_2} G(t, x(t), u(t))dt.$$

(See Section 1.1 of [5].) Pairs $(x, u) \in X(T_1, T_2)$ are called trajectory–control pairs.

Let $T_1 \ge 0$. Denoted by $X(T_1, \infty)$ the set of all pairs of functions $x : [T_1, \infty) \to E$, $u : [T_1, \infty) \to F$ such that for every $T_2 > T_1$, the pair $x : [T_1, T_2] \to E$, $u : [T_1, T_2] \to F$ belongs to $X(T_1, T_2)$. Pairs $(x, u) \in X(T_1, \infty)$ are called trajectory–control pairs.

A function $x : I \to E$, where I is either $[T_1, T_2]$ or $[T_1, \infty)$ $(0 \le T_1 < T_2)$, is called a trajectory if there exists a Lebesgue measurable function $u : I \to F$ (referred to as a control) such that $(x, u) \in X(T_1, T_2)$ or $(x, u) \in X(T_1, \infty)$, respectively).

For this class of problem for each $(t, x, u) \in \mathcal{M}$, set

$$\psi_1(t, x, u) = \max\{\psi(\rho_E(x, \theta_0)),\ \psi(\rho_F(u, \theta_1)),$$

$$\psi((G(t, x, u) - a_0\rho_E(x, \theta_0))_+)(G(t, x, u) - a_0\rho_E(x, \theta_0))_+\}.$$

For each pair of numbers $T_2 > T_1 \geq 0$, each $(x, u) \in X(T_1, T_2)$, and each $g \in \mathfrak{M}_\psi$, define

$$I^g(T_1, T_2, x, u) = \int_{T_1}^{T_2} g(t, x(t), u(t))dt \in (-\infty, \infty].$$

In Chap. 8, we show now that assumption (A0) holds for this class of problems.

Let us consider our general class of problems.

We say that the integrand f possesses the turnpike property (or TP for short) if for each $\epsilon > 0$ and each $M > 0$ there exist $\delta > 0$ and $L > 0$ such that for each $T_1 \geq 0$, each $T_2 \geq T_1 + 2L$, and each $(x, u) \in X(T_1, T_2)$ that satisfy

$$I^f(T_1, T_2, x, u) \leq \min\{U^f(T_1, T_2) + M,\ U^f(T_1, T_2, x(T_1), x(T_2)) + \delta\}$$

there exist $\tau_1, \tau_2 \in [0, L]$ such that

$$\rho_E(x(t), x_f(t)) \leq \epsilon \text{ for all } t \in [T_1 + \tau_1, T_2 - \tau_2].$$

Moreover, if $\rho_E(x(T_2), x_f(T_2)) \leq \delta$, then $\tau_2 = 0$, and if $T_1 \geq L$ and $\rho_E(x(T_1), x_f(T_1)) \leq \delta$, then $\tau_1 = 0$.

The next theorem is the main result of Chap. 8. It is proved in Sect. 8.13.

Theorem 1.10 *f has TP if and only if the following properties hold:*

(P1) For each (f)-good pair $(x, u) \in X(0, \infty)$,

$$\lim_{t\to\infty} \rho_E(x(t), x_f(t)) = 0.$$

(P2) For each $\epsilon > 0$ and each $M > 0$, there exist $\delta > 0$ and $L > 0$ such that for each $T \geq 0$ and each $(x, u) \in X(T, T + L)$ that satisfy

$$I^f(T, T + L, x, u) \leq \min\{U^f(T, T + L, x(T), x(T + L)) + \delta,$$

$$I^f(T, T + L, x_f, u_f) + M\}$$

there exists $s \in [T, T + L]$ such that $\rho_E(x(s), x_f(s)) \leq \epsilon$.

In Chap. 9, we study the stability of the turnpike phenomenon. In particular, it is shown that if TP holds, then it is stable under small perturbations of the objective function.

Chapter 2
Differential Inclusions

In this chapter, we discuss Banach space-valued functions, set-valued mappings in infinite dimensional spaces, and related continuous-time dynamical systems. Some convergence results are obtained.

2.1 Banach Space-Valued Functions

Let $(X, \|\cdot\|)$ be a Banach space and $a < b$ be real numbers. For any set $E \subset R^1$, define

$$\chi_E(t) = 1 \text{ for all } t \in E \text{ and } \chi_E(t) = 0 \text{ for all } t \in R^1 \setminus E.$$

If a set $E \subset R^1$ is Lebesgue measurable, then its Lebesgue measure is denoted by $|E|$ or by $\mathrm{mes}(E)$.

A function $f : [a, b] \to X$ is called a simple function if there exists a finite collection of Lebesgue measurable sets $E_i \subset [a, b]$, $i \in I$, mutually disjoint, and $x_i \in X$, $i \in I$ such that

$$f(t) = \sum_{i \in I} \chi_{E_i}(t) x_i, \ \ t \in [a, b].$$

A function $f : [a, b] \to X$ is strongly measurable if there exists a sequence of simple functions $\phi_k : [a, b] \to X$, $k = 1, 2, \ldots$, such that

$$\lim_{k \to \infty} \|\phi_k(t) - f(t)\| = 0, \ \ t \in [a, b] \text{ almost everywhere (a.e.).} \tag{2.1}$$

For every simple function $f(\cdot) = \sum_{i \in I} \chi_{E_i}(\cdot) x_i$, where the set I is finite, define its Bochner integral by

A. J. Zaslavski, *Turnpike Phenomenon in Metric Spaces*, Springer Optimization and Its Applications 201, https://doi.org/10.1007/978-3-031-27208-0_2

$$\int_a^b f(t)dt = \sum_{i \in I} |E_i| x_i.$$

Let $f : [a, b] \to X$ be a strongly measurable function. We say that f is Bochner integrable if there exists a sequence of simple functions $\phi_k : [a, b] \to X$, $k = 1, 2, \ldots$, such that (2.1) holds and the sequence $\{\int_a^b \phi_k(t)dt\}_{k=1}^{\infty}$ strongly converges in X. In this case, we define the Bochner integral of the function f by

$$\int_a^b f(t)dt = \lim_{k \to \infty} \int_a^b \phi_k(t)dt.$$

It is known that the integral defined above is independent of the choice of the sequence $\{\phi_k\}_{k=1}^{\infty}$ [73]. Similar to the Lebesgue integral, for any measurable set $E \subset [a, b]$, the Bochner integral of f over E is defined by

$$\int_E f(t)dt = \int_a^b \chi_E(t) f(t)dt.$$

The following result is true (see Proposition 3.4, Chapter 2 of [73]).

Proposition 2.1 *Let $f : [a, b] \to X$ be a strongly measurable function. Then f is Bochner integrable if and only if the function $\|f(\cdot)\|$ is Lebesgue integrable. Moreover, in this case,*

$$\|\int_a^b f(t)dt\| \leq \int_a^b \|f(t)\|dt.$$

The Bochner integral possesses almost the same properties as the Lebesgue integral. If $f : [a, b] \to X$ is strongly measurable and $\|f(\cdot)\| \in L^p(a, b)$, for some $p \in [1, \infty)$, then we say that $f(\cdot)$ is L^p Bochner integrable. For every $p \geq 1$, the set of all L^p Bochner integrable functions is denoted by $L^p(a, b; X)$, and for every $f \in L^p(a, b; X)$,

$$\|f\|_{L^p(a,b;X)} = (\int_a^b \|f(t)\|^p dt)^{1/p}.$$

Clearly, the set of all Bochner integrable functions on $[a, b]$ is $L^1(a, b; X)$.

Let $a < b$ be real numbers. A function $x : [a, b] \to X$ is absolutely continuous (a.c.) on $[a, b]$ if for each $\epsilon > 0$ there exists $\delta > 0$ such that for each pair of sequences $\{t_n\}_{n=1}^q$, $\{s_n\}_{n=1}^q \subset [a, b]$ satisfying

$$t_n < s_n, \ n = 1, \ldots, q, \ \sum_{n=1}^q (s_n - t_n) \leq \delta,$$

$$(t_n, s_n) \cap (t_m, s_m) = \emptyset \text{ for all } m, n \in \{1, \ldots, q\} \text{ such that } m \neq n,$$

we have

$$\sum_{n=1}^{q} \|x(t_n) - x(s_m)\| \leq \epsilon.$$

The following result is true (see Theorem 1.124 of [21]).

Proposition 2.2 *Let X be a reflexive Banach space. Then every a.c. function $x : [a, b] \to X$ is a.e. differentiable on $[a, b]$ and*

$$x(t) = x(a) + \int_a^t (dx/dt)(s)ds, \ t \in [a, b],$$

where $dx/dt \in L^1(a, b; X)$ is the strong derivative of x.

Let $-\infty < \tau_1 < \tau_2 < \infty$. Denote by $W^{1,1}(\tau_1, \tau_2; X)$ (or $W^{1,1}(\tau_1, \tau_2)$ if the space X is understood) the set of all functions $x : [\tau_1, \tau_2] \to X$ for which there exists a Bochner integrable function $u : [\tau_1, \tau_2] \to X$ such that for all $t \in (\tau_1, \tau_2]$,

$$x(t) = x(\tau_1) + \int_{\tau_1}^t u(s)ds.$$

For each $x \in X$ and each $r > 0$, set

$$B(x, r) = \{y \in X : \|x - y\| \leq r\} \text{ and } B^0(x, r) = \{y \in X : \|x - y\| < r\}.$$

For each nonempty set Y and each function $h : Y \to R^1$, set

$$\inf(h; Y) = \inf\{h(y) : y \in Y\}.$$

Let Y be a topological space and $Z \subset Y$. Then Z is equipped with the relative topology. The set Z is relatively compact if its closure is compact.

In the sequel, we use the following result (see Theorem 4 on page 13 of [12]).

Theorem 2.3 *Let $-\infty < a < b < \infty$ $(X, \|\cdot\|)$ be a reflexive Banach space, and let for every integer $k \geq 1$, $x_k \in W^{1,1}(a, b; X)$. Assume that for each real number $t \in [a, b]$, the sequence $\{x_k(t)\}_{k=1}^{\infty}$ is a relatively compact subset of X and that there exists a positive function $c \in L^1(a, b; X)$ such that*

$$\|x_k'(t)\| \leq c(t)$$

for almost every (a.e.) $t \in [a, b]$ and every integer $k \geq 1$. Then there are a subsequence $\{x_{k_i}\}_{i=1}^{\infty}$ and $x \in W^{1,1}(a, b; X)$ such that x_{k_i} converges to the function x uniformly on $[a, b]$ and that the function x_{k_i}' converges weakly to the function x' in $L^1(a, b; X)$ as $i \to \infty$.

2.2 Set-Valued Mappings and a Convergence Result

Let X be a Hausdorff topological space. Suppose that $a+\infty = \infty$ and $a-\infty = -\infty$ for every $a \in R^1$.

Let $f : X \to [-\infty, \infty]$ and $x_0 \in X$. We say that f is upper semicontinuous at x_0 if for every $\epsilon > 0$, there exists a neighborhood V of x_0 in X such that for every $x \in V$,

$$f(x) \le f(x_0) + \epsilon.$$

Let X, Y be nonempty sets and $F : X \to 2^Y$. Define the domain of F

$$\mathrm{Dom}(F) = \{x \in X : \ F(x) \neq \emptyset\}$$

and the graph

$$\mathrm{graph}(F) = \{(x, y) \in X \times Y : \ y \in F(x)\}.$$

The mapping F is proper if $\mathrm{Dom}(F) \neq \emptyset$.

Assume that X, Y are Hausdorff topological spaces. We say that F is upper semicontinuous (u.s.c.) at $x_0 \in X$ if for any open set $N \subset Y$ containing $F(x_0)$ there exists a neighborhood M of x_0 in X such that $F(M) \subset N$.

We say that F is upper semicontinuous if it is upper semicontinuous at every $x_0 \in X$.

For the proof of the following result, see Proposition 2 on page 41 of [12]).

Proposition 2.4 *If F is upper semicontinuous and $F(z)$ is closed for every $z \in X$, then graph(F) is closed.*

Assume that X is a Hausdorff topological space and Y is a locally convex Hausdorff space. Denote by Y^* its dual. The bilinear form $\langle p, y\rangle = p(y)$ is a duality pairing. We supply the space Y with the weak topology $\sigma(Y, Y^*)$. For every $K \subset Y$, define

$$\sigma_K(p) = \sigma(K, p) = \sup\{p(y) : \ y \in K\} \in [-\infty, \infty], \ \ p \in Y^*.$$

(Here we assume that the supremum over empty set is $-\infty$.) We say that F is upper hemicontinuous (u.h.c.) at $x_0 \in X$ if for every $p \in Y^*$ the function $x \to \sigma(F(x), p), \ x \in X$ is upper semicontinuous at x_0. We say that F is upper hemicontinuous if it is upper hemicontinuous at every $x_0 \in X$.

Now we present the proof of the following result (see Theorem 1 on page 60 of [12]).

Theorem 2.5 *Let $(X, \|\cdot\|)$ be a normed space, $(Y, \|\cdot\|)$ be a Banach space, $F : X \to 2^Y$ be a proper hemicontinuous mapping such that $F(x)$ is closed and convex for every $x \in X$, $-\infty < a < b < \infty$, and let for every integer $k \ge 1$, $x_k : [a, b] \to$*

X and $y_k : [a, b] \to Y$ be strongly measurable functions such that for almost every $t \in [a, b]$ and for every neighborhood $\mathcal{N}$ of zero in $X \times Y$ there exists an integer $k_0 = k_0(t, \mathcal{N})$ such that for each integer $k \geq k_0$,

$$(x_k(t), y_k(t)) \in graph(F) + \mathcal{N}.$$

Assume that x_k converges almost everywhere to a function $x : [a, b] \to X$ as $k \to \infty$ and that $y_k \in L^1(a, b; Y)$ converges weakly to y in $L^1(a, b; Y)$ as $k \to \infty$. Then for almost every $t \in [a, b]$, $y(t) \in F(x(t))$.

Proof For a convex subset of a Banach space (here $L^1(a, b; Y)$), the strong closure coincides with the weak closure (here in the weak topology $\sigma(L^1, L^\infty)$). This fact implies that for every integer $h \geq 1$, y belongs to the strong closure of the convex hull co($\{y_k : k \geq h$ is an integer $\}$). Hence, for any integer $h \geq 1$, there exist an integer $k(h) \geq h$, numbers $\alpha_h^k \geq 0$, $k = h, \ldots, k(h)$ satisfying

$$\sum_{k=h}^{k(h)} \alpha_h^k = 1$$

such that

$$v_h = \sum_{k=h}^{k(h)} \alpha_h^k y_k$$

converges to y as $h \to \infty$ in $L^1(a, b; Y)$. This implies that there exists a subsequence (again denoted by v_h, $h = 1, 2, \ldots$) that converges to y for almost every $t \in [a, b]$.

Let $t \in [a, b]$, $x_h(t)$ converge to $x(t)$ in X, and $v_h(t)$ converge to $y(t)$ in Y as $h \to \infty$. In order to complete the proof of the theorem, it is sufficient to show that

$$y(t) \in F(x(t)).$$

Assume the contrary. Then

$$y(t) \notin F(x(t)),$$

and there exists $p \in Y^*$ such that

$$p(y(t)) > \sup\{p(z) : z \in F(x(t))\} = \sigma(F(x(t), p). \tag{2.2}$$

(If $F(x(t)) = \emptyset$, then the right-hand side of (2.2) is $-\infty$.)

Fix $\lambda > \sigma(F(x(t)), p)$ and $\eta > 0$. Since the mapping F is upper hemicontinuous, there exists a neighborhood $\mathcal{M}_0$ of zero in X such that

$$\sigma(F(u), p) \le \lambda \text{ for all } u \in x(t) + \mathcal{M}_0.$$

There exists an integer $k_1 \ge 1$ such that

$$x_k(t) \in x(t) + 2^{-1}\mathcal{M}_0 \text{ for all integers } k \ge k_1.$$

In view of our assumptions, there exists an integer $k_0 \ge k_1$ such that for each integer $k \ge k_0$ there exists

$$(u_k, w_k) \in \text{graph}(F)$$

such that

$$u_k \in x_k(t) + 2^{-1}\mathcal{M}_0, \ \|y_k(t) - w_k\| \le \eta.$$

Therefore, for every integer $k \ge k_0$,

$$u_k \in x(t) + \mathcal{M}_0,$$

$$p(y_k(t)) \le p(w_k) + \eta\|p\| \le \sigma(F(u_k), p) + \eta\|p\| \le \lambda + \eta\|p\|,$$

and

$$p(y_k(t)) \le \lambda + \eta\|p\| \text{ for all integers } k \ge k_0.$$

Let $h \ge k_0$ be a fixed integer. By multiplying the above inequality by α_h^k and summing them, we deduce that

$$p(v_h(t)) \le \lambda + \eta\|p\|.$$

Letting $h \to \infty$, we obtain that

$$p(y(t)) \le \lambda + \|p\|\eta.$$

Since η is any positive number and λ is an arbitrary number satisfying $\lambda > \sigma(F(x), p)$, we conclude that

$$p(y(t)) \le \sigma(F(x), p).$$

This contradicts (2.2). The contradiction we have reached completes the proof of Theorem 2.5.

2.3 Dynamical Systems

Let $(X, \langle \cdot, \cdot \rangle)$ be a Hilbert space equipped with an inner product $\langle \cdot, \cdot \rangle$ that induces the norm

$$\|x\| = \langle x, x \rangle^{1/2}, \ x \in X.$$

For each $x \in X$ and each pair of nonempty sets $A, B \subset X$, define

$$\rho(x, B) = \inf\{\|x - y\| : \ y \in B\}$$

and

$$\mathrm{dist}(A, B) = \sup\{\rho(x, B) : \ x \in A\}.$$

Assume that a mapping $F : X \to 2^X$ is proper ($\mathrm{Dom}(F) \neq \emptyset$) and satisfies the following assumptions:

(A1) The mapping F is upper hemicontinuous, and $F(x)$ is a closed, convex set in X for every point $x \in X$.
(A2) For each $x \in X$, there exist $r, M > 0$ such that

$$F(y) \subset B(0, M) \text{ for each } y \in B(x, r).$$

Let $-\infty < T_1 < T_2 < \infty$. A function $x \in W^{1,1}(T_1, T_2; X)$ is called a trajectory if

$$x'(t) \in F(x(t)), \ t \in [T_1, T_2] \text{ almost everywhere (a.e.).} \tag{2.3}$$

Denote by $Y(T_1, T_2)$ the collection of all trajectories $x : [T_1, T_2] \to X$.

Let $T_1 \in R^1$. A function $x : [T_1, \infty) \to X$ is a trajectory if for every $T_2 > T_1$ its restriction to the interval $[T_1, T_2]$ belongs to $Y(T_1, T_2)$. Denote by $Y(T_1, \infty)$ the set of all trajectories $x : [T_1, \infty) \to X$.

A function $x : R^1 \to X$ is a trajectory if for each pair of numbers $T_2 > T_1$ its restriction to the interval $[T_1, T_2]$ belongs to $Y(T_1, T_2)$. Denote by $Y(-\infty, \infty)$ the collection of all trajectories $x : R^1 \to X$.

Suppose that the function $\phi : X \to [0, \infty)$ is locally Lipschitz and satisfies

$$\lim_{\|z\| \to \infty} \phi(z) = \infty \tag{2.4}$$

and that the following assumption holds:

(A3) For each $x \in X$ and each $h \in F(x)$, if there exists

$$\lim_{t \to 0} t^{-1}(\phi(x + th) - \phi(x)),$$

then this limit is non-positive.

Proposition 2.6 *Assume that* $-\infty < T_1 < T_2 < \infty$ *and that* $x \in Y(T_1, T_2)$. *Then the functions* x *and* $\phi \circ x$ *are Lipschitz, the function* $\phi \circ x$ *is decreasing on* $[T_1, T_2]$, *and for a.e.* $\tau \in [T_1, T_2]$, *there exist* $x'(\tau)$ *and* $(\phi \circ x)'(\tau)$ *and*

$$(\phi \circ x)'(\tau) = \lim_{h \to 0} h^{-1}[\phi(x(\tau) + hx'(\tau)) - \phi(x(\tau))] \le 0.$$

Proof In view of (A2), for every $\tau \in [T_1, T_2]$, there exist $M_\tau, r_\tau > 0$ such that

$$F(x(t)) \subset B(0, M_\tau) \text{ for each } y \in B(x(\tau), r_\tau). \tag{2.5}$$

Clearly, the set $x([T_1, T_2])$ is compact. In view of (2.5),

$$x([T_1, T_2]) \subset \cup\{B^0(x(\tau), r_\tau) : \tau \in [T_1, T_2]\}.$$

Therefore, there exist $\tau_i \in [T_1, T_2]$, $i = 1, \dots, q$ such that

$$x([T_1, T_2]) \subset \cup_{i=1}^{q} B^0(x(\tau_i), r_{\tau_i}). \tag{2.6}$$

Set

$$M = \max\{M_{\tau_i} : i = 1, \dots, q\}.$$

By (2.3), (2.5) and (2.6), for a.e. $\tau \in [T_1, T_2]$, there exists $j \in \{1, \dots, q\}$ such that

$$x(\tau) \in B^0(x(\tau_j), r_{\tau_j}),$$

$$x'(\tau) \in F(x(\tau)) \subset B(0, M_{\tau_j}) \subset B(0, M).$$

Thus the function $x : [T_1, T_2] \to X$ is Lipschitz. Clearly, $\phi \circ x$ is locally Lipschitz on the compact set $[T_1, T_2]$. Therefore, the function $\phi \circ x$ is Lipschitz on $[T_1, T_2]$ too, and for a.e. $t \in [T_1, T_2]$, there exist $x'(t)$ and $(\phi \circ x)'(t)$.

Let $\tau \in (0, T)$, and there exist $x'(\tau)$ and $(\phi \circ x)'(\tau)$. The function ϕ is Lipschitz in a neighborhood of $x(\tau)$ with a Lipschitz constant $L > 0$. Therefore,

$$x'(\tau) = \lim_{h \to 0} h^{-1}[x(\tau + h) - x(\tau)]$$

and

$$h^{-1}|[(\phi \circ x)(\tau + h) - (\phi \circ x)(\tau)] - [\phi(x(\tau) + hx'(\tau)) - \phi(x(\tau))]|$$

$$= h^{-1}|(\phi \circ x)(\tau + h) - \phi(x(\tau) + hx'(\tau))|$$

$$\le Lh^{-1}\|x(\tau + h) - x(\tau) - hx'(\tau)\| \to 0 \text{ as } h \to 0.$$

Together with (A3), this completes the proof of Proposition 2.6.

Assume that X_0 is nonempty, closed subset of X such that for each $r > 0$ the set

$$X_0 \cap B(x, r) \text{ is compact} \tag{2.7}$$

and that the following assumption holds:

(A4) For each bounded nonempty set B and each $\epsilon > 0$, there exists a natural number $T(B, \epsilon)$ such that for each $T \geq T(B, \epsilon)$,

$$\text{dist}(\{x(T) : \ x \in Y(0, T) \text{ and } x(0) \in B\}, X_0) \leq \epsilon.$$

In Sect. 2.5, we prove the following result.

Proposition 2.7 *The following properties are equivalent:*

(1) There exists $x \in Y(0, \infty)$ such that $x(t) \in X_0$ for every $t \geq 0$.
(2) There exists $x \in Y(0, \infty)$.
(3) There exist $M > 0$ and $x_k \in Y(0, T_k)$, $k = 1, 2, \ldots$, such that $\lim_{k\to\infty} T_k \to \infty$ and $\phi(x_k(0)) \leq M$, $k = 1, 2, \ldots$.

In this chapter, in the sequel, we suppose that there exists $x \in Y(0, \infty)$. Define

$$\Omega(F) = \{z \in X : \ \text{for every positive number } \epsilon \text{ there is a function}$$

$$x \in Y(0, \infty) \text{ for which } \liminf_{t\to\infty} \|z - x(t)\| \leq \epsilon\}. \tag{2.8}$$

Evidently, $\Omega(F)$ is a closed subset of X.

In Sect. 2.6, we prove the following result.

Proposition 2.8 *$\Omega(F) \neq \emptyset$ and $\Omega(F) \subset X_0$.*

It is not difficult to see that the following two results hold.

Proposition 2.9 *For every function $x \in Y(0, \infty)$,*

$$\lim_{t\to\infty} \rho(x(t), \Omega(F)) = 0.$$

Proposition 2.10 *Assume that $B \subset X$ be a nonempty, closed set such that for every $x \in Y(0, \infty)$,*

$$\lim_{t\to\infty} \rho(x(t), B) = 0.$$

Then $\Omega(F) \subset B$.

The following theorem will be proved in Sect. 2.7.

Theorem 2.11 *Let ϵ, M be positive real numbers. Then there is a positive number T such that for every function $x \in Y(0, T)$ that satisfies*

$$\phi(x(0)) \leq M$$

the inequality

$$\inf\{\rho(x(t), \Omega(F)) : t \in [0, T]\} \leq \epsilon$$

holds.

Note that the dynamical system considered here was studied in [150, 151] when the space X was finite dimensional.

2.4 Auxiliary Results

Lemma 2.12 *Let $M > 0$. Then there exist $r, L > 0$ such that for each*

$$z \in \cup\{B(x, r) : x \in X_0 \cap B(0, M)\}$$

the inclusion $F(z) \subset B(0, L)$ holds.

Proof In view of (A2), for each $x \in X_0 \cap B(0, M)$, there exist $L_x, r_x > 0$ such that

$$F(y) \subset B(0, L_x) \text{ for each } y \in B(x, 2r_x). \tag{2.9}$$

Clearly,

$$X_0 \cap B(0, M) \subset \cup\{B^0(x, r_x) : x \in X_0 \cap B(0, M)\}. \tag{2.10}$$

Since the set $X_0 \cap B(0, M)$ is compact, it follows from (2.10) that there exists a finite set

$$x_i \in X_0 \cap B(0, M), \ i = 1, \dots, q,$$

such that

$$X_0 \cap B(0, M) \subset \cup_{i=1}^q B^0(x_i, r_{x_i}). \tag{2.11}$$

Set

$$L = \max\{L_{x_i} : i = 1, \dots, q\}, \tag{2.12}$$

$$r = \min\{r_{x_i} : i = 1, \dots, q\}. \tag{2.13}$$

Assume that

$$z \in \cup\{B(x, r) : \ x \in X_0 \cap B(0, M)\}.$$

By the inclusion above, there exists

$$x \in X_0 \cap B(0, M) \cap B(z, r). \tag{2.14}$$

Equations (2.11) and (2.14) imply that there exists

$$i \in \{1, \dots, q\} \tag{2.15}$$

such that

$$x \in B^0(x_i, r_{x_i}). \tag{2.16}$$

It follows from (2.13), (2.14), and (2.16) that

$$z \in B(x_i, 2r_{x_i}). \tag{2.17}$$

In view of (2.9), (2.13), and (2.17),

$$F(z) \subset B(0, L_{x_i}) \subset B(0, L).$$

Lemma 2.12 is proved.

Lemma 2.13 *Let $M > 0$. Then there exist $r, L > 0$ such that for each*

$$z_1, z_2 \in \cup\{B(x, r) : \ x \in X_0 \cap B(0, M)\} \tag{2.18}$$

satisfying

$$\|z_1 - z_2\| \le r \tag{2.19}$$

the inequality

$$|\phi(z_1) - \phi(z_2)| \le L\|z_1 - z_2\|$$

holds.

Proof Since the function ϕ is locally Lipschitz for each $x \in X_0 \cap B(0, M)$, there exist $L_x, r_x > 0$ such that

$$|\phi(z_1) - \phi(z_2)| \le L_x\|z_1 - z_2\| \text{ for all } z_1, z_2 \in B(x, 4r_x). \tag{2.20}$$

Since the set $X_0 \cap B(0, M)$ is compact, there exists a finite set

$$x_i \in X_0 \cap B(0, M), \ i = 1, \dots, q$$

such that

$$X_0 \cap B(0, M) \subset \cup_{i=1}^{q} B^0(x_i, r_{x_i}). \tag{2.21}$$

Set

$$L = \max\{L_{x_i} : i = 1, \dots, q\}, \tag{2.22}$$

$$r = \min\{r_{x_i} : i = 1, \dots, q\}. \tag{2.23}$$

Assume that $z_1, z_2 \in X$ satisfy (2.18) and that

$$\|z_1 - z_2\| \le r. \tag{2.24}$$

By (2.18), there exists

$$x \in X_0 \cap B(0, M) \cap B(z_1, r). \tag{2.25}$$

It follows from (2.24) and (2.25) that

$$\|x - z_1\| \le r, \ \|z_2 - x\| \le \|z_2 - z_1\| + \|z_1 - x\| \le 2r. \tag{2.26}$$

In view of (2.25), there exists $p \in \{1, \dots, q\}$ such that

$$x \in B^0(x_p, r_{x_p}). \tag{2.27}$$

Equations (2.23), (2.26), and (2.27) imply that

$$\|z_i - x_p\| \le \|z_i - x\| + \|x - x_p\|$$

$$\le 2r + r_{x_p} \le 3r_{x_p}. \tag{2.28}$$

By (2.20), (2.22), and (2.28),

$$|\phi(z_1) - \phi(z_2)| \le L_{x_p}\|z_1 - z_2\| \le L\|z_1 - z_2\|.$$

Lemma 2.13 is proved.

Theorem 2.14 *Let* $-\infty < a < b < \infty$, $M, L, r > 0$, *and*

$$F(z) \subset (0, L) \tag{2.29}$$

for each

$$z \in E := \cup\{B(x, z) : x \in X_0 \cap B(0, M)\}.$$

Assume that for each integer $k \geq 1$, $x_k \in Y(a, b)$ *satisfies*

$$x_k(t) \in E, \ t \in [a, b] \tag{2.30}$$

and that for each $t \in [a, b]$,

$$\lim_{k\to\infty} \rho(x_k(t), X_0) = 0. \tag{2.31}$$

Then there exist a subsequence $\{x_{k_i}\}_{i=1}^{\infty}$ *and* $x \in Y(a, b)$ *such that*

$$x_{k_i} \text{ converges to } x \text{ uniformly on } [a, b] \tag{2.32}$$

and

$$x'_{k_i} \text{ converges to } x' \text{ weakly in } L^1(a, b; X) \tag{2.33}$$

as $i \to \infty$.

Proof By (2.3), (2.29), and (2.30), for each integer $k \geq 1$ and almost every $t \in [a, b]$,

$$\|x'_k(t)\| \leq L. \tag{2.34}$$

It follows from (2.30), (2.31), and compactness of bounded, closed subsets of X_0 that for every $t \in [a, b]$, the sequence $\{x_k(t)\}_{k=1}^{\infty}$ is relatively compact in X. Together with (2.34) and Theorem 2.3, this implies that there exist a subsequence $\{x_{k_i}\}_{i=1}^{\infty}$ and $x \in W^{1,1}(a, b; X)$ such that (2.32) and (2.33) hold. Theorem 2.5 implies that $x'(t) \in F(x(t))$ for almost every $t \in [a, b]$. Thus $x \in Y(a, b)$. Theorem 2.14 is proved.

2.5 Proof of Proposition 2.7

Clearly, (1) implies (2) and (2) implies (3). In order to complete the proof of the proposition, it is sufficient to show that (3) implies (1).

Assume that

$$x_k \in Y(0, T_k), \ k = 1, 2, \ldots, \ \lim_{k\to\infty} T_k = \infty,$$

$$\phi(x_k(0)) \leq M, \ k = 1, 2, \ldots. \tag{2.35}$$

We may assume that

$$T_k > k, \ k = 1, 2, \ldots. \tag{2.36}$$

Proposition 2.6 and (2.35) imply that

$$\phi(x_k(t)) \le \phi(x_k(0)) \le M \text{ for all integers } k \ge 1 \text{ and all } t \in [0, T_k]. \tag{2.37}$$

Set

$$C = \{x_k(t) : \ k = 1, 2, \ldots, \ t \in [0, T_k]\}. \tag{2.38}$$

In view of (2.4) and (2.37), the set C is bounded, and there exists $M_0 > 1$ such that

$$C \subset B(0, M_0).$$

Lemmas 2.12 and 2.13 imply that there exist $L > 1, r \in (0, 1)$ such that the following properties hold:

(i) For each

$$z \in \cup\{B(x, r) : \ x \in X_0 \cap B(0, M_0 + 1)\},$$

we have

$$F(z) \subset B(0, L).$$

(ii) For each

$$z_1, z_2 \in \cup\{B(x, r) : \ x \in X_0 \cap B(0, M_0 + 1)\}$$

satisfying

$$\|z_1 - z_2\| \le r,$$

we have

$$|\phi(z_1) - \phi(z_2)| \le L\|z_1 - z_2\|.$$

Assumption (A4) and Eqs. (2.36) and (2.38) imply that for every natural number k there exists an integer $n(k) > k$ such that for each integer $n \ge n(k)$ and each $t \in [n(k), T_n]$,

$$\rho(x_n(t), X_0) \le (2k)^{-1}r. \tag{2.39}$$

For each integer $k \geq 1$, define $y_k \in Y(0, n(k))$ by

$$y_k(t) = x_{2n(k)}(t + n_k), \ t \in [0, n(k)]. \tag{2.40}$$

It follows from (2.38)–(2.40) that for each integer $k \geq 1$ and each $t \in [0, n_k]$,

$$\rho(y_k(t), X_0) = \rho(x_{2n(k)}(n(k) + t), X_0) \leq (2k)^{-1} r,$$

$$\|y_k(t)\| \leq M_0,$$

and

$$\rho(y_k(t), X_0 \cap B(0, M_0 + 2^{-1})) \leq (2k)^{-1} r. \tag{2.41}$$

Property (i) and (2.41) imply that for each $t \geq 0$,

$$\lim_{k \to \infty} \rho(y_k(t), X_0 \cap B(0, M_0 + 1)) = 0. \tag{2.42}$$

By Theorem 2.14, property (i), (2.41) and (2.42), extracting subsequences and using the diagonalization process, we obtain that there exist a strictly increasing sequence of natural numbers $\{k_j\}_{j=1}^{\infty}$ and $x \in Y(0, \infty)$ such that y_{k_j} converges to x uniformly on $[0, p]$ as $j \to \infty$ for any integer $p \geq 1$. In view of (2.42), $x(t) \in X_0$ for all $t \geq 0$. Thus (1) holds and Proposition 2.7 is proved.

2.6 Proof of Proposition 2.8

Let $x \in Y(-\infty, \infty)$. In view of (A4),

$$\lim_{t \to \infty} \rho(x(t), X_0) = 0. \tag{2.43}$$

Proposition 2.6 implies that

$$\phi(x(t)) \leq \phi(x(0)), \ t \in [0, \infty). \tag{2.44}$$

From Equations (2.4) and (2.44), there exists $M_0 > 0$ such that

$$\|x(t)\| \leq M_0, \ t \in [0, \infty). \tag{2.45}$$

By (2.43) and (2.45),

$$\lim_{t \to \infty} \rho(x(t), X_0 \cap B(0, M_0 + 1)) = 0. \tag{2.46}$$

In view of (2.46), for each integer $k \geq 1$, there exists

$$z_k \in X_0 \cap B(0, M_0 + 1) \tag{2.47}$$

and a number $t_k \geq k$ such that

$$\|x(t_k) - z_k\| \leq k^{-1}. \tag{2.48}$$

Since the set $X_0 \cap B(0M_0 + 1)$ is compact, there exists a convergent subsequence $\{z_{k_i}\}_{i=1}^{\infty}$. By (2.48),

$$\lim_{i\to\infty} x(t_{k_i}) = \lim_{i\to\infty} z_{k_i} \in \Omega(F)$$

and

$$\Omega(F) \neq \emptyset.$$

Let $z \in \Omega(F)$ and $\epsilon > 0$. There exists $x \in Y(0, \infty)$ such that

$$\liminf_{t\to\infty} \|z - x(t)\| \leq \epsilon. \tag{2.49}$$

Assumption (A4) implies that

$$\lim_{t\to\infty} \rho(x(t), X_0) = 0.$$

Together with (2.49), this implies that

$$\rho(z, X_0) \leq 2\epsilon.$$

Since the set X_0 is closed and ϵ is an arbitrary positive number, we conclude that $z \in X_0$ and $\Omega(F) \subset X_0$. Proposition 2.8 is proved.

2.7 Proof of Theorem 2.11

Assume that the theorem does not hold. Then for each integer $k \geq 1$, there exists $x_k \in Y(0, k)$ such that

$$\phi(x_k(0)) \leq M, \tag{2.50}$$

$$\inf\{\rho(x(t), \Omega(F)) : t \in [0, k]\} > \epsilon. \tag{2.51}$$

Proposition 2.6 and (2.49) imply that

$$\phi(x_k(t)) \le M,\ t \in [0, k], k = 1, 2, \dots. \tag{2.52}$$

Equations (2.4) and (2.52) imply that there exists $M_1 > 0$ such that

$$\|x_k(t)\| \le M_1,\ t \in [0, k],\ k = 1, 2, \dots. \tag{2.53}$$

Lemma 2.12 implies that there exist $r \in (0, 1)$ and $L > 1$ such that the following property holds:

(i) For each $z \in \cup\{B(x, r) :\ x \in X_0 \cap B(0, M_1 + 1)\}$, we have

$$F(z) \subset B(0, L).$$

Assumption (A4) and (2.53) imply that for each integer $k \ge 1$ there exists an integer $n(k) > k$ such that for each integer $n \ge n(k)$ and each $t \in [n(k), n]$ we have

$$\rho(x_n(t), X_0) \le (2k)^{-1} r. \tag{2.54}$$

For each integer $k \ge 1$, define $y_k \in Y(0, n(k))$ by

$$y_k(t) = x_{2n(k)}(n(k) + t),\ t \in [0, n(k)]. \tag{2.55}$$

By (2.53)–(2.55), for each integer $k \ge 1$ and each $t \in [0, n_k]$,

$$\rho(y_k(t), X_0) = \rho(x_{2n(k)}(n(k) + t), X_0) \le (2k)^{-1} r, \tag{2.56}$$

$$\|y_k(t)\| \le M_1,$$

$$\rho(y_k(t), X_0 \cap B(0, M_1 + 2^{-1})) \le (2k)^{-1} r. \tag{2.57}$$

In view of (2.57), for each integer $t \ge 0$,

$$\lim_{k \to \infty} \rho(y_k(t), X_0 \cap B(0, M_1 + 1)) = 0. \tag{2.58}$$

By Theorem 2.14, property (i), (2.57) and (2.58), extracting subsequences and using the diagonalization process, we obtain that there exist a subsequence s $\{y_{k_i}\}_{i=1}^{\infty}$ and $y \in Y(0, \infty)$ such that

$$y_{k_i} \text{ converges to } y \text{ uniformly on } [0, p] \tag{2.59}$$

as $i \to \infty$ for any integer $p \ge 1$. It follows from (2.51), (2.55), and (2.59) that for every integer $t \ge 0$,

$$\rho(y(t), \Omega(F)) = \lim_{j\to\infty} \rho(y_{k_j}(t), \Omega(F)) \geq \epsilon. \tag{2.60}$$

Equations (2.58) and (2.59) imply that for each $t \geq 0$,

$$y(t) \in X_0 \cap B(0, M_1 + 1),$$

which is a compact set. Therefore, there exists a strictly increasing sequence of natural numbers $\{p_j\}_{j=1}^{\infty}$ such that there exists

$$\lim_{j\to\infty} y(p_j) \in \Omega(F).$$

This contradicts (2.60). The contradiction we have reached proves Theorem 2.11.

2.8 An Example

Assume that

$$X = R^n_+ := \{x = (x_1, \dots, x_n) \in R^n : \ x_i \geq 0, \ i = 1, \dots, n\},$$

$a : X \to 2^X \setminus \{\emptyset\}$, the graph of a is convex,

$$a(\lambda x) = \lambda a(x) \text{ for each } x \in X \text{ and each } \lambda \geq 0,$$

$$a(x) = (a(x) - R^n_+) \cap R^n_+, \ x \in X.$$

This set-valued map describes the von Neumann–Gale model of economic dynamics [75, 95, 97, 117] and is a convex process. It is known that there exists a triplet $(\widehat{\alpha}, \widehat{x}, \widehat{p})$ such that $\widehat{\alpha} \geq 0, \widehat{x}, \widehat{p} \in R^n_+$ satisfy

$$\|\widehat{x}\| = 1, \ \|\widehat{p}\| = 1, \ \widehat{\alpha}\widehat{x} \in a(\widehat{x}), \ \langle\widehat{\alpha}\widehat{p}, x\rangle \geq \widehat{p}, y\rangle$$

for each $x \in R^n_+$ and each $y \in a(x)$. The study of this mapping and the corresponding model can be reduced to the case when $\widehat{\alpha} = 1$. In this case,

$$F(x) = a(x) - x, \ x \in R^n_+, \ F(x) = \emptyset, x \in R^n \setminus R^n_+,$$

and this is a particular case of set-valued mappings and the corresponding differential inclusions considered in this chapter with

$$\Omega(F) = \{\lambda\widehat{x} : \ \lambda \in [0\infty)\},$$

$$\phi(x) = \sum_{i=1}^{n} |x_i|\widehat{p}_i, \ x \in X.$$

Chapter 3
Discrete-Time Dynamical Systems

In this chapter, we study a discrete-time dynamical system with a Lyapunov function in a metric space induced by a set-valued mapping. For this discrete-time dynamical system, we establish a turnpike result.

3.1 Preliminaries and Main Results

Assume that (X, ρ) is a metric space and $\mathcal{A}$ is a nonempty, closed subset of the metric space $X \times X$ equipped with the metric $\rho_1 : X \times X \to [0, \infty)$ defined by

$$\rho_1((x_1, x_2), (y_1, y_2)) = \rho(x_1, y_1) + \rho(x_2, y_2), \ \ x_1, x_2, y_1, y_2 \in X.$$

For each $x \in X$ and each $r > 0$, set

$$B(x, r) = \{y \in X : \ \rho(x, y) \le r\} \text{ and } B^0(x, r) = \{y \in X : \ \rho(x, y) < r\}.$$

Set

$$X_{\mathcal{A}} = \{x \in X : \ \{x\} \times X \cap \mathcal{A} \neq \emptyset\}. \tag{3.1}$$

For each $x \in X_{\mathcal{A}}$, set

$$a(x) = \{y \in X : \ (x, y) \in \mathcal{A}\}. \tag{3.2}$$

For each $x \in X$ and each pair of nonempty sets $A, B \subset X$, define

$$\rho(x, B) = \inf\{\rho(x, y) : \ y \in B\}$$

A. J. Zaslavski, *Turnpike Phenomenon in Metric Spaces*, Springer Optimization and Its Applications 201, https://doi.org/10.1007/978-3-031-27208-0_3

and

$$\mathrm{dist}(A, B) = \sup\{\rho(x, B) : x \in A\}. \tag{3.3}$$

For each nonempty subset $D \subset X$, set

$$a(D) = \cup\{a(x) : x \in D\}.$$

Fix $\widehat{\theta} \in X$. Assume that $\phi : X \to R^1$ is a continuous function such that

$$\phi(y) \leq \phi(x) \text{ for each } (x, y) \in \mathcal{A}. \tag{3.4}$$

Assume that X_0 is nonempty, closed subset of X such that for each $r > 0$ the set

$$X_0 \cap B(\widehat{\theta}, r) \text{ is compact .} \tag{3.5}$$

Let $T_2 > T_1$ be integers. A sequence $\{x_t\}_{t=T_1}^{T_2} \subset X$ is called a trajectory of a (or just a trajectory if the mapping a is understood) if $(x_t, x_{t+1}) \in \mathcal{A}$ for all integers $t \in \{T_1, \ldots, T_2 - 1\}$.

A sequence $\{x_t\}_{t=T_1}^{\infty} \subset X$ is called a trajectory of a (or just a trajectory if the mapping a is understood) if $(x_t, x_{t+1}) \in \mathcal{A}$ for all nonnegative integers $t \geq T_1$.

Denote by $Y(T_1, T_2)$ the collection of all trajectories $\{y_t\}_{t=T_1}^{T_2}$ and by $Y(T_1, \infty)$ the set of all trajectories $\{y_t\}_{t=T_1}^{\infty}$.

We assume that the following assumption holds:

(A) For each bounded, nonempty set C, there exists $M_C > 0$, and for each $\epsilon > 0$, there exists a natural number $n(C, \epsilon)$ such that for each integer $n \geq n(C, \epsilon)$,

$$\mathrm{dist}(\{x \in X : \text{ there is a trajectory } \{x_t\}_{t=0}^{n} \text{ such that } x_0 \text{ and } x_n = x\},$$

$$X_0 \cap B(\widehat{\theta}, M_C)\}) \leq \epsilon.$$

In Sect. 3.2, we prove the following result.

Proposition 3.1 *The following properties are equivalent:*

(1) There exists a trajectory $\{x_t\}_{t=0}^{\infty} \subset X_0$.
(2) There exists a trajectory $\{x_t\}_{t=0}^{\infty}$.
(3) There exists $M > 0$, *and for each integer* $n \geq 1$, *there exists a trajectory* $\{x_t\}_{t=0}^{n}$ *satisfying* $\rho(x_0, \widehat{\theta}) \leq M$.

In the sequel, we suppose that there exists a trajectory $\{x_t\}_{t=0}^{\infty}$.

Define

$$\Omega(a) = \{z \in X : \text{ for every } \epsilon > 0 \text{ there exists a trajectory } \{x_t\}_{t=0}^{\infty}$$

$$\text{for which } \liminf_{t \to \infty} \rho(z, x_t) \leq \epsilon\}. \tag{3.6}$$

Clearly, $\Omega(a)$ is a closed subset of the metric space (X, ρ).

In Sect. 3.3, we prove the following two results.

Proposition 3.2 *$\Omega(a) \neq \emptyset$ and $\Omega(a) \subset X_0$.*

Proposition 3.3 *For every trajectory $\{x_t\}_{t=0}^{\infty}$,*

$$\lim_{t\to\infty} \rho(x(t), \Omega(a)) = 0.$$

It is not difficult to see that the following result holds.

Proposition 3.4 *Assume that $B \subset X$ be a nonempty, closed set such that for every trajectory $\{x_t\}_{t=0}^{\infty}$,*

$$\lim_{t\to\infty} \rho(x(t), B) = 0.$$

Then $\Omega(a) \subset B$.

The following theorem will be proved in Sect. 3.4.

Theorem 3.5 *Let ϵ, M be positive real numbers. Then there is a positive integer T such that for every trajectory $\{x_t\}_{t=0}^{T}$ that satisfies*

$$\rho(x_0, \widehat{\theta}) \leq M$$

the inequality

$$\min\{\rho(x_t, \Omega(a)) : t = 0, \ldots, T\} \leq \epsilon$$

holds.

The next theorem is proved in Sect. 3.5.

Theorem 3.6 *The following properties are equivalent:*

(1) If a bounded sequence $\{x_t\}_{t=-\infty}^{\infty} \subset X$ satisfies

$$x_{t+1} \in a(x_t) \text{ and } \phi(x_{t+1}) = \phi(x_t)$$

for all integers t, then the inclusion $x_t \in \Omega(a)$ holds for all integers t.

(2) If a bounded sequence $\{x_t\}_{t=-\infty}^{\infty} \subset X_0$ satisfies

$$x_{t+1} \in a(x_t) \text{ and } \phi(x_{t+1}) = \phi(x_t)$$

for all integers t, then the inclusion $x_t \in \Omega(a)$ holds for all integers t.

(3) For each $M, \epsilon > 0$, there exist $\delta > 0$ and a natural number L such that for each integer $T > 2L$ and each trajectory $\{x_t\}_{t=0}^{T}$ that satisfy

$$\rho(x_0, \widehat{\theta}) \le M \text{ and } \phi(x_0) - \phi(x_T) \le \delta$$

the inequality

$$\rho(x_t, \Omega(a)) \le \epsilon$$

holds for all integers $t = L, \dots, T - L$.

Assume that properties of Theorem 3.6 hold. The next theorem is a weak turnpike result. It is proved in Sect. 3.6.

Theorem 3.7 *Let* $\epsilon, M > 0$. *Then there exist natural numbers* L, Q *such that for every natural number* $T > L$ *and every trajectory* $\{x_t\}_{t=0}^T$ *that satisfy*

$$\rho(x_0, \widehat{\theta}) \le M,$$

there exist an integer $q \in [1, Q]$ *and finite sequences of integers* $\{a_i\}_{i=1}^q, \{b_i\}_{i=1}^q \subset [0, T]$ *such that*

$$a_i \le b_i,\ i = 1, \dots, q,\ a_{i+1} > b_i,\ i \in \{1, \dots, q\} \setminus \{q\},$$

$$\rho(x_t, \Omega(a)) \le \epsilon,\ t \in \{a_i, \dots, b_i\},\ i = 1, \dots, q,$$

and

$$Card(\{0, \dots, T\} \setminus \cup_{i=1}^q \{a_i, \dots, b_i\}) \le L.$$

It is not difficult to see that the set-valued mapping a from Sect. 2.8 and the discrete-time dynamical system induced by a satisfy all the assumptions posed in this section.

It should be mentioned that a particular case of our system, when a is a self-mapping of a compact metric space, was introduced in [96]. Several results related to this particular case are presented in [149].

3.2 Proof of Proposition 3.1

In the proof of Proposition 3.1, we use the following simple auxiliary result that is proved easily.

Lemma 3.8 *Assume that* C *is a nonempty, compact set in* X. *Then every sequence* $\{x_i\}_{i=0}^\infty \subset X$, *which satisfies* $\lim_{i\to\infty} \rho(x_i, C) = 0$, *has a subsequence that converges to a point of* C.

Proof of Proposition 3.1 Evidently, (1) implies (2) and (2) implies (3). Assume that (3) holds. There exist $M > 0$ and trajectories $\{x_t^{(n)}\}_{t=0}^n \subset X$, $n = 1, 2, \dots$, such that

$$\rho(\widehat{\theta}, x_0^{(n)}) \le M, \ n = 1, 2, \dots. \tag{3.7}$$

By (3.7) and assumption (A), there exists $M_0 > 0$ such that for every natural number k there exists an integer $n(k) > k$ such that for each integer $n \ge n(k)$ and each integer $t \in [n(k), n]$ we have

$$\rho(x_t^{(n)}, X_0 \cap B(\widehat{\theta}, M_0)) \le k^{-1}. \tag{3.8}$$

For each integer $k \ge 1$, define a trajectory

$$y_t^{(k)} = x_{t+n(k)}^{2n(k)}, \ t = 0, \dots, n(k). \tag{3.9}$$

Equations (3.8) and (3.9) imply that for each integer $k \ge 1$ and each $t \in \{0, \dots, n(k)\}$,

$$\rho(y_t^{(k)}, X_0 \cap B(\widehat{\theta}, M_0)) = \rho(x_{t+n(k)}^{2n(k)}, X_0 \cap B(\widehat{\theta}, M_0)) \le k^{-1}. \tag{3.10}$$

In view of (3.10), for each integer $t \ge 0$,

$$\lim_{t \to \infty} \rho(y_t^{(k)}, X_0 \cap B(\widehat{\theta}, M_0)) = 0. \tag{3.11}$$

By Lemma 3.8 and (3.11), extracting subsequences and using the diagonalization process, we obtain that there exist a strictly increasing sequence of natural numbers $\{k_j\}_{j=1}^\infty$ such that for each integer $t \ge 0$ there exists

$$x_t = \lim_{j \to \infty} y_t^{(k_j)} \in X_0 \cap B(\widehat{\theta}, M_0).$$

Since the set $\mathcal{A}$ is closed, Eq. (3.9) implies that $\{x_t\}_{t=0}^\infty$ is a trajectory. Thus property (1) holds and (3) implies (1). This completes the proof of Proposition 3.1.

3.3 Proofs of Propositions 3.2 and 3.3

Let $\{x_t\}_{t=0}^\infty$ be a trajectory. Assumption (A) implies that there exists $M_0 > 0$ such that

$$\lim_{t \to \infty} \rho(x_t, X_0 \cap B(\widehat{\theta}, M_0)) = 0. \tag{3.12}$$

Lemma 3.8 and (3.5) and (3.12) imply that the sequence $\{x_t\}_{t=0}^{\infty}$ has a subsequence that converges to a point of $\Omega(a)$. Hence,

$$\Omega(a) \neq \emptyset.$$

Let $z \in \Omega(a)$ and $\epsilon > 0$. There exists a trajectory $\{x_t\}_{t=0}^{\infty}$ such that

$$\liminf_{t\to\infty} \rho(z, x_t) \leq \epsilon.. \tag{3.13}$$

In view of (A),

$$\lim_{t\to\infty} \rho(x_t, X_0) = 0.$$

By the equation above and (3.13),

$$\rho(z.X_0) \leq \epsilon.$$

Since the set X_0 is closed and ϵ is an arbitrary positive number, we conclude that $z \in X_0$ and $\Omega(a) \subset X_0$. Proposition 3.2 is proved.

Proposition 3.3 follows from Lemma 3.8 and assumption (A).

3.4 Proof of Theorem 3.5

Assume that Theorem 3.5 does not hold. Then for every integer n, there exists a trajectory $\{x_t^{(n)}\}_{t=0}^{n}$ such that

$$\rho(\widehat{\theta}, x_0^{(n)}) \leq M, \tag{3.14}$$

$$\rho(x_t^{(n)}, \Omega(a)) > \epsilon,\ t = 0, , \ldots, n. \tag{3.15}$$

Assumption (A) and (3.14) imply that there exists $M_1 > 0$ such that for every integer $k \geq 1$ there exists an integer $n(k) \geq k$ such that for each integer $n \geq n(k)$,

$$\rho(x_t^{(n)}, X_0 \cap B(\theta, M_1)) \leq k^{-1},\ t = n(k), \ldots, n. \tag{3.16}$$

For each integer $k \geq 1$, define a trajectory

$$y_t^{(k)} = x_{t+n(k)}^{(2n(k))},\ t = 0, \ldots, n(k). \tag{3.17}$$

It follows from (3.15)–(3.17) that for each integer $k \geq 1$ and $t = 0, \ldots, n(k)$,

$$\rho(y_t^{(k)}, \Omega(a)) > \epsilon, \tag{3.18}$$

$$\rho((y_t^{(k)}, X_0 \cap B(\widehat{\theta}, M_1)) \leq k^{-1}. \tag{3.19}$$

By Lemma 3.8 and (3.19), extracting subsequences and using the diagonalization process we obtain that there exists a strictly increasing sequence of natural numbers $\{k_j\}_{j=1}^{\infty}$ such that for each integer $t \geq 0$ there exists

$$y_t = \lim_{j\to\infty} y_t^{(k_j)}. \tag{3.20}$$

Since the set $\mathcal{A}$ is closed, it follows from (3.20) that $\{y_t\}_{t=0}^{\infty}$ is a trajectory. Equations (3.18)–(3.20) imply that for every integer $t \geq 0$,

$$\rho(y_t, \widehat{\theta}) \leq M_1, \ \rho(y_t, \Omega(a)) \geq \epsilon, \tag{3.21}$$

$$y_t \in X_0 \cap B(\widehat{\theta}, M_1).$$

Since in view of (3.5), the set $X_0 \cap B(\widehat{\theta}, M_1)$ is compact, there exists a subsequence $\{y_{t_i}\}_{i=1}^{\infty}$ that converges to a point of $\Omega(a)$. This contradicts (3.21). The contradiction we have reached proves Theorem 3.5.

3.5 Proof of Theorem 3.6

Clearly, (1) implies (2) and (3) implies (1). Therefore, in order to prove the theorem, it is sufficient to show that (2) implies (3).

Assume that (2) holds and (3) does not hold. Then there exist ϵ, $M > 0$ such that for each integer $k \geq 1$ there exist an integer $T_k > 2k$ and a trajectory $\{x_t^{(k)}\}_{t=0}^{T_k}$ such that

$$\rho(x_0^{(k)}, \widehat{\theta}) \leq M, \tag{3.22}$$

$$|\phi(x_0^{(k)}) - \phi(x_{T_k}^{(k)})| \leq k^{-1} \tag{3.23}$$

and

$$\max\{\rho(x_t^{(k)}, \Omega(a)) : \ t = k, \ldots, T_k - k\} > \epsilon. \tag{3.24}$$

By (A) and (3.22), there exists $M_1 > 0$ such that for each integer $p \geq 1$ there exists an integer $k(p) > p$ such that for each integer $k \geq k(p)$ and all $t = k(p), \ldots, T_k$ we have

$$\rho(x_t^{(k)}, X_0 \cap B(\theta, M_1)) \le p^{-1}. \quad (3.25)$$

Let $p \ge 1$ be an integer. In view of (3.24), there exists an integer

$$\tau_p \in \{2k(p), \ldots, T_{2k(p)} - 2k(p)\}$$

such that

$$\rho(x_{\tau_p}^{(2k(p))}, \Omega(a)) > \epsilon. \quad (3.26)$$

Define a trajectory

$$y_t^{(p)} = x_{t+\tau_p}^{(2k(p))}, \; t = -\tau_p, \ldots, T_{2k(p)} - \tau_p. \quad (3.27)$$

Equations (3.23) and (3.25)–(3.27) imply that

$$\rho(y_0^{(p)}, \Omega(a)) = \rho(x_{\tau_p}^{(2k(p))}, \Omega(a)) > \epsilon \quad (3.28)$$

and that for all $t = -\tau_p, \ldots, T_{2k(p)} - \tau_p$,

$$\rho(y_t^{(p)}, X_0 \cap B(\widehat{\theta}, M_1)) = \rho(x_{t+\tau_p}^{(2k(p))}, X_0 \cap B(\widehat{\theta}, M_1)) \le p^{-1}, \quad (3.29)$$

$$\phi(y_{-\tau_p}^{(p)}) - \phi(y_{T_{2k(p)}-\tau_p}^{(p)}) \le (2k(p))^{-1} \le (2p)^{-1}. \quad (3.30)$$

In view of (3.29), for each integer t,

$$\lim_{p\to\infty} \rho(y_t^{(p)}, X_0 \cap B(\widehat{\theta}, M_1)) = 0. \quad (3.31)$$

By Lemma 3.8, (3.5) and (3.31), extracting subsequences and using the diagonalization process, we obtain that there exist a strictly increasing sequence of natural numbers $\{p_i\}_{i=1}^{\infty}$ such that for each integer t there exists

$$y_t = \lim_{i\to\infty} y_t^{(p_i)}. \quad (3.32)$$

Since the set $\mathcal{A}$ is closed, it follows from (3.30), (3.31), and (3.32) that for each integer t we have

$$\phi(y_t) = \phi(y_{t+1}), \; y_{t+1} \in a(y_t), \; y_t \in X_0.$$

Property (2) implies that

$$y_t \in \Omega(a) \text{ for each integer } t.$$

On the other hand, it follows from (3.28) and (3.32) that

$$\rho(y_0, \Omega(a)) \geq \epsilon.$$

The contradiction we have reached completes the proof of Theorem 3.6.

3.6 Proof of Theorem 3.7

We may assume without loss of generality that $\epsilon < 1$ and $M > 0$. Assumption (A) implies that there exist $M_1 > M$ and an integer $L_1 \geq 1$ such that the following property holds:

(a) For each integer $n \geq L_1$ and each trajectory $\{x_t\}_{t=0}^n$ satisfying $\rho(x_0, \widehat{\theta}) \leq M$, we have

$$\rho(x_t, X_0 \cap B(\widehat{\theta}, M_1)) \leq 1, \ |\phi(x_t)| \leq M_1$$

for all $t = L_1, \dots, n$.

Property (3) of Theorem 3.6 implies there exist $\delta \in (0, 1)$ and a natural number L_0 such that the following property holds:

(b) For each integer $T > 2L_0$ and each trajectory $\{x_t\}_{t=0}^T$ that satisfy

$$\rho(x_0, \widehat{\theta}) \leq M_1 + 2 \text{ and } \phi(x_0) - \phi(x_T) \leq \delta$$

the inequality

$$\rho(x_t, \Omega(a)) \leq \epsilon$$

holds for all integers $t = L_0, \dots, T - L_0$.

Choose integers

$$Q > 2M_1\delta^{-1} + 2 \tag{3.33}$$

and

$$L > (4L_0 + 8)(2\delta^{-1}M_1 + 2) + L_1 + 4. \tag{3.34}$$

Suppose that $T > L$ is a natural number and that a trajectory $\{x_t\}_{t=0}^T$ satisfies

$$\rho(x_0, \widehat{\theta}) \leq M, \ \phi(x_0) \leq M. \tag{3.35}$$

Property (a) and (3.35) imply that

$$\rho(x_t, \widehat{\theta}) \leq M_1 + 2, \; |\phi(x_t)| \leq M_1 \tag{3.36}$$

for all $t = L_1, \dots, T$.

By induction, we define a strictly increasing finite sequence $t_i \in \{0, \dots, T\}$, $i = 0, \dots, q$. Set

$$t_0 = L_1.$$

If

$$\phi(x_T) \geq \phi(x_{t_0}) - \delta,$$

then set $t_1 = T$ and complete to construct the sequence.

Assume that

$$\phi(x_T) < \phi(x_{t_0}) - \delta.$$

Evidently, there is an integer $t_1 \in (t_0, T]$ satisfying

$$\phi(x_{t_1}) < \phi(x_0) - \delta$$

and that if an integer S satisfies $t_0 < S < t_1$, then

$$\phi(x_S) \geq \phi(x_0) - \delta.$$

If $t_1 = T$, then we complete to construct the sequence.

Assume that $k \geq 1$ is an integer and that we defined a strictly increasing sequence $t_0, \dots, t_k \in \{0, 1, \dots, T\}$ such that

$$t_0 = L_1, \; t_k \leq T$$

and that for each $i \in \{0, \dots, k-1\}$,

$$\phi(x_{t_{i+1}}) < \phi(x_{t_i}) - \delta,$$

and if an integer S satisfies $t_i < S < t_{i+1}$, then

$$\phi(x_S) \geq \phi(x_{t_i}) - \delta.$$

(It is not difficult to see that the assumption is true with $k = 1$).

If $t_k = T$, then we complete to construct the sequence. Assume that $t_k < T$. If

$$\phi(x_T) \geq \phi(x_{t_k}) - \delta,$$

then we set $t_{k+1} = T$ and complete to construct the sequence.

Assume that

$$\phi(x_T) < \phi(x_{t_k}) - \delta.$$

Evidently, there is a natural number $t_{k+1} \in (t_k, T]$ for which

$$\phi(x_{t_{k+1}}) < \phi(x_{t_k}) - \delta$$

and that if an integer S satisfies $t_k < S < t_{k+1}$, then

$$\phi(x_S) \geq \phi(x_{t_k}) - \delta.$$

Evidently, the assumption made for k is true for $k + 1$ too. Therefore by induction, we constructed the strictly increasing finite sequence of integers $t_i \in [0, T]$, $i = 0, \ldots, q$, such that

$$t_0 = L_1, \ t_q = T \tag{3.37}$$

and that for every i satisfying $0 \leq i < q - 1$,

$$\phi(x_{t_{i+1}}) < \phi(x_{t_i}) - \delta \tag{3.38}$$

and for each $i \in \{0, \ldots, q - 1\}$ and each integer S satisfying $t_i < S < t_{i+1}$, we have

$$\phi(x_S) \geq \phi(x_{t_i}) - \delta. \tag{3.39}$$

By (3.4), (3.33),(3.35), (3.36), and (3.38),

$$2M_1 \geq \phi(x_{t_0}) - \phi(x_{t_{q-1}})$$

$$\geq \sum \{\phi(x_{t_i}) - \phi(x_{t_{i+1}}) : \ i \text{ is an integer}, \ 0 \leq i \leq q - 2\} \geq \delta(q - 1)$$

and

$$q \leq 1 + 2\delta^{-1}M_1 < Q. \tag{3.40}$$

Define

$$E = \{i \in \{0, \ldots, q - 1\} : \ t_{i+1} - t_i \geq 2L_0 + 4\}. \tag{3.41}$$

Let

$$i \in E. \tag{3.42}$$

By (3.41) and (3.42),

$$t_{i+1} - 1 - t_i \geq 2L_0 + 3. \tag{3.43}$$

Equations (4.43) and (4.47) imply that

$$\phi(x_{t_{i+1}-1}) \geq \phi(x_{t_i}) - \delta. \tag{3.44}$$

Equations (3.36), (3.37), (3.43), (3.44) and property (b) applied to the trajectory $\{x_t\}_{t=t_i}^{t_{i+1}-1}$ imply that

$$\rho(x_t, \Omega(a)) \leq \epsilon,\ t = t_i + L_0, \ldots, t_{i+1} - 1 - L_0. \tag{3.45}$$

Set

$$a_i = t_i + L_0,\ b_i = t_{i+1} - L_0 - 1.$$

By (3.37) and (3.45),

$$\{0, \ldots, T\} \setminus \cup_{i \in E}\{a_i, \ldots, b_i\}$$

$$\subset \cup\{\{t_i, \ldots, t_{i+1}\} : i \in \{0, \ldots, q-1\} \setminus E\} \cup \{0, \ldots, L_1 - 1\}$$

$$\cup \{\{t_i, \ldots, t_i + L_0 - 1\} \cup \{t_{i+1} - L_0, \ldots, t_{i+1}\} : i \in E\}. \tag{3.46}$$

It follows from (3.40), (3.41), (3.44), and (3.46) that

$$\mathrm{Card}(\{0, \ldots, T\} \setminus \cup_{i \in E}\{a_i, \ldots, b_i\})$$

$$\leq q(2L_0 + 5) + (2L_0 + 2)q + L_1 = q(4L_0 + 7) + L_1$$

$$\leq (4L_0 + 7)(1 + 2M_1\delta^{-1}) + L_1 < L.$$

Theorem 3.7 is proved.

Chapter 4
Continuous-Time Dynamical Systems

In this chapter, we study a class of continuous-time dynamical systems that is an analog of the class of discrete-time dynamical systems considered in Chap. 3. This class also contains the class of dynamical systems studied in Chap. 2.

4.1 Preliminaries and Main Results

Assume that (X, ρ) is a metric space. For each $x \in X$ and each $r > 0$, set

$$B(x, r) = \{y \in X : \rho(x, y) \le r\} \text{ and } B^0(x, r) = \{y \in X : \rho(x, y) < r\}.$$

For each $x \in X$ and each pair of nonempty sets $A, B \subset X$, define

$$\rho(x, B) = \inf\{\rho(x, y) : y \in B\}$$

and

$$\mathrm{dist}(A, B) = \sup\{\rho(x, B) : x \in A\}.$$

Assume that for each pair of real numbers $T_1 < T_2$, $Y(T_1, T_2)$ is a nonempty set of functions $x : [T_1, T_2] \to X$ such that the following properties hold:

(A1) For each pair of numbers $T_1 < T_2$, each $S_1, S_2 \in [T_1, T_2]$ satisfying $S_1 < S_2$, and each $x \in Y(T_1, T_2)$, the restriction of x to $[S_1, S_2]$ belongs to $Y(S_1, S_2)$.

(A2) For each pair of real numbers $T_1 < T_2$, each $S \in R^1$, and each $x \in Y(T_1, T_2)$, the function $x(t + S)$, $t \in [T_1, T_2]$, belongs to $Y(T_1 + S, T_2 + S)$.

(A3) For each triplet of numbers $T_1 < T_2 < T_3$, each $x_1 \in Y(T_1, T_2)$, and each $x_2 \in Y(T_2, T_3)$ satisfying $x_1(T_2) = x_2(T_2)$, the function $x_3 : [T_1, T_3] \to X$ defined by

A. J. Zaslavski, *Turnpike Phenomenon in Metric Spaces*, Springer Optimization and Its Applications 201, https://doi.org/10.1007/978-3-031-27208-0_4

$$x_3(t) = x_1(t),\ t \in [T_1, T_2],\ x_3(t) = x_2(t),\ t \in (T_2, T_3]$$

belongs to $Y(T_1, T_3)$.

For each pair of numbers $T_1 < T_2$, elements of the set $Y(T_1, T_2)$ are called trajectories. Let $T_1 \in R^1$. A function $x : [T_1, \infty) \to X$ is a trajectory if for every $T_2 > T_1$ the restriction of x to $[T_1, T_2]$ belongs to $Y(T_1, T_2)$. The set of all trajectories $x : [T_1, \infty) \to X$ is denoted by $Y(T_1, \infty)$.

Denote by $Y(R^1) = Y(-\infty, \infty)$ the set of all functions $x : R^1 \to X$ such that for each pair of numbers $T_1 < T_2$ the restriction of x to $[T_1, T_2]$ belongs to $Y(T_1, T_2)$. The elements of $Y(R^1)$ are called trajectories too.

Fix $\widehat{\theta} \in X$.

Assume that $\phi : X \to R^1$ is a continuous function such that for each pair of numbers $T_1 < T_2$, each $x \in Y(T_1, T_2)$, and each $S_1, S_2 \in [T_1, T_2]$ satisfying $S_1 < S_2$,

$$\phi(x(S_2)) \leq \phi(x(S_1)). \tag{4.1}$$

Assume that X_0 is nonempty, closed subset of X such that for each $r > 0$ the set

$$X_0 \cap B(\widehat{\theta}, r) \text{ is compact} \tag{4.2}$$

and that the following assumptions hold:

(A4) For each pair of numbers $T_2 > T_1$, each $M > 0$, and each sequence $\{x_n\}_{n=1}^{\infty} \subset Y(T_1, T_2)$ that satisfy

$$\lim_{n\to\infty} \sup\{\rho(x_n(t), X_0 \cap B(\widehat{\theta}, M)) : t \in [T_1, T_2]\} = 0,$$

there exists subsequence $\{x_{n_k}\}_{k=1}^{\infty}$ that converges uniformly on $[T_1, T_2]$ to a trajectory $x \in Y(T_1, T_2)$ as $k \to \infty$.

(A5) For each bounded, nonempty set B, there exists $M > 0$, and for each $\epsilon > 0$, there exists a number $T(B, \epsilon) > 0$ such that for each $T \geq T(B, \epsilon)$ and each $x \in Y(0, T)$ satisfying $x(0) \in B$,

$$\rho(x(t), X_0 \cap B(\widehat{\theta}, M)) \leq \epsilon,\ t \in [T(B, \epsilon), T].$$

In Sect. 4.2, we prove the following result.

Proposition 4.1 *The following properties are equivalent:*

(1) There exist a trajectory $x \in Y(0, \infty)$ and $M_0 > 0$ such that for every $t \geq 0$,

$$x(t) \in X_0 \cap B(\widehat{\theta}, M_0).$$

(2) There exists $x \in Y(0, \infty)$.

(3) *There exists $M > 0$, and for each integer $k \geq 1$, there exists $x_k \in Y(0, T_k)$ such that*

$$\lim_{k\to\infty} T_k = \infty \text{ and } \rho(\widehat{\theta}, x_k(0)) \leq M, \ k = 1, 2, \ldots.$$

In this chapter, in the sequel, we suppose that $Y(0, \infty) \neq \emptyset$.

Define

$$\Omega = \{z \in X : \ \text{for very } \epsilon > 0 \text{ there exists } x \in Y(0, \infty)$$

$$\text{for which } \liminf_{t\to\infty} \rho(z, x(t)) \leq \epsilon\}. \tag{4.3}$$

Clearly, Ω is a closed subset of the metric space (X, ρ).

In Sect. 4.3, we prove the following result.

Proposition 4.2 *$\Omega \neq \emptyset$ and $\Omega \subset X_0$.*

Assumption (A4) implies the following proposition.

Proposition 4.3 *For every $x \in Y(0, \infty)$,*

$$\lim_{t\to\infty} \rho(x(t), \Omega) = 0.$$

It is not difficult to see that the following result holds.

Proposition 4.4 *Assume that $B \subset X$ be a nonempty, closed set such that for every $x \in Y(0, \infty)$,*

$$\lim_{t\to\infty} \rho(x(t), B) = 0.$$

Then $\Omega \subset B$.

The following theorem will be proved in Sect. 4.4.

Theorem 4.5 *Let ϵ, M be positive real numbers. Then there is a positive number T such that for every $x \in Y(0, T)$ that satisfies*

$$\rho(x(0), \widehat{\theta}) \leq M$$

the inequality

$$\inf\{\rho(x(t), \Omega) : \ t \in [0, T]\} \leq \epsilon$$

holds.

The next theorem is proved in Sect. 4.5.

Theorem 4.6 *The following properties are equivalent:*

(1) If a bounded function $x \in Y(-\infty, \infty)$ satisfies

$$\phi(x(t)) = \phi(x(0))$$

for all $t \in R^1$, then the inclusion $x(t) \in \Omega$ holds for all $t \in R^1$.

(2) If a bounded function $x \in Y(-\infty, \infty)$ satisfies

$$x(t) \in X_0 \text{ and } \phi(x(t)) = \phi(x(0))$$

for all $t \in R^1$, then the inclusion $x(t) \in \Omega$ holds for all real numbers t.

(3) For each $M, \epsilon > 0$, there exist $\delta > 0$ and $L \geq 1$ such that for each $T > 2L$ and each $x \in Y(0, T)$ that satisfy

$$B(x(0), \widehat{\theta}) \leq M \text{ and } \phi(x(0)) - \phi(x(T)) \leq \delta$$

the inequality

$$\rho(x(t), \Omega) \leq \epsilon$$

holds for all $t \in [L, T - L]$.

In Sect. 4.6, we prove the following turnpike result.

Theorem 4.7 *Assume that properties of Theorem 4.6 hold. Let $\epsilon, M > 0$. Then there exist natural numbers L, Q such that for every number $T > L$ and every trajectory $x \in Y(0, T)$ that satisfy*

$$\rho(x(0), \widehat{\theta}) \leq M$$

there exist an integer $q \in [1, Q]$ and finite sequences of integers $\{a_i\}_{i=1}^q, \{b_i\}_{i=1}^q \subset [0, T]$ such that

$$a_i \leq b_i, \ i = 1, \ldots, q, \ a_{i+1} > b_i, \ i \in \{1, \ldots, q\} \setminus \{q\},$$

$$\rho(x(t), \Omega) \leq \epsilon, \ t \in [a_i, b_i], \ i = 1, \ldots, q,$$

and

$$mes([0, T] \setminus \cup_{i=1}^q [a_i, b_i]) \leq L.$$

4.2 Proof of Proposition 4.1

Evidently, (1) implies (2) and (2) implies (3). Assume that (3) holds. We show that this implies (1). There exist $M > 0$ and $x_n \in Y(0, n)$, $n = 1, 2, \ldots$, such that

$$\rho(\widehat{\theta}, x_n(0)) \leq M,, \ n = 1, 2, \ldots. \tag{4.4}$$

By (4.10) and assumption (A5), there exists $M_0 > 0$ such that for every natural number k there exists an integer $n(k) > k$ such that for each integer $n \geq n(k)$ and each $t \in [n(k), n]$, we have

$$\rho(x_n(t), X_0 \cap B(\widehat{\theta}, M_0)) \leq k^{-1}. \tag{4.5}$$

For each integer $k \geq 1$, define

$$y_k(t) = x_{2n(k)}(t + n(k)), \ t \in [0, n(k)]. \tag{4.6}$$

In view of (4.6),

$$y_k \in Y(0, n(k)), \ k = 1, 2, \ldots.$$

Equations (4.5) and (4.6) imply that for each integer $k \geq 1$ and each $t \in [0, n(k)]$,

$$\rho(y_k(t), X_0 \cap B(\widehat{\theta}, M_0)) = \rho(x_{2n(k)}(t + n(k)), X_0 \cap B(\widehat{\theta}, M_0)) \leq k^{-1}. \tag{4.7}$$

In view of (4.7), for each integer $m \geq 1$,

$$\lim_{k \to \infty} \sup\{\rho(y_k(t), X_0 \cap B(\widehat{\theta}, M_0)) : \ t \in [0, m]\} = 0. \tag{4.8}$$

By assumption (A4) and (4.8), extracting subsequences and using the diagonalization process, we obtain that there exist a strictly increasing sequence of natural numbers $\{k_j\}_{j=1}^{\infty}$ and $x \in Y(0, \infty)$ such that for each integer $m \geq 1$,

$$y_{k_j}(t) \to x(t) \text{ as } j \to \infty \text{ uniformly on } [0, m].$$

Together with (4.8), this implies that

$$x(t) \in X_0 \cap B(\widehat{\theta}, M_0), \ t \in [0, \infty)$$

and (1) holds. This completes the proof of Proposition 4.1.

4.3 Proof of Proposition 4.2

Let $x \in Y(0, \infty)$. Assumption (A5) implies that there exists $M_0 > 0$ and that for each integer $k \geq 1$ there exists an integer $n(k) > k$ such that

$$\rho(x(t), X_0 \cap B(\widehat{\theta}, M_0)) \leq k^{-1}, \ t \in [n(k), \infty). \tag{4.9}$$

Equation (4.9) implies that the sequence $\{x(n)\}_{n=1}^{\infty}$ (here n is a natural number) has a subsequence that converges to a point of $\Omega \cap B(\widehat{\theta}, M_0)$. Hence,

$$\Omega \neq \emptyset.$$

Let $z \in \Omega$ and $\epsilon > 0$. There exists $x \in Y(0, \infty)$ such that

$$\liminf_{t \to \infty} \rho(z, (x(t)) \leq \epsilon.. \tag{4.10}$$

In view of (A5),

$$\lim_{t \to \infty} \rho(x(t), X_0) = 0.$$

By the equation above and (4.10),

$$\rho(z, X_0) \leq \epsilon.$$

Since the set X_0 is closed and ϵ is an arbitrary positive number, we conclude that $z \in X_0$ and $\Omega \subset X_0$. Proposition 4.2 is proved.

4.4 Proof of Theorem 4.5

Assume that Theorem 4.5 does not hold. Then for every integer $n \geq 1$ there exists $x_n \in Y(0, n)$ such that

$$\rho(\widehat{\theta}, x_n(0)) \leq M, \tag{4.11}$$

$$\rho(x_n(t)), \Omega) > \epsilon, \ t \in [0, n]. \tag{4.12}$$

Assumption (A5) implies that there exists $M_1 > 0$ such that for every integer $k \geq 1$ there exists an integer $n(k) \geq k$ such that for each integer $n \geq n(k)$ and each $t \in [n(k), n]$,

$$\rho(x_n(t), X_0 \cap B(\theta, M_1)) \leq k^{-1}. \tag{4.13}$$

For each integer $k \geq 1$, define

$$y_k(t) = x_{2n(k)}(t + n(k)), \ t \in [0, n(k)]. \tag{4.14}$$

It follows from (4.12) and (4.14) that for each integer $k \geq 1$ and each $t \in [0, n(k)]$,

$$\rho(y_k(t), \Omega) > \epsilon, \tag{4.15}$$

$$\rho((y_k(t), X_0 \cap B(\widehat{\theta}, M_1)) \le k^{-1}. \tag{4.16}$$

By assumption (A4) and (4.16), extracting subsequences and using the diagonalization process, we obtain that there exist a strictly increasing sequence of natural numbers $\{k_j\}_{j=1}^{\infty}$ and $y \in Y(0, \infty)$ such that for each integer $m \ge 1$,

$$y_{k_j}(t) \to y(t) \text{ as } j \to \infty \text{ uniformly on } [0, m].$$

Together with (4.15) and (4.16), this implies that for every $t \in [0, \infty)$,

$$y(t) \in X_0 \cap B(\widehat{\theta}, M_1) \text{ and } \rho(y(t), \Omega) \ge \epsilon.$$

On the other hand, by Proposition 4.3,

$$\lim_{t \to \infty} \rho(y(t), \Omega) = 0.$$

The contradiction we have reached proves Theorem 4.5.

4.5 Proof of Theorem 4.6

Clearly, (1) implies (2) and (3) implies (1). Therefore, in order to prove the theorem, it is sufficient to show that (2) implies (3).

Assume that (2) holds and (3) does not hold. Then there exist $\epsilon, M > 0$ such that for each integer $k \ge 1$ there exist

$$T_k > 2k, \ x_k \in Y(0, T_k) \tag{4.17}$$

such that

$$\rho(x_k(0), \widehat{\theta}) \le M, \tag{4.18}$$

$$|\phi(x_k(0)) - \phi(x_k(T_k))| \le k^{-1} \tag{4.19}$$

and

$$\sup\{\rho(x_k(t)), \Omega) : t \in [k, T_k - k]\} > \epsilon. \tag{4.20}$$

By (A5), (4.17), and (4.18), there exists $M_1 > 0$, and for each integer $p \ge 1$, there exists an integer $k(p) > p$ such that for each integer $k \ge k(p)$ and each $t \in [k(p), T_k]$ we have

$$\rho(x_k(t), X_0 \cap B(\theta, M_1)) \le p^{-1}. \tag{4.21}$$

Let $p \geq 1$ be an integer. In view of (4.20), there exists

$$\tau_p \in [2k(p), T_{2k(p)} - 2k(p)] \tag{4.22}$$

such that

$$\rho(x_{2k(p)}(\tau_p), \Omega) > \epsilon. \tag{4.23}$$

Define

$$y_p(t) = x_{2k(p)}(t + \tau_p), \ t \in [-\tau_p, T_{2k(p)} - \tau_p]. \tag{4.24}$$

Clearly,

$$y_p \in Y(-\tau_p, T_{2k(p)} - \tau_p).$$

Equations (4.23) and (4.24) imply that

$$\rho(y_p(0), \Omega) = \rho(x_{2k(p)}(\tau_p), \Omega) > \epsilon. \tag{4.25}$$

By (4.21), (4.22), and (4.24), for all $t \in [k(p) - \tau_p, T_{2k(p)} - \tau_p]$,

$$\rho(y_p(t), \Omega \cap B(\widehat{\theta}, M_1)) = \rho(x_{2k(p)}(t + \tau_p), \Omega \cap B(\widehat{\theta}, M_1)) \leq p^{-1}. \tag{4.26}$$

Equations (4.14) and (4.24) imply that

$$\phi(y_p(-\tau_p)) - \phi(y_p(T_{2k(p)} - \tau_p)) \leq (2k(p))^{-1} \leq (2p)^{-1}. \tag{4.27}$$

In view of (4.22) and (4.26), for each integer $m \geq 1$,

$$\lim_{p \to \infty} \sup\{\rho(y_p(t), X_0 \cap B(\widehat{\theta}, M_1)) : \ t \in [-m, m]\} = 0. \tag{4.28}$$

By assumption (A4) and (4.28), extracting subsequences and using the diagonalization process, we obtain that there exists a strictly increasing sequence of natural numbers $\{p_i\}_{i=1}^{\infty}$ and $x \in Y(R^1)$ such that for each integer $m \geq 1$,

$$y_{p_i} \to x \text{ as } i \to \infty \text{ uniformly on } [-m, m]. \tag{4.29}$$

It follows from (4.27) and (4.29) that for every $t \in R^1$,

$$x(t) = \lim_{i \to \infty} y_{p_i}(t) \in X_0 \cap B(\widehat{\theta}, M_1), \tag{4.30}$$

$$\phi(x(t)) = \phi(x(0)). \tag{4.31}$$

Property (2) and Eqs. (4.30) and (4.31) imply that

$$x(t) \in \Omega,\ t \in R^1.$$

Together with (4.30), this implies that for all sufficiently large natural numbers i,

$$\rho(y_{p_i}(0), \Omega) \leq \epsilon/2.$$

This contradicts (4.25). The contradiction we have reached proves that (3) holds and completes the proof of Theorem 4.6.

Since bounded, closed subsets of X_0 are compact, it is not difficult to see that the following result is true.

Lemma 4.8 *Let $M_1 > 0$. Then there exist $M_0, r > 0$ such that for each $x \in X$ satisfying*

$$B(x, r_0) \cap B(\widehat{\theta}, M_1) \cap X_0 \neq\rightarrow \emptyset$$

the inequality $|\phi(x)| \leq M_0$ holds.

4.6 Proof of Theorem 4.7

We may assume without loss of generality that $\epsilon < 1$ and $M > 1$. Assumption (A5) and Lemma 4.8 imply that there exist $M_1 > M$ and an integer $L_1 \geq 1$ such that the following property holds:

(a) For each $T \geq L_1$ and each $x \in (0, T)$ satisfying $\rho(x(0), \widehat{\theta}) \leq M$, we have

$$\rho(x(t), X_0 \cap B(\widehat{\theta}, M_1)) \leq 1,\ |\phi(x(t))| \leq M_1$$

for all $t \in [L_1, n]$.

Property (3) of Theorem 4.6 implies there exist $\delta \in (0, 1)$ and $L_0 > 0$ such that the following property holds:

(b) For each $T > 2L_0$ and each $x \in Y(0, T)$ that satisfy

$$\rho(x(0), \widehat{\theta}) \leq M_1 + 2 \text{ and } \phi(x(0)) - \phi(x(T)) \leq \delta,$$

the inequality

$$\rho(x(t), \Omega) \leq \epsilon$$

holds for all $t \in [L_0, T - L_0]$.

Choose integers

$$Q > 2M_1\delta^{-1} + 2 \tag{4.32}$$

and

$$L > (4L_0 + 4L_1 + 8)(2\delta^{-1}M_1 + 2) + M + L_1 + 4. \tag{4.33}$$

Suppose that $T > L$ and that $x \in Y(0, T)$ satisfies

$$\rho(x(0), \widehat{\theta}) \leq M. \tag{4.34}$$

Property (a), (4.33), and (4.34) imply that

$$\rho(x(t), \widehat{\theta})) \leq M_1 + 2, \tag{4.35}$$

$$|\phi(x(t))| \leq M_1 \tag{4.36}$$

for all $t \in [L_1, T]$.

By induction, we define a strictly increasing finite sequence $t_i \in [0, T]$, $i = 0, \dots, q$. Set

$$t_0 = L_1.$$

If

$$\lim_{t \to T^-} \phi(x(t)) \geq \lim_{t \to t_0^+} \phi(x(t)) - \delta,$$

then set $t_1 = T$ and complete to construct the sequence.

Assume that

$$\lim_{t \to T^-} \phi(x(t)) < \lim_{t \to t_0^+} \phi(x(t)) - \delta.$$

Set

$$t_1 = \sup\{t \in (t_0, T] : \phi(x(t)) \geq \lim_{t \to t_0^+} \phi(x(t)) - \delta\}.$$

It is not difficult to see that t_1 is well-defined and

$$\lim_{t \to t_1^-} \phi(x(t)) \geq \lim_{t \to t_0^+} \phi(x(t)) - \delta,$$

$$\lim_{t \to t_1^+} \phi(x(t)) \le \lim_{t \to t_0^+} \phi(x(t)) - \delta.$$

If $t_1 = T$, then we complete to construct the sequence.

Assume that $k \ge 1$ is an integer, that we defined a finite strictly increasing sequence $t_0, \dots, t_k \in [0, T]$ such that

$$t_0 = L_1, \ t_k \le T,$$

and that for each $i \in \{0, \dots, k-1\}$,

$$\lim_{t \to t_{i+1}^-} \phi(x(t)) \ge \lim_{t \to t_i^+} \phi(x(t)) - \delta,$$

$$\lim_{t \to t_{i+1}^+} \phi(x(t)) \le \lim_{t \to t_i^+} \phi(x(t)) - \delta.$$

(It is not difficult to see that the assumption is true with $k = 1$.)

If $t_k = T$, then we complete to construct the sequence. Assume that $t_k < T$. If

$$\lim_{t \to T^-} \phi(x(t)) \ge \lim_{t \to t_k^+} \phi(x(t)) - \delta,$$

then we set $t_{k+1} = T$ and complete to construct the sequence.

Assume that

$$\lim_{t \to T^-} \phi(x(t)) < \lim_{t \to t_k^+} \phi(x(t)) - \delta.$$

Set

$$t_{k+1} = \sup\{t \in (t_k, T] : \ \phi(x(t)) \ge \lim_{t \to t_k^+} \phi(x(t)) - \delta\}.$$

It is not difficult to see that t_{k+1} is well-defined,

$$\lim_{t \to t_{k+1}^-} \phi(x(t)) \ge \lim_{t \to t_k^+} \phi(x(t)) - \delta,$$

$$\lim_{t \to t_{k+1}^+} \phi(x(t)) \le \lim_{t \to t_k^+} \phi(x(t)) - \delta.$$

Thus, the assumption made for k is true for $k + 1$ too. Therefore by induction, we constructed the strictly increasing finite sequence $t_i \in [0, T]$, $i = 0, \dots, q$, such that

$$t_0 = L_1, \ t_q = T \tag{4.37}$$

and that for every i satisfying $0 \le i < q - 1$,

$$\lim_{t \to t_{i+1}^+} \phi(x(t)) \le \lim_{t \to t_i^+} \phi(x(t)) - \delta \le \lim_{t \to t_{i+1}^+} \phi(x(t)), \tag{4.38}$$

$$\lim_{t \to T^-} \phi(x(t)) \ge \lim_{t \to t_{q-1}^+} \phi(x(t)) - \delta. \tag{4.39}$$

By (4.33) and (4.36)–(4.38),

$$2M_1 \ge M_1 + M \ge \lim_{t \to t_0^+} \phi(x(t)) - \lim_{t \to t_{q-1}^+} \phi(x(t))$$

$$\ge \sum \{ \lim_{t \to t_i^+} \phi(x(t)) - \lim_{t \to t_{i+1}^+} \phi(x(t)) :$$

$$i \text{ is an integer, } 0 \le i \le q - 2\} \ge \delta(q - 1)$$

and

$$q \le 1 + 2\delta^{-1} M_1 < Q. \tag{4.40}$$

Define

$$E = \{i \in \{0, \dots, q - 1\} : t_{i+1} - t_i \ge 2L_0 + 4\}. \tag{4.41}$$

Let

$$i \in E. \tag{4.42}$$

By (4.41) and (4.42),

$$t_{i+1} - 1 - (t_i + 1) \ge 2L_0 + 2. \tag{4.43}$$

Equations (4.37)–(4.39) imply that

$$\phi(x(t_i + 1)) - \phi(x(t_{i+1} - 1)) \le \delta. \tag{4.44}$$

Equations (4.35), (4.43), (4.44) and property (b) applied to the restriction of x to the interval $[t_i + 1, t_{i+1} - 1]$ imply that

$$\rho(x(t), \Omega) \le \epsilon, \ t \in [t_i + L_0 + 1, t_{i+1} - 1 - L_0]. \tag{4.45}$$

Set

$$a_i = t_i + L_0 + 1, \ b_i = t_{i+1} - L_0 - 1. \tag{4.46}$$

By (4.37) and (4.46),

$$[0, T] \setminus \cup_{i \in E}[a_i, b_i]$$

$$\subset \cup\{[t_i, t_{i+1}] : i \in \{0, \dots, q-1\} \setminus E\} \cup [0, L_1]$$

$$\cup \{[t_i, t_i + L_0 + 1] \cup [t_{i+1} - L_0 - 1, t_{i+1}] : i \in E\}. \tag{4.47}$$

It follows from (4.33), (4.41), (4.44), and (4.47) that

$$\text{mes}([0, T] \setminus \cup_{i \in E}[a_i, b_i])$$

$$\leq q(2L_0 + 4) + (2L_0 + 2)q + L_1 = q(4L_0 + 6) + L_1$$

$$\leq (4L_0 + 6)(1 + 2M_1\delta^{-1}) + L_1 < L.$$

Theorem 4.7 is proved.

Chapter 5
General Dynamical Systems with a Lyapunov Function

In this chapter we develop a turnpike theory for a class of general dynamical systems in a metric space with a Lyapunov function. This class of dynamical systems contains as particular cases the classes of dynamical systems studied in Chaps. 3 and 4. We obtain several turnpike results and show that the turnpike phenomenon is stable under small perturbations of objective functions.

5.1 Preliminaries and Two Turnpike Results

Assume that (X, ρ) is a metric space. For each $x \in X$ and each $r > 0$, set

$$B(x, r) = \{y \in X : \ \rho(x, y) \le r\} \text{ and } B^0(x, r) = \{y \in X : \ \rho(x, y) < r\}.$$

For each $x \in X$ and each pair of nonempty sets $A, B \subset X$, define

$$\rho(x, B) = \inf\{\rho(x, y) : \ y \in B\}$$

and

$$\mathrm{dist}(A, B) = \sup\{\rho(x, B) : \ x \in A\}.$$

Let Γ be either R^1 or the set of all integers $\mathcal{Z} = \{0, \pm 1, \pm 2, \dots\}$.

Assume that for each pair $T_1, T_2 \in \Gamma$ satisfying $T_1 < T_2$, $Y(T_1, T_2)$ is a nonempty set of functions $x : [T_1, T_2] \cap \Gamma \to X$ such that the following properties hold:

(A1) For each pair $T_1, T_2 \in \Gamma$, each pair $S_1, S_2 \in \Gamma$ satisfying $T_1 \le S_1 < S_2 \le T_2$, and each $x \in Y(T_1, T_2)$, the restriction of x to $[S_1, S_2]$ belongs to $Y(S_1, S_2)$.

A. J. Zaslavski, *Turnpike Phenomenon in Metric Spaces*, Springer Optimization and Its Applications 201, https://doi.org/10.1007/978-3-031-27208-0_5

$$\pi(x) = \lim_{T\to\infty} \sup\{\phi(x(T)) : x \in Y(0,T) \text{ and } x(0) = x\}. \tag{5.6}$$

(We assume that the supremum over an empty set is $-\infty$.)
It is clear that

$$\pi(x) \le \phi(x),\ x \in X. \tag{5.7}$$

Evidently, the following result is true:

Proposition 5.1 *Let $T > 0$ and $x \in Y(0,T)$. Then $\pi(x(s)) \le \pi(x(0))$ for all $s \in [0,T] \cap \Gamma$.*

The following result is proved in Sect. 5.3.

Proposition 5.2 *Let $x \in X$. The value $\pi(x)$ is finite if and only if for every $T \in \Gamma \cap (0,\infty)$ there exists $y \in Y(0,T)$ satisfying $y(0) = x$.*

In Sect. 5.4, we prove the following result.

Proposition 5.3 *Let $M_0 > 0$. Then there exists $M > 0$ such that for each $x \in B(\widehat{\theta}, M_0)$ satisfying $\pi(x) > -\infty$,*

$$B(\widehat{\theta}, M) \cap \Omega \cap \phi^{-1}(\pi(x)) \ne \emptyset.$$

The next theorem is our first turnpike result. It is proved in Sect. 5.5.

Theorem 5.4 *Let $\epsilon > 0$ and $\widehat{x} \in X$ satisfy $\pi(\widehat{x}) > -\infty$. Then there exist $\delta > 0$ and $L > 0$ such that for every $T \in \Gamma \cap (2L,\infty)$ and every $x \in Y(0,T)$ which satisfies*

$$x(0) = \widehat{x} \text{ and } \phi(x(T)) \ge \pi(\widehat{x}) - \delta$$

the inequality

$$\rho(x(t), \Omega) \le \epsilon$$

holds for all $t \in [L, T-L] \cap \Gamma$.

Clearly, Theorem 5.4 establishes the turnpike property for approximate optimal trajectories which have a fixed starting point with respect to the objective function ϕ. The turnpike is the set Ω and δ and L depend on ϵ and the initial point $\widehat{x}$.
For each $x \in X$ satisfying $\pi(x) > -\infty$, set

$$F(x) = \Omega \cap \phi^{-1}(\pi(x)),$$

which is nonempty in view of Proposition 5.3.
The following result which is a generalization of the previous theorem is proved in Sect. 5.7. According to Theorem 5.4, for all $t \in [L, T-L] \cap \Gamma$, the distance

between $x(t)$ and the turnpike Ω does not exceed ϵ. According to our next result, for all $t \in [L, T - L] \cap \Gamma$, the distance between $x(t)$ and the set $F(\widehat{x}) \cap B(\widehat{\theta}, M)$ does not exceed ϵ, where a positive constant M depends on ϵ and the initial point $\widehat{x}$ and $F(\widehat{x})$ is the set of all points of the turnpike Ω where the value of the Lyapunov function ϕ is $\pi(\widehat{x})$.

Theorem 5.5 *Let $\epsilon > 0$ and $\widehat{x} \in X$ satisfy $\pi(\widehat{x}) > -\infty$. Then there exist $M, L, \delta > 0$ such that for every $T \in \Gamma \cap (2L, \infty)$ and every $x \in Y(0, T)$ which satisfies*

$$x(0) = \widehat{x} \text{ and } \phi(x(T)) \geq \pi(\widehat{x}) - \delta$$

the inequality

$$\rho(x(t), F(\widehat{x}) \cap B(\widehat{\theta}, M)) \leq \epsilon$$

holds for all $t \in [L, T - L] \cap \Gamma$.

The next two results are proved in Sect. 5.8.

Proposition 5.6 *Assume that $X_0 = X$, $\widehat{x} \in X$ satisfies $\pi(\widehat{x}) > -\infty$, and that for each $S \in \Gamma$, the set*

$$\{z(t) : z \in Y(0, S), \; z(0) = \widehat{x}, \; t \in [0, S]\}$$

is bounded. Then there exists $x \in Y(0, \infty)$ such that $x(0) = \widehat{x}$ and

$$\pi(\widehat{x}) = \lim_{t \to \infty} \phi(x(t)).$$

Proposition 5.7 *Assume that $X_0 = X$, $x \in X$, and there exists $r > 0$ such that for each $S \in \Gamma \cap (0, \infty)$ the set*

$$\{z(t) : z \in Y(0, S), \; \rho(z(0), x) \leq r, \; t \in [0, S]\}$$

is bounded. Then the function $\pi : X \to R^1 \cup \{-\infty\}$ is upper semicontinuous at x.

5.2 Two Auxiliary Results

Proposition 5.8 *Let $x \in \Omega$. Then there exists a bounded function $y \in Y(\Gamma)$ such that*

$$y(0) = x, \; \phi(y(t)) = \phi(y(0)), \; t \in \Gamma$$

and $y(t) \in \Omega$, $t \in \Gamma$.

Set

$$x(t) = y(t), \ t \in \Gamma \cap (-\infty, 0).$$

Clearly, $x \in Y(\Gamma)$ and $y(\Gamma)$ is bounded,

$$\phi(x(t)) = \phi(x(0)), \ t \in \Gamma.$$

Since y is bounded, it follows from (A5) that x is bounded on Γ too. Proposition 5.8 implies that $x(t) \in \Gamma, t \in \Gamma$. Proposition 5.10 is proved.

5.3 Proof of Proposition 5.2

Assume that for every integer $n \geq 1$, there exists $x_n \in Y(0, n)$ such that

$$x_n(0) = x.$$

It is sufficient to show that $\pi(x)$ is finite. By (A5), there exists $M > 0$, and for each integer $p \geq 1$, there exists an integer $n(p) \geq p$ such that for each integer $n \geq n(p)$ and each $t \in \Gamma \cap [n(p), n]$,

$$\rho(x_n(t), B(\widehat{\theta}, M)) \leq p^{-1}. \tag{5.23}$$

Lemma 4.8 implies that there exist $M_0, r_0 > 0$ such that for each $\xi \in X$ satisfying

$$B(\xi, r_0) \cap B(\widehat{\theta}, M) \cap X_0 \neq \emptyset,$$

we have

$$|\phi(\xi)| \leq M_0. \tag{5.24}$$

It follows from (5.23) and (5.24) that for each integer p satisfying $p^{-1} < r_0/2$, each integer $n \geq n(p)$, and each $t \in \Gamma \cap [n(p), n]$,

$$|\phi(x_n(t))| \leq M_0$$

and

$$\sup\{\phi(y) : \ y \in Y(0, n), \ y(0) = x\} \geq -M_0.$$

Thus $\pi(x) \geq -M_0$. On the other hand, $\pi(x) \leq \phi(x)$. Proposition 5.2 is proved.

5.4 Proof of Proposition 5.3

Assumption (A5) implies that there exists $M_1 > M_0$, and for each integer $k \geq 1$, there exists an integer $n(k) > k$ such that the following property holds:

(i) For each integer $n \geq n(k)$, each $y \in Y(0, n)$ satisfying

$$\rho(y(0), \widehat{\theta}) \leq M_0,$$

and each $t \in \Gamma \cap [n(k), n]$,

$$\rho(y(t), X_0 \cap B(\widehat{\theta}, M)) \leq k^{-1}.$$

Let

$$x \in B(\widehat{\theta}, M_0)$$

and $\pi(x)$ be finite. Proposition 5.2 implies that for each integer $n \geq 1$ there exists $x_n \in Y(0, n)$ such that

$$x_n(0) = x, \tag{5.25}$$

$$\begin{aligned} \phi(x_n(n)) &\geq \sup\{\phi(\xi(n)) : \xi \in Y(0, n),\ \xi(0) = x\} - n^{-1} \\ &\geq \pi(x) - n^{-1}. \end{aligned} \tag{5.26}$$

For each integer $k \geq 1$, define $y_k \in Y(0, n(k))$ by

$$y_k(t) = x_{2n(k)}(t + n(k)),\ t \in [0, n(k)] \cap \Gamma. \tag{5.27}$$

Property (i) and (5.27) imply that for each integer $k \geq 1$ and each $t \in [0, n(k)] \cap \Gamma$,

$$\rho(y_k(t), X_0 \cap B(\widehat{\theta}, M)) = \rho(x_{2n(k)}(t + n(k)), X_0 \cap B(\widehat{\theta}, M)) \leq k^{-1}. \tag{5.28}$$

In view of (28), for each integer $m \geq 1$,

$$\lim_{k \to \infty} \sup\{\rho(y_k(t), X_0 \cap B(\widehat{\theta}, M)) : t \in \Gamma \cap [0, m]\} = 0. \tag{5.29}$$

By assumption (A4) and (5.29), extracting subsequences, and using the diagonalization process, we obtain that there exist a strictly increasing sequence of natural numbers $\{k_i\}_{i=1}^{\infty}$ and $y \in Y(0, \infty)$ such that for each integer $m \geq 1$,

$$y_{k_j} \to y \text{ as } j \to \infty \text{ uniformly on } [0, m] \cap \Gamma. \tag{5.30}$$

It follows from (5.27) and (5.30) that for every $t \in [0, \infty) \cap \Gamma$,

$$y(t) = \lim_{j \to \infty} y_{k_j}(t) = \lim_{j \to \infty} x_{2n(k_j)}(t + n(k_j)). \tag{5.31}$$

Let $t \in [0, \infty) \cap \Gamma$. By (5.1) and (5.26), for each integer $j \geq 1$ and each $t \in [0, n(k_j)] \cap \Gamma$,

$$\phi(x_{2n(k_j)}(t + n(k_j)) \geq \phi(x_{2n(k_j)}(2n(k_j))) \geq \pi(x) - (2n(k_j))^{-1}. \tag{5.32}$$

In view of (5.31) and (5.32),

$$\phi(y(t)) \geq \pi(x). \tag{5.33}$$

By (5.6), for each integer $j \geq 1$ and each $t \in [0, n(k_j)] \cap \Gamma$,

$$\phi(x_{2n(k_j)}(t + n(k_j)) \leq \sup\{\phi(\xi) : \xi \in Y(0, t + n(k_j)), \ \xi(0) = x\} \to \pi(x)$$

as $j \to \infty$ and

$$\limsup_{j \to \infty} \phi(x_{2n(k_j)}(t + n(k_j)) \leq \pi(x).$$

Together with (5.31), this implies that

$$\phi(y(t)) \leq \pi(x).$$

Combined with (5.33), this implies that

$$\phi(y(t)) = \pi(x), \ t \in [0, \infty) \cap \Gamma. \tag{5.34}$$

In view of (5.29) and (5.30),

$$y(t) \in X_0 \cap B(\widehat{\theta}, M), \ t \in [0, \infty) \cap \Gamma. \tag{5.35}$$

In view of (5.35), the sequence $\{y(n)\}_{n=1}^{\infty}$ has a limit point

$$\xi \in \Omega \cap X_0 \cap B(\widehat{\theta}, M).$$

By (5.34), $\phi(\xi) = \pi(x)$. Proposition 5.3 is proved.

5.5 Proof of Theorem 5.4

Assume that Theorem 5.4 does not hold. Then, for each natural number k, there exist

$$T_k \in (2k, \infty) \cap \Gamma, \ \ x_k \in Y(0, T_k) \tag{5.36}$$

such that

$$x_k(0) = \widehat{x}, \tag{5.37}$$

$$\phi(x_k(T_k)) \geq \pi(\widehat{x}) - k^{-1}, \tag{5.38}$$

and

$$\sup\{\rho(x_k(t), \Omega) : t \in [k, T_k - k] \cap \Gamma\} > \epsilon. \tag{5.39}$$

In view of (5.39), for each integer $k \geq 1$, there exists

$$\tau_k \in [k, T_k - k] \cap \Gamma \tag{5.40}$$

such that

$$\rho(x_k(\tau_k), \Omega) > \epsilon. \tag{5.41}$$

Equations (5.1), (5.37), and (5.38) imply that for each integer $k \geq 1$ and each $t \in [0, T_k] \cap \Gamma$,

$$\phi(\widehat{x}) \geq \phi(x_k(t)) \geq \pi(\widehat{x}) - 1/k. \tag{5.42}$$

Let $k \geq 1$ be an integer. Set

$$y_k(t) = x_k(t + \tau_k),\ t \in [-\tau_k, T_k - \tau_k] \cap \Gamma. \tag{5.43}$$

In view of (5.36), (5.41), and (5.43),

$$y_k \in Y(-\tau_k, T_k - \tau_k),\ \rho(y_k(0), \Omega) > \epsilon. \tag{5.44}$$

Assumption (A5) implies that there exists $M_0 > 0$ such that the following property holds:

(i) For each integer $p \geq 1$, there exists an integer $k(p) \geq p$ such that for each integer $k \geq k(p)$ and each $t \in \Gamma \cap [k(p), T_k]$,

$$\rho(x_k(t), X_0 \cap B(\widehat{\theta}, M_0)) \leq p^{-1}.$$

Let $p \geq 1$ be an integer. Property (i) and (5.43) imply that for each integer $k \geq k(p)$ and each $t \in [k(p) - \tau_k, T_k - \tau_k] \cap \Gamma$

$$\rho(y_k(t), X_0 \cap B(\widehat{\theta}, M_0)) = \rho(x_k(t + \tau_k), X_0 \cap B(\widehat{\theta}, M_0)) \leq p^{-1}. \tag{5.45}$$

It follows from (5.40) and (5.45) that for each $t \in \Gamma$ and each integer $k \geq k(p)+|t|$,

$$\rho(y_k(t), X_0 \cap B(\widehat{\theta}, M_0)) \leq p^{-1}.$$

This implies that for each integer $m \geq 1$,

$$\lim_{k\to\infty} \sup\{\rho(y_k(t), X_0 \cap B(\widehat{\theta}, M_0)) : t \in [-m, m] \cap \Gamma\} = 0. \tag{5.46}$$

By assumption (A4) and (5.46), extracting subsequences, and using the diagonalization process, we obtain that there exist a strictly increasing sequence of natural numbers $\{k_j\}_{j=1}^{\infty}$ and $y \in Y(\Gamma)$ such that for each $t \in \Gamma$,

$$y(t) = \lim_{j\to\infty} y_{k_j}(t) \in X_0 \cap B(\widehat{\theta}, M_0) \tag{5.47}$$

uniformly on $[-m, m] \cap \Gamma$ for every integer $m \geq 1$. It follows from (5.44) and (5.47) that

$$\rho(y(0), \Omega) \geq \epsilon. \tag{5.48}$$

Let $t \in \Gamma$. Equations (5.38), (5.40), (5.43), and (5.47) imply that

$$\phi(y(t)) = \lim_{j\to\infty} \phi(y_{k_j}(t)) = \lim_{j\to\infty} \phi(x_{k_j}(t + \tau_{k_j}))$$

$$\geq \lim_{j\to\infty} (\pi(\widehat{x}) - k_j^{-1}) = \pi(\widehat{x}).$$

On the other hand, Eqs. (5.6), (5.43), and (5.47) imply that

$$\phi(y(t)) = \lim_{j\to\infty} \phi(x_{k_j}(t + \tau_{k_j}))$$

$$= \lim_{j\to\infty} \sup\{\phi(z(t + k_j)) : z \in Y(0, t + k_j),\ z(0) = \widehat{x}\} = \pi(\widehat{x}).$$

In view of the equation above,

$$\phi(x(t)) = \pi(\widehat{x}),\ t \in \Gamma. \tag{5.49}$$

Assumption (A6), (5.47), and (5.49) imply that $y(t) \in \Omega$, $t \in \Gamma$, and in particular $y(0) \in \Omega$. This contradicts (5.48). The contradiction we have reached completes the proof of Theorem 5.4.

5.6 An Auxiliary Result for Theorem 5.5

Lemma 5.11 *Assume that $K \subset U \subset X$, where the set U is open and the set K is nonempty and compact, $g : U \to R^1$, every point of K is a continuity point of g, and that $\epsilon > 0$. Then there exists $\delta > 0$ such that for each $z_1, z_2 \in X$ which satisfy*

$$\rho(z_1, z_2) \le \delta, \ \rho(z_i, K) \le \delta, \ i = 1, 2,$$

the inclusion $z_1, z_2 \in U$ and the inequality

$$|g(z_1) - g(z_2)| \le \epsilon$$

hold.

Proof For each $x \in K$, there exists $r_x \in (0, 1)$ such that

$$B(x, 4r_x) \subset U, \tag{5.50}$$

$$|g(z_1) - g(z_2)| \le \epsilon/4 \text{ for all } z_1, z_2 \in B(x, 4r_x). \tag{5.51}$$

In view of (5.50),

$$K \subset \cup\{B^0(x, r_x) : \ x \in K\}. \tag{5.52}$$

Since the set K is compact, (5.52) implies that there exist an integer $q \ge 1$ and

$$x_i \in K, \ i = 1, \dots, q, \tag{5.53}$$

such that

$$K \subset \cup_{i=1}^{q} B^0(x_i, r_{x_i}). \tag{5.54}$$

Set

$$\delta = 2^{-1} \min\{r_{x_i} \ : \ i = 1, \dots, q\}. \tag{5.55}$$

Assume that $z_1, z_2 \in X$ satisfy

$$\rho(z_1, z_2) \le \delta, \ \rho(z_i, K) \le \delta, \ i = 1, 2. \tag{5.56}$$

In view of (5.56), there exists $y \in K$ such that

$$\rho(z_1, y) \le \delta, \ \rho(z_2, y) \le 2\delta. \tag{5.57}$$

By (5.54), there exists $j \in \{1, \dots, q\}$ such that

$$\rho(y, x_j) \le r_{x_j}. \tag{5.58}$$

It follows from (5.55), (5.57), and (5.58) that for $i = 1, 2$,

$$\rho(z_i, x_j) \le \rho(z_i, y) + \rho(y, x_j) \le 2\delta + r_{x_j} \le 2r_{x_j}. \tag{5.59}$$

By (5.50), (5.51), and (5.59),

$$z_1, z_2 \in U,$$

$$|g(z_1) - g(z_2)| \le \epsilon/4.$$

Lemma 5.11 is proved.

5.7 Proof of Theorem 5.5

We may assume that $\epsilon < 1$. Assumption (A5) implies that there exists $M > 2$, and for each integer $p \ge 1$, there exists an integer $L_p > p$ such that the following property holds:

(i) For each $T \in \Gamma \cap [L_p, \infty)$ and each $x \in Y(0, T)$ satisfying $x(0) = \widehat{x}$, we have

$$\rho(x(t), X_0 \cap B(\widehat{\theta}, M - 2)) \le p^{-1},\ t \in \Gamma \cap [L_p, T].$$

Proposition 5.3 implies that

$$F(\widehat{x}) = \Omega \cap \phi^{-1}(\pi(\widehat{x})) \ne \emptyset. \tag{5.60}$$

We show that there exists $\delta_1 > 0$ such that the following property holds:

(ii) For each $z \in \Omega \cap B(\widehat{\theta}, M)$ satisfying

$$|\phi(z) - \pi(\widehat{x})| \le 2\delta_1,$$

the inequality

$$\rho(z, F(\widehat{x}) \cap B(\widehat{\theta}, M)) \le \epsilon/4$$

is true.

Assume the contrary. Then there exists a sequence

$$\{z_i\}_{i=1}^{\infty} \subset \Omega \cap B(\widehat{\theta}, M), \tag{5.61}$$

$$\lim_{i\to\infty} \phi(z_i) = \pi(\widehat{x}), \tag{5.62}$$

$$\rho(z_i, F(\widehat{x}) \cap B(\widehat{\theta}, M)) > \epsilon/4. \tag{5.63}$$

Since the set $\Omega \cap B(\widehat{\theta}, M + 1)$ is compact, we can assume without loss of generality that there exists $\lim_{i\to\infty} z_i$. Equations (5.61) and (5.62) imply that

$$\lim_{i\to\infty} z_i \in \Omega \cap B(\widehat{\theta}, M), \ \phi(\lim_{i\to\infty} z_i) = \pi(\widehat{x}). \tag{5.64}$$

By (5.63),

$$\rho(\lim_{i\to\infty} z_i, F(\widehat{x}) \cap B(\widehat{\theta}, M)) \geq \epsilon/4.$$

This contradicts (5.64). The contradiction we have reached proves that there exists $\delta_1 > 0$ such that property (ii) holds.

Lemma 5.11 implies that there exists $\epsilon_1 \in (0, \epsilon/4)$ such that the following property holds:

(iii)

$$|\phi(z_1) - \phi(z_2)| \leq \delta_1/4$$

for each $z_1, z_2 \in X$ which satisfy

$$\rho(z_1, z_2) \leq 4\epsilon_1, \ \rho(z_i, \Omega \cap B(\widehat{\theta}, M + 1)) \leq 4\epsilon_1, \ i = 1, 2.$$

By definition (5.6), there exists a natural number L_0 such that

$$|\sup\{\phi(z(L_0)) : \ z \in Y(0, L_0) \text{ and } z(0) = \widehat{x}\} - \pi(\widehat{x})| \leq \delta_1/2. \tag{5.65}$$

Theorem 5.4 implies that there exist $\delta \in (0, \delta_1)$ and an integer $L > 2L_0 + 2L_1$ such that the following property holds:

(iv) For every $T \in \Gamma \cap (2L, \infty)$ and every $x \in Y(0, T)$ which satisfies

$$x(0) = \widehat{x} \text{ and } \phi(x(T)) \geq \pi(\widehat{x}) - \delta,$$

the inequality

$$\rho(x(t), \Omega) \leq \epsilon_1$$

holds for all $t \in [L, T - L] \cap \Gamma$.

Assume that

$$T \in \Gamma \cap (2L, \infty), \ x \in Y(0, T), \tag{5.66}$$

$$x(0) = \widehat{x} \text{ and } \phi(x(T)) \geq \pi(\widehat{x}) - \delta. \tag{5.67}$$

Property (i) (with $p = 1$), (5.66), and (5.67) imply that

$$\rho(x(t), X_0 \cap B(\widehat{\theta}, M - 2)) \le 1, \ t \in \Gamma \cap [L_1, T]. \tag{5.68}$$

Property (iv), (5.67), and (5.68) imply that

$$\rho(x(t), \Omega) \le \epsilon_1, \ t \in [L, T - L] \cap \Gamma. \tag{5.69}$$

Assume that

$$t \in \Gamma \cap [L, T - L]. \tag{5.70}$$

In view of (5.69), there exists

$$z \in \Omega \tag{5.71}$$

such that

$$\rho(x(t), z) \le 2\epsilon_1. \tag{5.72}$$

By (5.1), (5.67), and the relation $L_0 < L \le t \le T$,

$$\phi(x(t)) \ge \phi(x(T)) \ge \pi(\widehat{x}) - \delta \ge \pi(\widehat{x}) - \delta_1. \tag{5.73}$$

In view of (5.1), (5.65), and the relation $L_0 < L \le t \le T$,

$$\phi(x(t)) \le \phi(x(L_0)) \le \pi(\widehat{x}) + \delta_1/2.$$

Together with (5.73), this implies that

$$|\phi(x(t)) - \pi(\widehat{x})| \le \delta_1. \tag{5.74}$$

It follows from (5.68), (5.70), and (5.72) that

$$x(t) \in B(\widehat{\theta}, M - 1) \tag{5.75}$$

and

$$z \in \Omega \cap B(\widehat{\theta}, M). \tag{5.76}$$

Property (iii), (5.72), and (5.76) imply that

$$|\phi(x(t)) - \phi(z)| \le \delta_1/4. \tag{5.77}$$

By (5.74) and (5.77),

$$|\phi(z) - \pi(\widehat{x})| \leq |\phi(z) - \phi(x(t))| + |\phi(x(t)) - \pi(\widehat{x})| \leq 3\delta_1/2. \tag{5.78}$$

Property (ii), (5.76), and (5.78) imply that

$$\rho(z, F(\widehat{x}) \cap B(\widehat{\theta}, M)) \leq \epsilon/4. \tag{5.79}$$

It follows from (5.72) and (5.79) that

$$\rho(x(t), F(\widehat{x}) \cap B(\widehat{\theta}, M)) \leq \epsilon/4 + 2\epsilon_1 < \epsilon/2.$$

Theorem 5.5 is proved.

5.8 Proofs of Propositions 5.6 and 5.7

Proof of Proposition 5.6 For each integer $n \geq 1$, there exists $x_n \in Y(0, n)$ such that

$$x_n(0) = \widehat{x}, \tag{5.79}$$

$$\phi(x_n(n)) \geq \sup\{\phi(z(n)) : \ z \in Y(0, n), \ z(0) = \widehat{x}\} - n^{-1}. \tag{5.80}$$

Assumption (A5) implies that there exist $M_0 > 0$ and an integer $L_0 > 0$ such that for each integer $n \geq L_0$,

$$\rho(x_n(t), \widehat{\theta}) \leq M_0, \ t \in [L_0, n] \cap \Gamma. \tag{5.81}$$

By our assumptions, there exists $M_1 > M_0$ such that

$$\{x_n(t) : \ t \in [0, L_0] \cap \Gamma, \ n = 1, 2, \dots\} \subset B(\widehat{\theta}, M_1). \tag{5.82}$$

By assumption (A4), (5.81), and (5.82), extracting subsequences, and using the diagonalization process, we obtain that there exist a strictly increasing sequence of natural numbers $\{n_j\}_{j=1}^{\infty}$ and $x \in Y(0, \infty)$ such that for each integer $m \geq 1$,

$$x_{n_j} \to x \text{ as } j \to \infty \text{ uniformly on } [0, m] \cap \Gamma. \tag{5.83}$$

It follows from (5.79) and (5.83) that

$$x(0) = \widehat{x}. \tag{5.84}$$

In view of (5.84),

$$\lim_{t\to\infty} \phi(x(t)) \le \pi(x). \tag{5.85}$$

It follows from (5.1), (5.80), and (5.83) that for each $t \in \Gamma$,

$$\phi(x(t)) = \lim_{j\to\infty} \phi(x_{n_j}(t)) \ge \limsup_{j\to\infty} \phi(x_{n_j}(n_j))$$

$$\ge \lim_{j\to\infty} (\pi(\widehat{x}) - n_j^{-1}) = \pi(\widehat{x}).$$

Combined with (5.85), this implies that

$$\lim_{t\to\infty} \phi(x(t)) = \pi(x).$$

Proposition 5.6 is proved.

Proof of Proposition 5.7 Assume that $\{x_n\}_{n=1}^{\infty} \subset X$ and

$$\lim_{n\to\infty} x_n = x. \tag{5.86}$$

We show that

$$\pi(x) \ge \limsup_{n\to\infty} \pi(x_n).$$

We may assume without loss of generality that there exists

$$\lim_{n\to\infty} \pi(x_n) > -\infty \tag{5.87}$$

and

$$\pi(x_n) > -\infty \text{ for any integer } n \ge 1. \tag{5.88}$$

Assumption (A5) and (5.86) imply that there exist $M_0 > 0$ and an integer $L_0 > 0$ such that the following property holds:

(a) For each integer $n \ge 1$, each $T \in \Gamma \cap [L_0, \infty)$, and each $y \in Y(0, T)$ satisfying $y(0) = x_n$,

$$\rho(y(t), \widehat{\theta}) \le M_0, \ t \in [L_0, T] \cap \Gamma.$$

Let $r > 0$ be as guaranteed by the assumptions of the proposition. In view of (5.86), we may assume without loss of generality that

$$\rho(x_n, \widehat{x}) \leq r, \; n = 1, 2, \ldots. \tag{5.89}$$

It follows from (5.88), (5.89), the assumptions of our proposition, the choice of r, and Proposition 5.6 that for each integer $n \geq 1$ there exists $y_n \in Y(0, \infty)$ such that

$$y_n(0) = x_n, \; \lim_{t \to \infty} \phi(y_n(t)) = \pi(x_n). \tag{5.90}$$

Property (a) implies that for each integer $n \geq 1$ and each $t \in [L_0, \infty) \cap \Gamma$,

$$\rho(y_n(t), \widehat{\theta}) \leq M_0. \tag{5.91}$$

By the assumptions of the proposition and the choice of r, there exists $M_1 > M_0$ such that

$$\{z(t) : \; z \in Y(0, L_0), \; \rho(z(0), x) \leq r, \; t \in [0, L_0] \cap \Gamma\} \subset B(\widehat{\theta}, M_1). \tag{5.92}$$

Equations (5.89), (5.91), and (5.92) imply that

$$\{y_n(t) : \; t \in \Gamma, \; n = 1, 2, \ldots\} \subset B(\widehat{\theta}, M_1). \tag{5.93}$$

By assumption (A4) and Eq. (5.93), extracting subsequences, using the diagonalization process, and re-indexing, we obtain that there exist a strictly increasing sequence of natural numbers $\{n_j\}_{j=1}^{\infty}$ and $y \in Y(0, \infty)$ such that for each integer $m \geq 1$,

$$y_{n_j} \to y \text{ as } j \to \infty \text{ uniformly on } [0, m] \cap \Gamma. \tag{5.94}$$

It follows from (5.86), (5.90), and (5.94) that

$$y(0) = x.$$

By (5.87), (5.90), and (5.94), for each $t \in \Gamma \cap [0, \infty)$,

$$\phi(y(t)) = \lim_{j \to \infty} \phi(y_{n_j}(t))$$

$$\geq \limsup_{j \to \infty} \pi(x_{n_j}) = \lim_{n \to \infty} \pi(x_n)$$

$$\pi(y(t)) \geq \lim_{t \to \infty} \phi(y(t)) \geq \lim_{n \to \infty} \pi(x_n).$$

Proposition 5.7 is proved.

5.9 A Weak Turnpike Result

In the sequel we use the following property.

(P1) If $x_1, x_2 \in \Omega$ satisfy $\phi(x_1) = \phi(x_2)$, then $x_1 = x_2$.

This property holds for models of economic dynamics which are prototype of our dynamical system. When (P1) holds, we obtain the following weak turnpike result.

Theorem 5.12 *Assume that (P1) holds and that $\epsilon \in (0, 1)$, $M > 0$. Then there exist natural numbers L and Q such that for every number $T \in \Gamma \cap (L, \infty)$ and every trajectory $x \in Y(0, T)$ which satisfies*

$$\rho(x(0), \widehat{\theta}) \leq M,$$

there exist an integer $q \in [1, Q]$ and finite sequences of integers $\{a_i\}_{i=1}^q, \{b_i\}_{i=1}^q \subset \Gamma \cap [0, T]$ such that

$$a_i \leq b_i,\ i = 1, \ldots, q,\ a_{i+1} > b_i,\ i \in \{1, \ldots, q\} \setminus \{q\}$$

and $z_i \in \Omega$, $i = 1, \ldots, q$ such that for each $i \in \{1, \ldots, q\}$,

$$\rho(x(t), z_i) \leq \epsilon,\ t \in [a_i, b_i] \cap \Gamma$$

and

$$mes(\Gamma \cap [0, T] \setminus \cup_{i=1}^q [a_i, b_i]) \leq L.$$

For the proof of Theorem 5.12, the following two lemmas are used. The proof of the first one is obvious.

Lemma 5.13 *Let $M, \epsilon > 0$. Then there exists $\delta > 0$ such that if*

$$x_1, x_2 \in \Omega \cap B(\widehat{\theta}, M)$$

satisfy $|\phi(x_1) - \phi(x_2)| \leq \delta$, then $\rho(x_1, x_2) \leq \epsilon$.

Lemma 5.14 *Assume that property (P1) holds and that $\epsilon \in (0, 1]$, $M > 0$. Then there exist $\delta > 0$ and an integer $L \geq 1$ such that for every $T \in \Gamma \cap (2L, \infty)$ and every $x \in Y(0, T)$ satisfying*

$$\rho(x(0), \widehat{\theta}) \leq M \tag{5.95}$$

and

$$\phi(x(0)) \leq \phi(x(T)) + \delta, \tag{5.96}$$

there exists $z \in \Omega$ such that the inequality $\rho(x(t), z) \le \epsilon$ holds for all $t \in [L, T - L] \cap \Gamma$.

Proof Assumption (A5) implies that there exists $M_0 > M$, and for each integer $p \ge 1$, there exists an integer $k(p) \ge p$ such that the following property holds:

(a) For each $T \in \Gamma \cap [k(p), \infty)$ and each $z \in Y(0, T)$ satisfying $\rho(z(0), \widehat{\theta}) \le M$, we have

$$\rho(z(t), X_0 \cap B(\widehat{\theta}, M_0)) \le 1/p, \ t \in [k(p), T] \cap \Gamma.$$

Lemma 5.13 implies that there exists $\epsilon_0 \in (0, \epsilon/4)$ such that the following property holds:

(b) For each $z_1, z_2 \in \Omega \cap B(\widehat{\theta}, M_0 + 1)$ satisfying $|\phi(x_1) - \phi(x_2)| \le \epsilon_0$, we have $\rho(z_1, z_2) \le \epsilon/4$.

Lemma 5.11 implies that there exists $\epsilon_1 \in (0, \epsilon_0/2)$ such that the following property holds:

(c) For each $y_1, y_2 \in X$ which satisfy

$$\rho(y_1, y_2) \le \epsilon_1, \ \rho(y_i, \Omega \cap B(\widehat{\theta}, M_0 + 1)) \le \epsilon_1, \ i = 1, 2,$$

the inequality

$$|\phi(y_1) - \phi(y_2)| \le \epsilon_0/4.$$

holds.

Theorems 3.6 (for $\Gamma = \mathcal{Z}$) and 4.6 (for $\Gamma = R^1$) imply that there exist $\delta \in (0, \epsilon_1/4)$ and an integer $L \in \Gamma \cap [1, \infty)$ such that the following property holds:

(d) For every $T \in \Gamma \cap (2L, \infty)$ and every $x \in Y(0, T)$ which satisfies (5.95) and (5.96), the inequality

$$\rho(x(t), \Omega) \le \epsilon_1$$

holds for all $t \in [L, T - L] \cap \Gamma$.

We may assume without loss of generality that

$$L > k(2). \tag{5.97}$$

Assume that

$$T \in \Gamma \cap (2L, \infty), \ x \in Y(0, T) \tag{5.98}$$

and that (5.95) and (5.96) hold. Property (d) and Eqs. (5.95)–(5.98) imply that

$$\rho(x(t), \Omega) \le \epsilon_1, \ t \in [L, T - L] \cap \Gamma. \tag{5.99}$$

Property (a) (with $p = 2$) and Eqs. (5.95), (5.97), and (5.98) imply that

$$\rho(x(t), \widehat{\theta}) \le M_0 + 2^{-1}, \ t \in [L, T] \cap \Gamma. \tag{5.100}$$

Fix

$$t_0 \in [L, T - L] \cap \Gamma \tag{5.101}$$

and

$$t \in [L, T - L] \cap \Gamma \tag{5.102}$$

By (5.99)–(5.102), there exist

$$z_0, z \in \Omega \tag{5.103}$$

such that

$$\rho(x(t_0), z_0) \le \epsilon_1, \tag{5.104}$$

$$\rho(x(t), z) \le \epsilon_1. \tag{5.105}$$

In view of (5.100), (5.102), (5.104), and (5.105),

$$z, z_0 \in B(\widehat{\theta}, M_0 + 1). \tag{5.106}$$

Property (c) and Eqs. (5.103)–(5.106) imply that

$$|\phi(x(t_0)) - \phi(z_0)| \le \epsilon_0/4, \tag{5.107}$$

$$|\phi(x(t)) - \phi(z)| \le \epsilon_0/4. \tag{5.108}$$

By (5.1) and (5.96),

$$|\phi(x(t_0)) - \phi(x(t))| \le \delta \le \epsilon_0/4. \tag{5.109}$$

It follows from (5.107)–(5.109) that

$$|\phi(z_0) - \phi(z)| \le |\phi(z_0) - \phi(x(t_0))| + |\phi(x(t_0)) - \phi(x(t))|$$

$$+ |\phi(x(t)) - \phi(z)| \le \epsilon_0/4 + \epsilon_0/4 + \epsilon_0/4.$$

The equation above, property (b), and Eqs. (5.103) and (5.106) imply that

$$\rho(z_0, z) \le \epsilon/4.$$

Together with (5.105), this implies that

$$\rho(z_0, x(t)) \leq \rho(z_0, z) + \rho(z, x(t)) \leq \epsilon/4 + \epsilon/4 < \epsilon.$$

Thus

$$z_0 \in \Omega$$

and

$$\rho(z_0, x(t)) < \epsilon, \ t \in [L, T - L] \cap \Gamma.$$

Lemma 5.14 is proved.

Proof of Theorem 5.12 Assumption (A5) and Lemma 5.11 imply that there exist $M_1 > M$, $M_2 > 0$, and for each integer $p \geq 1$, there exists an integer $L_p \geq p$ such that the following property holds:

(a) For each $T \in \Gamma \cap [L_p, \infty)$ and each $y \in Y(0, T)$ satisfying $\rho(y(0), \widehat{\theta}) \leq M$, we have

$$\rho(y(t), X_0 \cap B(\widehat{\theta}, M_1)) \leq p^{-1}, \ t \in \Gamma \cap [L_p, T],$$

$$|\phi(y(t))| \leq M_2, \ t \in [L_p, T].$$

In this section, for a trajectory x, we use the notation $\lim_{t \to S^+} \phi(x(t))$ and $\lim_{t \to S^-} \phi(x(t))$ in their usual sense if $\Gamma = R^1$ and if $\Gamma = \mathcal{Z}$, then $\lim_{t \to S^+} \phi(x(t)) = \phi(x(S + 1))$ and $\lim_{t \to S^-} \phi(x(t)) = \phi(x(S - 1))$.
Lemma 5.14 implies that there exist $\delta \in (0, \epsilon)$ and an integer $L_0 \geq 1$ such that the following property holds:

(b) For every $T \in \Gamma \cap (2L_0, \infty)$ and every $x \in Y(0, T)$ satisfying $\rho(x(0), \widehat{\theta}) \leq M_1$ and $\phi(x(0)) \leq \phi(x(T)) + \delta$, there exists $z \in \Omega$ such that the inequality $\rho(x(t), z) \leq \epsilon$ holds for all $t \in [L_0, T - L_0] \cap \Gamma$.
Choose integers

$$Q > 2(M_1 + M_2)\delta^{-1} + 1 \tag{5.110}$$

and

$$L > 2L_0 + 2 + (4L_0 + 8L_1 + 8)(2\delta^{-1}(M_1 + M_2) + 1). \tag{5.111}$$

Suppose that $T \in \Gamma \cap [L, \infty)$ and that $x \in Y(0, T)$ satisfies

$$\rho(x(0), \widehat{\theta}) \leq M. \tag{5.112}$$

Property (a), (5.111), and (5.112) imply that for every $t \in [L_1, T] \cap \Gamma$.

$$\rho(x(t), \widehat{\theta})) \leq M_1 + 1, \ |\phi(x(t))| \leq M_2. \tag{5.113}$$

By induction, we define a strictly increasing finite sequence $t_i \in \Gamma \cap [0, T]$, $i = 0, \ldots, q$. Set

$$t_0 = L_1.$$

If

$$\lim_{t \to T^-} \phi(x(t)) \geq \lim_{t \to t_0^+} \phi(x(t)) - \delta,$$

then set $t_1 = T$ and complete to construct the sequence.
Assume that

$$\lim_{t \to T^-} \phi(x(t)) < \lim_{t \to t_0^+} \phi(x(t)) - \delta.$$

Set

$$t_1 = \inf\{t \in (t_0, T] \cap \Gamma : \phi(x(t)) < \lim_{t \to L_0^+} \phi(x(t)) - \delta\}.$$

It is not difficult to see that t_1 is well-defined, in the case $\Gamma = R^1$,

$$\lim_{t \to t_1^+} \phi(x(t)) \leq \lim_{t \to t_0^+} \phi(x(t)) - \delta \leq \lim_{t \to t_1 -} \phi(x(t))$$

and in the case $\Gamma = \mathcal{Z}$,

$$\phi(x(t_1)) < \phi(x(t_0)) - \delta$$

and for each $s \in \Gamma$ satisfying $t_0 < s < t_1$,

$$\phi(x(s)) \geq \phi(x(t_0)) - \delta.$$

If $t_1 = T$, then we complete to construct the sequence.
Assume that $k \geq 1$ is an integer and that we defined a finite strictly increasing sequence $t_0, \ldots, t_k \in [0, T] \cap \Gamma$ such that

$$t_0 = L_1, \ t_k \leq T$$

and that for each $i \in \{0, \ldots, k-1\}$, if $\Gamma = R^1_+$, then

$$\lim_{t \to t_{i+1}^+} \phi(x(t)) \leq \lim_{t \to t_i^+} \phi(x(t)) - \delta \leq \lim_{t \to t_{i+1}^-} \phi(x(t)), \tag{5.114}$$

if $\Gamma = \mathcal{Z}$, then

$$\phi(x(t_{i+1})) < \phi(x(t_i)) - \delta, \tag{5.115}$$

and if $s \in \Gamma$ satisfies $t_i < s < t_{i+1}$, then

$$\phi(x(s)) \geq \phi(x(t_i)) - \delta. \tag{5.116}$$

(It is not difficult to see that the assumption is true with $k = 1$.) If $t_k = T$, then we complete to construct the sequence. Assume that $t_k < T$. If

$$\lim_{t \to T^-} \phi(x(t)) \geq \lim_{t \to t_k^+} \phi(x(t)) - \delta,$$

then we set $t_{k+1} = T$ and complete to construct the sequenced.
Assume that

$$\lim_{t \to T^-} \phi(x(t)) < \lim_{t \to t_k^+} \phi(x(t)) - \delta.$$

Set

$$t_{k+1} = \inf\{t \in \Gamma \cap (t_k, T] : \phi(x(t)) < \lim_{t \to t_k^+} \phi(x(t)) - \delta\}.$$

It is not difficult to see that t_{k+1} is well-defined, $t_{k+1} > t_k$, if $\Gamma = R^1$, then

$$\lim_{t \to t_{k+1}^+} \phi(x(t)) \leq \lim_{t \to t_k^+} \phi(x(t)) - \delta \leq \lim_{t \to t_{k+1}^-} \phi(x(t)),$$

if $\Gamma = \mathcal{Z}$, then

$$\phi(x(t_{k+1})) < \phi(x(t_k)) - \delta,$$

and for each $s \in \Gamma \cap (t_k, t_{k+1})$,

$$\phi(x(s)) \geq \phi(x(t_k)) - \delta.$$

Thus, the assumption made for k is true for $k+1$ too. Therefore, by induction, we constructed the strictly increasing finite sequence $t_i \in [0, T] \cap \Gamma$, $i = 0, \ldots, q$ such that

$$t_0 = L_1, \ t_q = T,$$

for every integer i satisfying $0 \leq i < q - 1$, if $\Gamma = R^1$, then (5.114) holds, if $\Gamma = \mathcal{Z}$, then (5.115) holds, and if $s \in \Gamma$ satisfies $t_i < s < t_{i+1}$, then (5.116) holds. If $\Gamma = R^1$, then

$$\lim_{t\to T^-} \phi(x(t)) \geq \lim_{t\to t_{q-1}^+} \phi(x(t)) - \delta,$$

and if $\Gamma = \mathcal{Z}$, then for all integers s satisfying $t_{q-1} < s < T$,

$$\phi(x(s)) > \phi(x(t_{q-1})) - \delta.$$

By the relation $t_0 = L_1$, (5.110), (5.113), and (5.114),

$$2M_2 \geq \lim_{t\to t_0^+} \phi(x(t)) - \lim_{t\to t_{q-1}^+} \phi(x(t))$$

$$= \sum \{ \lim_{t\to t_i^+} \phi(x(t)) - \lim_{t\to t_{i+1}^+} \phi(x(t)) :$$

$$i \text{ is an integer, } 0 \leq i \leq q-2\} \geq \delta(q-1)$$

and

$$q \leq 1 + 2\delta^{-1} M_2 < Q. \tag{5.117}$$

Define

$$E = \{i \in \{0, \ldots, q-1\} : \; t_{i+1} - t_i \geq 2L_0 + 4\}. \tag{5.118}$$

Let

$$i \in E. \tag{5.119}$$

By (5.118) and (5.119),

$$t_{i+1} - 1 - (t_i + 1) \geq 2L_0 + 2. \tag{5.120}$$

Equations (5.114) and (5.120) imply that

$$\phi(x(t_{i+1} - 1)) \geq \phi(x(t_i)) - \delta. \tag{5.121}$$

Equations (5.113), (5.120), and (5.121) and property (b) applied to the restriction of x to the interval $[t_i + 1, t_{i+1} - 1]$ imply that there exists

$$z_i \in \Omega$$

such that

$$\rho(x(t), z_i) \le \epsilon,\ t \in [t_i + L_0 + 1, t_{i+1} - 1 - L_0] \cap \Gamma. \tag{5.122}$$

Set

$$a_i = t_i + L_0 + 1,\ b_i = t_{i+1} - L_0 - 1. \tag{5.123}$$

By (5.111), (5.117), (5.118), and (5.123),

$$\text{mes}([0, T] \cap \Gamma \setminus \cup_{i \in E}[a_i, b_i])$$

$$\le L_1 + \text{mes}(\cup\{[t_i, t_{i+1}] \cap \Gamma : i \in \{0, \dots, q-1\} \setminus E\})$$

$$+ \text{mes}(\cup\{[t_i, t_i + L_0 + 1] \cap \Gamma \cup [t_{i+1} - L_0 - 1, t_{i+1}] \cap \Gamma : i \in E\})$$

$$\le L_1 + (2L_0 + 5)(1 + 2M_2\delta^{-1}) < L.$$

Theorem 5.12 is proved.

5.10 Turnpike Results

Theorems 5.4 and 5.5 establish the turnpike property for approximate optimal trajectories which have a fixed starting point with respect to the objective function ϕ. On the other hand, there are many turnpike results where the turnpike property holds for approximate optimal trajectories with an initial point belonging to a given bounded set [117, 129, 148]. In this chapter we prove the following three turnpike results for approximate optimal trajectories with an initial point belonging to a given bounded set assuming that the function π is continuous at every point of Ω. This assumption holds for models of economic dynamics which are prototype of our dynamical system.

Theorem 5.15 *Assume that the function π is continuous at every point of Ω. Let $\epsilon \in (0, 1)$, $M > 0$. Then there exist $\delta > 0$ and $L \in \Gamma \cap [1, \infty)$ such that for every $T \in \Gamma \cap (2L, \infty)$ and every trajectory $x \in Y(0, T)$ which satisfies*

$$\rho(x(0), \widehat{\theta}) \le M \text{ and } \phi(x(T)) \ge \pi(x(0)) - \delta,$$

the value $\pi(x(0))$ is finite and

$$\rho(x(t), \Omega) \le \epsilon$$

for all $t \in [L, T - L] \cap \Gamma$.

The turnpike is the set Ω and δ and L depend on ϵ, M.

The following result is a generalization of the previous theorem. According to Theorem 5.15, for all $t \in [L, T - L] \cap \Gamma$, the distance between $\rho(x(t))$ and the turnpike Ω does not exceed ϵ. According to our next result, for all $t \in [L, T - L] \cap \Gamma$, the distance between $\rho(x(t))$ and the set $\phi^{-1}(\lambda) \cap B(\widehat{\theta}, M_1) \cap \Omega$ does not exceed ϵ, where a positive constant λ belongs to $\phi(B(\widehat{\theta}, M_1) \cap \Omega)$ and is close to $\pi(x(0))$.

Theorem 5.16 *Assume that the function π is continuous at every point of Ω. Let $\epsilon \in (0, 1)$, $M > 0$. Then there exist $\delta, M_1 > 0$ and $L \in \Gamma \cap [1, \infty)$ such that for every $T \in \Gamma \cap (2L, \infty)$ and every trajectory $x \in Y(0, T)$ which satisfies*

$$\rho(x(0), \widehat{\theta}) \leq M \text{ and } \phi(x(T)) \geq \pi(x(0)) - \delta,$$

the value $\pi(x(0))$ is finite and there exists

$$\lambda \in \phi(B(\widehat{\theta}, M_1) \cap \Omega) \cap [\pi(x(0)) - \epsilon, \pi(x(0)) + \epsilon]$$

such that for all $t \in [L, T - L] \cap \Gamma$,

$$\rho(x(t), \phi^{-1}(\lambda) \cap B(\widehat{\theta}, M_1) \cap \Omega) \leq \epsilon$$

and $|\phi(x(t)) - \lambda| \leq \epsilon$.

In our final result, we assume the property (P1) holds and obtain for all $t \in [L, T - L] \cap \Gamma$,

$$\rho(x(t), F(x(0))) \leq \epsilon$$

for approximate optimal trajectories with an initial point belonging to a given bounded set.

Theorem 5.17 *Assume that property (P1) holds and that the function π is continuous at every point of Ω. Let $\epsilon \in (0, 1)$, $M > 0$. Then there exist $\delta > 0$ and $L \in \Gamma \cap [1, \infty)$ such that for every $T \in \Gamma \cap (2L, \infty)$ and every trajectory $x \in Y(0, T)$ which satisfies*

$$\rho(x(0), \widehat{\theta}) \leq M \text{ and } \phi(x(T)) \geq \pi(x(0)) - \delta,$$

the value $\pi(x(0))$ is finite and for all $t \in [L, T - L] \cap \Gamma$,

$$\rho(x(t), F(x(0))) \leq \epsilon \text{ and } |\phi(x(t)) - \pi(x(0))| \leq \epsilon.$$

5.11 Auxiliary Results

In the previous section, we stated three turnpike results under the assumption that the function π is continuous at every point of Ω. In this section $\mathcal{A}$ and a are defined as in Sect. 5.1. For the discrete-time case when $\Gamma = \mathcal{Z}$, we show when the continuity of π holds.

Proposition 5.18 *Let $\Gamma = \mathcal{Z}$ $\xi \in \Omega$. Assume that for each $\epsilon > 0$ there exist $\delta > 0$ and $\xi_1 \in \Omega$ such that $\rho(\xi, \xi_1) \le \epsilon$ and that for each $x \in B(\xi, \delta)$ the inclusion $a(\xi_1) \subset a(x)$ holds. Then the function π is lower semicontinuous at the point ξ.*

Proof Let $\epsilon > 0$. There exists $\epsilon_1 \in (0, \epsilon)$ such that

$$|\phi(z) - \phi(\xi)| \le \epsilon/4 \tag{5.124}$$

for each $z \in B(\xi, \epsilon_1)$. There exist $\delta \in (0, \epsilon_1)$ and $\xi_1 \in \Omega$ such that

$$\rho(\xi, \xi_1) \le \epsilon_1 \tag{5.125}$$

and that the following property holds:

For each $x \in B(\xi, \delta)$,

$$a(\xi_1) \subset a(x). \tag{5.126}$$

Assume that

$$x \in B(\xi, \delta).$$

Then the inclusion (5.126) holds. Corollary 5.9, the inclusion ξ, $\xi_1 \in \Omega$, and (5.126) imply that

$$\pi(x) \ge \pi(\xi_1) = \phi(\xi_1) \ge \phi(\xi) - \epsilon/4 = \pi(\xi) - \epsilon/4.$$

Thus

$$\pi(x) \ge \pi(\xi) - \epsilon/4, \ x \in B(\xi, \delta).$$

Proposition 5.18 is proved.

Proposition 5.19 *Let $\Gamma = \mathcal{Z}$ $\xi \in \Omega$. Assume that for each $\epsilon > 0$ there exist $\delta > 0$ and $\xi_1 \in \Omega$ such that $\rho(\xi, \xi_1) \le \epsilon$ and that for each $x \in B(\xi, \delta)$ the inclusion $a(x) \subset a(\xi)$ holds. Then the function π is upper semicontinuous at the point ξ.*

Proof Let $\epsilon > 0$. There exists $\epsilon_1 \in (0, \epsilon)$ such that

$$|\phi(z) - \phi(\xi)| \le \epsilon/4 \tag{5.127}$$

for each $z \in B(\xi, \epsilon_1)$. There exist $\delta \in (0, \epsilon_1)$ and $\xi_1 \in \Omega$ such that

$$\rho(\xi, \xi_1) \le \epsilon_1 \tag{5.128}$$

and that for each $x \in B(\xi, \delta)$,

$$a(x) \subset a(\xi_1). \tag{5.129}$$

Assume that

$$x \in B(\xi, \delta). \tag{5.130}$$

In view of (5.130), the inclusion (5.129) holds. Corollary 5.9, (5.127)–(5.129), and the inclusion $\xi, \xi_1 \in \Omega$ imply that

$$\pi(x) \le \pi(\xi_1) = \phi(\xi_1) \le \phi(\xi) + \epsilon \le \pi(\xi) + \epsilon.$$

Proposition 5.19 is proved.

5.12 Proof of Theorem 5.15

Assumption (A5) implies that there exist $M_1 > M + 1$ and an integer $L_0 \ge 1$ such that the following property holds:

(i) For each $T \in \Gamma \cap [L_0, \infty)$ and each $y \in Y(0, T)$ satisfying $\rho(y(0), \widehat{\theta}) \le M$, we have

$$\rho(y(t), X_0 \cap B(\widehat{\theta}, M_1 - 1)) \le 1/2, \ t \in [L_0, T] \cap \Gamma.$$

Assumption (A6) and Theorems 3.6 (in the discrete case) and 4.6 (in the continuous case) imply that there exist $\epsilon_1 \in (0, \epsilon)$ and an integer $L_1 \ge 1$ such that the following property holds:

(ii) For every $T \in \Gamma \cap (2L_1, \infty)$ and every $y \in Y(0, T)$ which satisfies

$$\rho(y(0), \widehat{\theta}) \le M_1 \text{ and } \phi(y(0)) - \phi(y(T)) \le \epsilon_1,$$

the inequality

$$\rho(x(t), \Omega) \le \epsilon$$

holds for all $t \in [L_1, T - L_1] \cap \Gamma$.

In view of (5.2) and (5.4),

$$\Omega \cap B(\widehat{\theta}, M_1 + 1)$$

is a compact set. Corollary 5.9 implies that

$$\phi(x) = \pi(x),\ x \in \Omega. \tag{5.131}$$

Lemma 5.11 implies that there exists $\delta \in (0, \epsilon_1/4)$ such that the following property holds:

(iii) For each $z_1, z_2 \in X$ which satisfy

$$\rho(z_i, \Omega \cap B(\widehat{\theta}, M_1 + 1)) \leq \delta,\ i = 1, 2$$

and

$$\rho(z_1, z_2) \leq \delta,$$

we have

$$|\phi(z_1) - \phi(z_2)| \leq \epsilon_1/4,\ |\pi(z_1) - \pi(z_2)| \leq \epsilon_1/4.$$

Theorems 3.7 (in the discrete case) and 4.7 (in the continuous case) imply that there exists $L_3 \geq 1$ such that the following property holds:

(iv) For every number $T \in \Gamma \cap [L_3, \infty)$ and every trajectory $y \in Y(0, T)$ which satisfies

$$\rho(y(0), \widehat{\theta}) \leq M_1,$$

we have

$$\mathrm{mes}([0, T] \cap \Gamma :\ \rho(y(t), \Omega) > \delta\}) < L_3.$$

Choose an integer

$$L > 3(L_0 + L_1 + L_3). \tag{5.132}$$

Assume that

$$T \in (2L, \infty) \cap \Gamma,\ x \in Y(0, T), \tag{5.133}$$

$$\rho(x(0), \widehat{\theta}) \leq M, \tag{5.134}$$

$$\phi(x(T)) \geq \pi(x(0)) - \delta. \tag{5.135}$$

(Note that here we do not assume that $\pi(x(0))$ is finite.)

Property (i) and Eqs. (5.132)–(5.134) imply that for each $t \in [L_0, T] \cap \Gamma$,

$$\rho(x(t), X_0 \cap B(\widehat{\theta}, M_1 - 1)) \leq 2^{-1}. \tag{5.136}$$

Property (iv) and Eqs. (5.132) and (5.133) imply that there exists

$$\tau \in [L_0, L_0 + L_3] \cap \Gamma \tag{5.137}$$

such that

$$\rho(x(\tau), \Omega) \leq \delta. \tag{5.138}$$

In view of (5.138), there exists

$$z \in \Omega \tag{5.139}$$

such that

$$\rho(x(\tau), z) \leq \delta. \tag{5.140}$$

By (5.136) and (5.137),

$$\rho(x(\tau), \widehat{\theta}) \leq M_1. \tag{5.141}$$

Equations (5.140) and (5.141) imply that

$$\rho(z, \widehat{\theta}) \leq M_1 + 1. \tag{5.142}$$

Property (iii) and Eqs. (5.139), (5.140), and (5.142) imply that

$$|\phi(x(\tau)) - \phi(z)| \leq \epsilon_1/4, \tag{5.143}$$

$$|\pi(x(\tau)) - \pi(z)| \leq \epsilon_1/4. \tag{5.144}$$

In view of (5.131) and (5.139),

$$\phi(z) = \pi(z). \tag{5.145}$$

Proposition 5.1 and (5.1) 0imply that

$$\pi(x(\tau)) \leq \pi(x(0)) \leq \phi(x(0)). \tag{5.146}$$

Thus $\pi(x(0))$ is finite. By (5.135), (5.143), and (5.144)–(5.146),

$$\phi(x(\tau)) - \phi(x(T)) \leq \phi(z) + \epsilon_1/4 - \pi(x(0)) + \delta$$

$$\leq \phi(z)+\epsilon_1/4-\pi(x(\tau))+\delta \leq \phi(z)+\epsilon_1/4-\pi(z)+\epsilon_1/4+\delta < \epsilon_1. \qquad (5.147)$$

Equations (5.132), (5.133), (5.137), (5.141), and (5.147) and property (ii) applied to the restriction of x to $[\tau, T] \cap \Gamma$ imply that

$$\rho(x(t), \Omega) \leq \epsilon, \ t \in [\tau + L_1, T - L_1] \cap \Gamma.$$

Theorem 5.15 is proved.

5.13 Auxiliary Results for Theorem 5.16

Lemma 5.20 *Let $M, \epsilon > 0$ and let $\xi \in \Omega$ satisfy $\rho(\xi, \widehat{\theta}) \leq M$. Then there exists $\delta > 0$ such that for each $\eta \in \Omega \cap B(\widehat{\theta}, M)$ satisfying*

$$|\phi(\eta) - \phi(\xi)| \leq \delta,$$

the inequality

$$dist(\phi^{-1}(\phi(\eta)) \cap B(\widehat{\theta}, M) \cap \Omega, \phi^{-1}(\phi(\xi)) \cap B(\widehat{\theta}, M) \cap \Omega) \leq \epsilon$$

holds.

Proof Assume the contrary. Then, for each integer $i \geq 1$, there exists

$$\eta_i \in B(\widehat{\theta}, M) \cap \Omega$$

such that

$$|\phi(\eta_i) - \phi(\xi)| \leq 1/i, \qquad (5.148)$$

$$\rho(\eta_i, \phi^{-1}(\phi(\xi)) \cap B(\widehat{\theta}, M) \cap \Omega) > \epsilon. \qquad (5.149)$$

Since the set $\Omega \cap B(\widehat{\theta} M)$ is compact, we may assume without loss of generality that there exists

$$\eta = \lim_{i \to \infty} \eta_i.$$

Clearly,

$$\eta \in \Omega \cap B(\widehat{\theta}, M).$$

By (5.148),

$$\phi(\eta) = \lim_{i \to \infty} \phi(\eta_i) = \phi(\xi).$$

This implies that

$$\eta \in \phi^{-1}(\phi(\xi)) \cap B(\widehat{\theta}, M) \cap \Omega.$$

Therefore, for all integers $i \geq 1$,

$$\rho(\eta_i, \phi^{-1}(\phi(\xi)) \cap B(\widehat{\theta}, M) \cap \Omega) \leq \rho(\eta_i, \eta) \to 0$$

as $i \to \infty$. This contradicts (5.149). The contradiction we have reached completes the proof of Lemma 5.20.

Lemma 5.21 *Let $M, \epsilon > 0$. Then there exists $\delta > 0$ such that for each*

$$c \in \phi(B(\widehat{\theta}, M) \cap \Omega)$$

there exists

$$c_0 \in \phi(B(\widehat{\theta}, M) \cap \Omega)$$

such that

$$|c - c_0| \leq \epsilon,$$

and for each

$$\tilde{c} \in \phi(B(\widehat{\theta}, M) \cap \Omega) \cap [c - \delta, c + \delta],$$

the inequality

$$dist(\phi^{-1}(\tilde{c}) \cap B(\widehat{\theta}, M) \cap \Omega, \phi^{-1}(c_0) \cap B(\widehat{\theta}, M) \cap \Omega) \leq \epsilon$$

holds.

Proof Lemma 5.20 implies that for each

$$\xi \in \Omega \cap B(\widehat{\theta}, M),$$

there exists $\delta_\xi \in (0, \min\{\epsilon, 1\})$ such that the following property holds:

(i) For each $\eta \in \Omega \cap B(\widehat{\theta}, M)$ satisfying

$$|\phi(\eta) - \phi(\xi)| \leq \delta_\xi,$$

we have

$$\text{dist}(\phi^{-1}(\phi(\eta)) \cap B(\widehat{\theta}, M) \cap \Omega, \phi^{-1}(\phi(\xi)) \cap B(\widehat{\theta}, M) \cap \Omega) \leq \epsilon.$$

Clearly,

$$\phi(B(\widehat{\theta}, M) \cap \Omega) \subset \cup\{(\phi(\xi) - \delta_\xi/2, \phi(\xi) + \delta_\xi/2) : \xi \in B(\widehat{\theta}, M) \cap \Omega\},$$

and there exist $\xi_i \in B(\widehat{\theta}, M) \cap \Omega$, $i = 1, \dots, q$, such that

$$\phi(B(\widehat{\theta}, M) \cap \Omega) \subset \cup_{i=1}^{q}(\phi(\xi_i) - \delta_{\xi_i}/2, \phi(\xi_i) + \delta_{\xi_i}/2). \tag{5.150}$$

Set

$$\delta = \min\{\delta_{\xi_i} : i = 1, \dots, q\}/4. \tag{5.151}$$

Assume that

$$c \in \phi(B(\widehat{\theta}, M) \cap \Omega). \tag{5.152}$$

In view of (5.150) and (5.152), there exists $j \in \{1, \dots, q\}$ such that

$$c \in (\phi(\xi_j) - \delta_{\xi_j}/2, \phi(\xi_j) + \delta_{\xi_j}/2). \tag{5.153}$$

Set

$$c_j = \phi(\xi_j). \tag{5.154}$$

By (5.153) and (5.154),

$$|c - c_j| < 2^{-1}\delta_{\xi_j} < \epsilon.$$

Assume that

$$\tilde{c} \in \phi(B(\widehat{\theta}, M) \cap \Omega) \cap [c - \delta, c + \delta]. \tag{5.155}$$

It follows from (5.151), (5.153), and (5.154) that

$$|\tilde{c} - \phi(\xi_j)| \le 2^{-1}\delta_{\xi_j} + \delta < \delta_{\xi_j}. \tag{5.156}$$

Equations (5.154)–(5.156) and property (i) applied with $\xi = \xi_j$ imply that

$$\mathrm{dist}(\phi^{-1}(\tilde{c}) \cap B(\widehat{\theta}, M) \cap \Omega, \phi^{-1}(c_j) \cap B(\widehat{\theta}, M) \cap \Omega) \le \epsilon.$$

Lemma 5.21 is proved.

5.14 Proofs of Theorems 5.16 and 5.17

Proof of Theorem 5.16 Assumption (A5) implies that there exist $M_1 > M + 1$ and an integer $L_0 \geq 1$ such that the following property holds:

(i) For each $T \in \Gamma \cap [L_0, \infty)$ and each $y \in Y(0, T)$ satisfying $\rho(y(0), \widehat{\theta}) \leq M$, we have

$$\rho(y(t), X_0 \cap B(\widehat{\theta}, M_1 - 1)) \leq 1/2, \; t \in [L_0, T] \cap \Gamma.$$

Lemma 5.21 implies that there exists $\epsilon_0 \in (0, \epsilon/4)$ such that the following property holds:

(ii) For each

$$c \in \phi(B(\widehat{\theta}, M_1) \cap \Omega),$$

there exists

$$c_0 \in \phi(B(\widehat{\theta}, M_1) \cap \Omega)$$

such that

$$|c - c_0| \leq \epsilon/8,$$

and for each

$$\tilde{c} \in \phi(B(\widehat{\theta}, M_1) \cap \Omega) \cap [c - \epsilon_0, c + \epsilon_0],$$

the inequality

$$\mathrm{dist}(\phi^{-1}(\tilde{c}) \cap B(\widehat{\theta}, M_1) \cap \Omega, \phi^{-1}(c_0) \cap B(\widehat{\theta}, M_1) \cap \Omega) \leq \epsilon/4$$

is true.

Lemma 5.11 and the compactness of the set $\Omega \cap B(\widehat{\theta}, M_1 + 1)$ imply that there exists $\epsilon_1 \in (0, \epsilon_0/4)$ such that the following property holds:

(iii) For each $z_1, z_2 \in X$ which satisfy

$$\rho(z_i, \Omega \cap B(\widehat{\theta}, M_1 + 1)) \leq 2\epsilon_1, \; i = 1, 2$$

and

$$\rho(z_1, z_2) \leq 2\epsilon_1,$$

we have

$$|\phi(z_1) - \phi(z_2)| \leq \epsilon_0/4, \; |\pi(z_1) - \pi(z_2)| \leq \epsilon_0/4.$$

Assumption (A6) and Theorems 3.6 (in the discrete case) and 4.6 (in the continuous case) imply that there exist $\epsilon_2 \in (0, \epsilon_1/4)$ and an integer $L_1 \geq 1$ such that the following property holds:

(iv) For every $T \in \Gamma \cap (2L_1, \infty)$ and every $x \in Y(0, T)$ which satisfies

$$\rho(x(0), \widehat{\theta}) \leq M_1 \text{ and } \phi(x(0)) - \phi(x(T)) \leq \epsilon_2,$$

the inequality

$$\rho(x(t), \Omega) \leq \epsilon_1$$

holds for all $t \in [L_1, T - L_1] \cap \Gamma$.

Lemma 5.11 and the compactness of the set $\Omega \cap B(\widehat{\theta}, M_1 + 1)$ imply that there exists $\delta \in (0, \epsilon_2/4)$ such that the following property holds:

(v) For each $z_1, z_2 \in X$ which satisfy

$$\rho(z_i, \Omega \cap B(\widehat{\theta}, M_1 + 1)) \leq 2\delta, \ i = 1, 2$$

and

$$\rho(z_1, z_2) \leq 2\delta,$$

we have

$$|\phi(z_1) - \phi(z_2)| \leq \epsilon_2/4, \ |\pi(z_1) - \pi(z_2)| \leq \epsilon_2/4.$$

Theorems 3.7 (in the discrete case) and 4.7 (in the continuous case) imply that there exists an integer $L_2 > 2L_1$ such that the following property holds:

(vi) For every number $T \in \Gamma \cap [L_2, \infty)$ and every trajectory $x \in Y(0, T)$ which satisfies

$$\rho(x(0), \widehat{\theta}) \leq M,$$

we have

$$\text{mes}([0, T] \cap \Gamma : \ \rho(x(t), \Omega) > \delta\}) < L_2.$$

Choose an integer

$$L > 3(L_0 + L_1 + L_2). \tag{5.157}$$

Assume that

$$T \in (2L, \infty) \cap \Gamma, \ x \in Y(0, T), \tag{5.158}$$

$$\rho(x(0), \widehat{\theta}) \le M, \tag{5.159}$$

$$\phi(x(T)) \ge \pi(x(0)) - \delta. \tag{5.160}$$

(Note that here we do not assume that $\pi(x(0))$ is finite.)
Property (i) and Eqs. (5.157)–(5.159) imply that for each $t \in [L_0, T] \cap \Gamma$,

$$\rho(x(t), \widehat{\theta}) \le M_1 - 2^{-1}. \tag{5.161}$$

Property (vi) and Eq. (5.161) imply that there exists

$$\tau \in [L_0, L_0 + L_2] \cap \Gamma \tag{5.162}$$

such that

$$\rho(x(\tau), \Omega) \le \delta. \tag{5.163}$$

In view of (5.163), there exists

$$\xi \in \Omega \tag{5.164}$$

such that

$$\rho(x(\tau), \xi) \le \delta. \tag{5.165}$$

By (5.161), (5.162), and (5.165),

$$x(\tau) \in B(\widehat{\theta}, M_1 - 2^{-1}), \ \xi \in B(\widehat{\theta}, M_1). \tag{5.166}$$

Property (v) and Eqs. (5.164)–(5.166) imply that

$$|\phi(x(\tau)) - \phi(\xi)| \le \epsilon_2/4, \tag{5.167}$$

$$|\pi(x(\tau)) - \pi(\xi)| \le \epsilon_2/4. \tag{5.168}$$

Thus $\pi(x(0))$ is finite. Proposition 5.1, (5.1), and (5.6) imply that

$$\pi(x(\tau)) \le \pi(x(0)) \le \phi(x(0)). \tag{5.169}$$

Corollary 5.9, (5.1), (5.160), (5.164), and (5.167) imply that

$$\begin{aligned} \pi(\xi) = \phi(\xi) &\ge \phi(x(\tau)) - \epsilon_2/4 \\ &\ge \pi(x(0)) - \delta - \epsilon_2/4. \end{aligned} \tag{5.170}$$

It follows from (5.168) and (5.169) that

$$\pi(\xi) \le \epsilon_2/4 + \pi(x(\tau)) \le \epsilon_2/4 + \pi(x(0)). \tag{5.171}$$

Equations (5.160) and (5.167)–(5.170) imply that

$$\begin{aligned}\phi(x(T)) &\ge \pi(x(0)) - \delta \ge \pi(x(\tau)) - \delta \ge \pi(\xi) - \epsilon_2/4 - \delta \\ &\ge \phi(\xi) - \epsilon_2/4 - \epsilon_2/4 \ge \phi(x(\tau)) - 3\epsilon_2/4.\end{aligned} \tag{5.172}$$

In view of (5.157), (5.158), and (5.162),

$$T - \tau > L > 3L_1. \tag{5.173}$$

Equations (5.166), (5.172), and (5.173) and property (iv) applied to the restriction of x to $[\tau, T] \cap \Gamma$ imply that

$$\rho(x(t), \Omega) \le \epsilon_1,\ t \in [\tau + L_1, T - L_1] \cap \Gamma. \tag{5.174}$$

Let

$$t \in [\tau + L_1, T - L_1] \cap \Gamma. \tag{5.175}$$

In view of (5.174) and (5.175), there exists

$$\xi(t) \in \Omega \tag{5.176}$$

such that

$$\rho(x(t), \xi(t)) \le \epsilon_1. \tag{5.177}$$

By (5.161), (5.162), (5.174), and (5.177),

$$\rho(\xi(t), \widehat{\theta}) \le M_1. \tag{5.178}$$

Property (iii) and Eqs. (5.176)–(5.178) imply that

$$|\phi(x(t)) - \phi(\xi(t))| \le \epsilon_0/4, \tag{5.179}$$

$$|\pi(x(t)) - \pi(\xi(t))| \le \epsilon_0/4. \tag{5.180}$$

Corollary 5.9 and (5.176) imply that

$$\phi(\xi(t)) = \pi(\xi(t)). \tag{5.181}$$

It follows from (5.1), (5.172), and (5.175) that

$$\phi(x(t)) \le \phi(x(\tau)) \le \phi(x(T)) + \epsilon_2 \le \phi(x(t)) + \epsilon_2$$

and

$$|\phi(x(t)) - \phi(x(\tau))| \le \epsilon_2. \tag{5.182}$$

By (5.167), (5.179), and (5.182),

$$|\phi(\xi) - \phi(\xi(t))| \le |\phi(\xi) - \phi(x(\tau))| + |\phi(x(\tau)) - \phi(x(t))|$$

$$+ |\phi(x(t)) - \phi(\xi(t))| \le \epsilon_2/4 + \epsilon_2 + \epsilon_0/4 < \epsilon_0. \tag{5.183}$$

In view of (5.170) and (5.171),

$$|\pi(\xi) - \pi(x(0))| \le \epsilon_2/4. \tag{5.184}$$

By Eqs. (5.179) and (5.183),

$$|\phi(x(t)) - \phi(\xi)| \le \epsilon. \tag{5.185}$$

It follows from (5.164), (5.166), (5.176), (5.178), (5.183), and property (ii) (applied with $c = \phi(\xi)$, $\tilde{c} = \phi(\xi(t))$, $c_0 = \phi(\widehat{\xi})$) that there exists

$$\widehat{\xi} \in B(\widehat{\theta}, M_1) \cap \Omega$$

such that

$$|\phi(\widehat{\xi}) - \phi(\xi)| \le \epsilon/8,$$

and for each $t \in \Gamma \cap [\tau + L_1, T - L_1]$,

$$\rho(\xi(t), \phi^{-1}(\phi(\widehat{\xi})) \cap B(\widehat{\theta}, M_1) \cap \Omega) \le \epsilon/4.$$

Together with (5.177), this implies that

$$\rho(x(t), \phi^{-1}(\phi(\widehat{\xi})) \cap B(\widehat{\theta}, M_1) \cap \Omega) \le \epsilon.$$

Setting $\lambda = \phi(\widehat{\xi})$, we complete the proof of Theorem 5.16.

Proof of Theorem 5.17 It is not difficult to see that the following proposition is true.

Proposition 5.22 *Assume that property (P1) holds and that* $x \in Y(0, \infty)$*. Then there exists* $\lim_{t\to\infty} x(t) \in \Omega$.

Property (P1), assumption (A5), Proposition 5.3, and (5.5) imply that if (P1) holds, $x \in X$, and $\pi(x)$ is finite, then $F(x) \in \Omega$ is a singleton and that

$$F : \{x \in X : \pi(x) > -\infty\} \to \Omega.$$

Since all bounded, closed subsets of Ω are compact, property (P1) implies the next result.

Proposition 5.23 *Assume that (P1) holds and $M, \epsilon > 0$. Then there exists $\delta > 0$ such that for each*

$$z_1, z_2 \in \Omega \cap B(\widehat{\theta}, M)$$

satisfying

$$|\phi(z_1) - \phi(z_2)| \le \delta,$$

the inequality $\rho(z_1, z_2) \le \epsilon$ holds.

Now it is easy to see that Theorem 5.17 follows from Theorem 5.16 and Propositions 5.3 and 5.23.

5.15 Extensions of Theorems 5.16 and 5.17

In the previous sections we showed that approximate optimal trajectories are closed to the turnpike except in regions close to the endpoints. In this chapter we prove the following result which shows under certain additional assumptions that if an initial point of approximate optimal trajectory is close to the turnpike, then the trajectory is close to the turnpike in the region containing the left endpoint.

Theorem 5.24 *Assume that the following assumptions hold:*

(B1) For each $M > 0$, there exists $M_0 > 0$ such that for each $T \in \Gamma \cap (0, \infty)$ and each $y \in Y(0, T)$ satisfying $y(0) \in B(\widehat{\theta}, M)$ the inclusion $y(t) \in B(\widehat{\theta}, M_0)$ holds for all $t \in [0, T] \cap \Gamma$.

(B2) For each $\epsilon, M > 0$, there exists $\delta > 0$ such that for each $T \in (0, \infty) \cap \Gamma$ and each $y \in Y(0, T)$ satisfying

$$\rho(y(0), X_0 \cap B(\widehat{\theta}, M)) \le \delta,$$

the inequality

$$\rho(y(t), X_0) \le \epsilon, \ t \in [0, T] \cap \Gamma$$

is true.

Assume that the function π is continuous at every point of Ω. Let $\epsilon \in (0, 1)$, $M > 0$. Then there exist $\delta, M_1 > 0$ and $L \in \Gamma \cap [1, \infty)$ such that for every $T \in \Gamma \cap (2L, \infty)$ and every trajectory $x \in Y(0, T)$ which satisfies

$$\rho(x(0), \widehat{\theta}) \le M \text{ and } \phi(x(T)) \ge \pi(x(0)) - \delta,$$

the value $\pi(x(0))$ is finite and there exist

$$\lambda \in \phi(B(\widehat{\theta}, M_1) \cap \Omega) \cap [\pi(x(0)) - \epsilon, \pi(x(0)) + \epsilon]$$

and $\tau \in [0, L] \cap \Gamma$ such that for each $t \in [\tau, T - L] \cap \Gamma$,

$$\rho(x(t), \phi^{-1}(\lambda) \cap B(\widehat{\theta}, M_1) \cap \Omega) \le \epsilon$$

and $|\phi(x(t)) - \lambda| \le \epsilon$. Moreover, if $\rho(x(0), \Omega) \le \delta$, then $\tau = 0$.

The following theorem easily follows from Theorem 5.24 and Propositions 5.3 and 5.23.

Theorem 5.25 *Assume that assumptions (B1) and (B2) and property (P1) hold and that the function π is continuous at every point of Ω. Let $\epsilon \in (0, 1)$, $M > 0$. Then there exist $\delta, M_1 > 0$ and $L \in \Gamma \cap [1, \infty)$ such that for every $T \in \Gamma \cap (2L, \infty)$ and every trajectory $x \in Y(0, T)$ which satisfies*

$$\rho(x(0), \widehat{\theta}) \le M \text{ and } \phi(x(T)) \ge \pi(x(0)) - \delta,$$

the value $\pi(x(0))$ is finite and there exists $\tau \in [0, L] \cap \Gamma$ such that for each $t \in [\tau, T - L] \cap \Gamma$,

$$\rho(x(t), F(x(0)) \le \epsilon \text{ and } |\phi(x(t)) - \pi(x(0))| \le \epsilon.$$

Moreover, if $\rho(x(0), \Omega) \le \delta$, then $\tau = 0$.

5.16 An Auxiliary Result for Theorem 5.24

Lemma 5.26 *Assume that (B1) and (B2) hold. Let $\epsilon, M > 0$ and L be a natural number. Then there exist $\delta > 0$ and an integer $L_1 \ge L$ such that for every $T \in [L_1, \infty) \cap \Gamma$ and every $x \in Y(0, T)$ satisfying*

$$\rho(x(0), \widehat{\theta}) \le M, \ \rho(x(0), \Omega) \le \delta, \ \phi(x(T)) \ge \phi(x(0)) - \delta$$

the inequality $\rho(x(t), \Omega) \le \epsilon$ holds for all $t \in [0, L] \cap \Gamma$.

Proof Assume that the lemma is not true. Then, for each integer $k \geq 1$, there exist $T_k \in [L + k, \infty) \cap \Gamma$ and $x_k \in Y(0, T_k)$ such that

$$\rho(x_k(0), \widehat{\theta}) \leq M, \tag{5.186}$$

$$\rho(x_k(0), \Omega) \leq 1/k, \tag{5.187}$$

$$\phi(x_k(T_k)) \geq \phi(x_k(0)) - 1/k, \tag{5.188}$$

$$\sup\{\rho(x_k(t), \Omega) : t \in [0, L] \cap \Gamma\} > \epsilon. \tag{5.189}$$

In view of (B1) and (5.186), there exists $M_0 > M$ such that

$$\phi(x_k(t), \widehat{\theta}) \leq M_0, \ t \in [0, T_k] \cap \Gamma, \ k = 1, 2, \ldots. \tag{5.190}$$

It follows from (B2), (5.187), and (5.190) that

$$\lim_{k \to \infty} \sup\{\rho(x_k(t), X_0) : t \in [0, T_k] \cap \Gamma\} = 0. \tag{5.191}$$

By (A4), (5.190) and (5.191), extracting subsequence, using diagonalization process, and re-indexing, we obtain that there exist a subsequence $\{x_{k_j}\}_{j=1}^{\infty}$ and $x \in Y(0, \infty)$ such that for every integer $m \geq 1$,

$$x_{k_j} \text{ converge to } x \text{ as } j \to \infty \text{ uniformly on } [0, m]. \tag{5.192}$$

Equations (5.187) and (5.190)–(5.192) imply that

$$x(t) \in X_0 \cap B(\widehat{\theta}, M_0), \ t \in \Gamma,$$

$$x(0) \in \Omega. \tag{5.193}$$

In view of (5.188) and (5.192),

$$\phi(x(t)) = \phi(x(0)), \ t \in [0, \infty) \cap \Gamma. \tag{5.194}$$

Proposition 5.10, (5.193), and (5.194) imply that

$$x(t) \in \Omega, \ t \in [0, \infty) \cap \Gamma.$$

By the equation above, (5.192), for all sufficiently large natural numbers j,

$$\rho(x_{k_j}(t), \Omega) < \epsilon/2, \ t \in [0, L] \cap \Gamma.$$

This contradicts (5.189). The contradiction we have reached completes the proof of Lemma 5.26.

5.17 Proof of Theorem 5.24

Theorem 5.24 follows from Theorem 5.16 and the following result.

Proposition 5.27 *Assume that (B1) and (B2) hold and that the function π is continuous at every point of Ω. Let $\epsilon \in (0, 1)$, $M > 0$. Then there exist $\delta, M_1 > 0$ and $L \in \Gamma \cap [1, \infty)$ such that for every $T \in \Gamma \cap (2L, \infty)$ and every trajectory $x \in Y(0, T)$ which satisfies*

$$\rho(x(0), \widehat{\theta}) \leq M, \ \rho(x(0), \Omega) \leq \delta \text{ and } \phi(x(T)) \geq \pi(x(0)) - \delta,$$

the value $\pi(x(0))$ is finite and there exists

$$\lambda \in \phi(B(\widehat{\theta}, M_1 + 1) \cap \Omega) \cap [\pi(x(0)) - \epsilon, \pi(x(0)) + \epsilon]$$

such that for all $t \in [0, T - L] \cap \Gamma$,

$$\rho(x(t), \phi^{-1}(\lambda) \cap B(\widehat{\theta}, M_1 + 1) \cap \Omega) \leq \epsilon$$

and $|\phi(x(t)) - \lambda| \leq \epsilon$.

Proof Assumption (B1) implies that there exists $M_1 > M + 1$ such that the following property holds:

(i) For each $T \in \Gamma \cap (0, \infty)$ and each $y \in Y(0, T)$ satisfying $y(0) \in B(\widehat{\theta}, M)$, the inclusion $y(t) \in B(\widehat{\theta}, M_1)$ holds for all $t \in [0, T] \cap \Gamma$.
Lemma 5.21 implies that there exists $\epsilon_0 \in (0, \epsilon/4)$ such that the following property holds:

(ii) For each

$$c \in \phi(B(\widehat{\theta}, M_1 + 1) \cap \Omega)$$

there exists

$$c_0 \in \phi(B(\widehat{\theta}, M_1 + 1) \cap \Omega)$$

such that

$$|c - c_0| \leq \epsilon/8,$$

and for each

$$\tilde{c} \in \phi(B(\widehat{\theta}, M_1 + 1) \cap \Omega) \cap [c - \epsilon_0, c + \epsilon_0],$$

the inequality

$$\text{dist}(\phi^{-1}(\tilde{c}) \cap B(\widehat{\theta}, M_1 + 1) \cap \Omega, \phi^{-1}(c_0) \cap B(\widehat{\theta}, M_1 + 1) \cap \Omega) \leq \epsilon/4$$

is true.

Lemma 5.11 and the compactness of the set $\Omega \cap B(\widehat{\theta}, M_1+1)$ imply that there exists $\epsilon_1 \in (0, \epsilon_0/4)$ such that the following property holds:

(iii) For each $z_1, z_2 \in X$ which satisfy

$$\rho(z_i, \Omega \cap B(\widehat{\theta}, M_1 + 1)) \le 2\epsilon_1,\ i = 1, 2$$

and

$$\rho(z_1, z_2) \le 2\epsilon_1,$$

we have

$$|\phi(z_1) - \phi(z_2)| \le \epsilon_0/4,\ |\pi(z_1) - \pi(z_2)| \le \epsilon_0/4.$$

Assumption (A6) and Theorems 3.6 (in the discrete case) and 4.6 (in the continuous case) imply that there exist $\epsilon_2 \in (0, \epsilon_1/4)$ and an integer $L_1 \ge 1$ such that the following property holds:

(iv) For every $T \in \Gamma \cap (2L_1, \infty)$ and every $x \in Y(0, T)$ which satisfies

$$\rho(x(0), \widehat{\theta}) \le M_1 \text{ and } \phi(x(0)) - \phi(x(T)) \le \epsilon_2,$$

the inequality

$$\rho(x(t), \Omega) \le \epsilon_1$$

holds for all $t \in [L_1, T - L_1] \cap \Gamma$.

Lemma 5.26 implies that there exist $\epsilon_3 \in (0, \epsilon_2/2)$ and an integer $L \ge 2L_1$ such that the following property holds:

(v) For every $T \in [L, \infty) \cap \Gamma$ and every $x \in Y(0, T)$ satisfying

$$\rho(x(0), \widehat{\theta}) \le M,\ \rho(x(0), \Omega) \le \epsilon_3,\ \phi(x(T)) \ge \phi(x(0)) - \epsilon_3,$$

the inequality $\rho(x(t), \Omega) \le \epsilon_1$ holds for all $t \in [0, L_1] \cap \Gamma$.

Lemma 5.11 and the compactness of the set $\Omega \cap B(\widehat{\theta}, M_1+1)$ imply that there exists $\delta \in (0, \epsilon_3/2)$ such that the following property holds:

(vi) For each $z_1, z_2 \in X$ which satisfy

$$\rho(z_i, \Omega \cap B(\widehat{\theta}, M_1 + 1)) \le 2\delta,\ i = 1, 2$$

and

$$\rho(z_1, z_2) \le 2\delta,$$

we have

$$|\phi(z_1) - \phi(z_2)| \le \epsilon_3/4, \ |\pi(z_1) - \pi(z_2)| \le \epsilon_3/4.$$

Assume that

$$T \in (L, \infty) \cap \Gamma, \ x \in Y(0, T), \tag{5.196}$$

$$\rho(x(0), \widehat{\theta}) \le M, \ \rho(x(0), \Omega) \le \delta, \tag{5.197}$$

$$\phi(x(T)) \ge \pi(x(0)) - \delta. \tag{5.198}$$

Property (i) and Eq. (5.197) imply that

$$\rho(x(t), \widehat{\theta}) \le M_1, \ t \in [0, T] \cap \Gamma. \tag{5.199}$$

In view of (5.197), there exists

$$z \in \Omega \tag{5.200}$$

such that

$$\rho(x(0), z) \le \delta. \tag{5.201}$$

By (5.197) and (5.201),

$$\rho(\widehat{\theta}, z) \le M + 1. \tag{5.202}$$

Property (vi) and Eqs. (5.200)–(5.202) imply that

$$|\pi(x(0)) - \pi(z)| \le \epsilon_3/4. \tag{5.203}$$

$$|\phi(x(0)) - \phi(z)| \le \epsilon_3/4. \tag{5.204}$$

Corollary 5.9 and (5.200) imply that

$$\phi(z) = \pi(z). \tag{5.205}$$

In view of (5.203)–(5.205),

$$|\phi(x(0)) - \pi(x(0))| \le \epsilon_3/2. \tag{5.206}$$

By (5.1), (5.198), and (5.206),

$$|\phi(x(T)) - \phi(x(0))| \le \epsilon_3. \tag{5.207}$$

Property (v), (5.196), (5.197), and (5.207) imply that

$$\rho(x(t), \Omega) \le \epsilon_1, \ t \in [0, L_1] \cap \Gamma. \tag{5.208}$$

By property (iv), (5.196), (5.197), and (5.207),

$$\rho(x(t), \Omega) \le \epsilon_1, \ t \in [L_1, T - L_1] \cap \Gamma.$$

It follows from the equation above and (5.208) that

$$\rho(x(t), \Omega) \le \epsilon_1, \ t \in [0, T - L_1] \cap \Gamma. \tag{5.209}$$

Let

$$t \in [0, T - L_1] \cap \Gamma. \tag{5.210}$$

In view of (5.209) and (5.210), there exists

$$z(t) \in \Omega \tag{5.211}$$

such that

$$\rho(x(t), z(t)) \le \epsilon_1. \tag{5.212}$$

By (5.199) and (5.212),

$$\rho(z(t), \widehat{\theta}) \le M_1 + 1. \tag{5.213}$$

Property (iii) and Eqs. (5.211)–(5.213) imply that

$$|\pi(z(t)) - \pi(x(t))| \le \epsilon_0/4, \ |\phi(z(t)) - \phi(x(t))| \le \epsilon_0/4. \tag{5.214}$$

By (5.204), (5.207), and (5.214),

$$|\phi(z) - \phi(z(t))| \le |\phi(z) - \phi(x(0))| + |\phi(x(0)) - \phi(x(t))|$$

$$+ |\phi(x(t)) - \phi(z(t))| \le \epsilon_3/4 + \epsilon_3 + \epsilon_0/4 < \epsilon_0. \tag{5.215}$$

It follows from (5.200), (5.210), (5.211), (5.213), (5.215), and property (ii) (applied with $c = \phi(z)$, $c_0 = \lambda$, $\tilde{c} = \phi(z(t))$) that there exists

$$\lambda \in \phi(B(\widehat{\theta}, M_1 + 1) \cap \Omega)$$

such that

$$|\phi(z) - \lambda| \le \epsilon/8,$$

and for each $t \in \Gamma \cap [0, T - L_1]$,

$$\rho(z(t), \phi^{-1}(\lambda) \cap B(\widehat{\theta}, M_1 + 1) \cap \Omega) \le \epsilon/4.$$

Together with (5.206) and (5.212), this implies that

$$\rho(x(t), \phi^{-1}(\lambda) \cap B(\widehat{\theta}, M_1 + 1) \cap \Omega) < \epsilon$$

and

$$|\lambda - \pi(x(0))| \le \epsilon.$$

Proposition 5.27 is proved.

5.18 Generalizations of Theorems 5.16 and 5.17

In the several turnpike results, we showed that the turnpike property holds when the optimality criterion is defined by the Lyapunov function ϕ. Now we prove the turnpike results where the optimality criterion is defined by an element of the large class of functions.

Let $\mathfrak{M}$ be a set of functions $\psi : X \to R^1$ such that $\phi \in \mathfrak{M}$ and the following assumptions hold:

(C1) For each $\epsilon > 0$ and each $M > 0$, there exists $\delta > 0$ such that for each $z \in X \cap B(\widehat{\theta}, M)$ satisfying $\rho(z, \Omega) \le \delta$ there exist $\xi_1, \xi_2 \in \Omega$ such that

$$\rho(z, \xi_i) \le \epsilon, \ i = 1, 2,$$

and for each $\psi \in \mathfrak{M}$ and each $T \in \Gamma \cap (0, \infty)$,

$$\sup\{\psi(x(T)) : x \in Y(0, T) \text{ and } x(0) = \xi_1\} - \epsilon$$

$$\le \sup\{\psi(x(T)) : x \in Y(0, T) \text{ and } x(0) = z\}$$

$$\le \sup\{\psi(x(T)) : x \in Y(0, T) \text{ and } x(0) = \xi_2\} + \epsilon$$

(note that by Proposition 5.8 all the values in the equations above are finite).

(C2) For each $\epsilon > 0$ and each $M > 0$, there exists $\delta > 0$ such that for each $\xi_1, \xi_2 \in \Omega \cap B(\widehat{\theta}, M)$ satisfying $\phi(\xi_1) \le \phi(\xi_2) - \epsilon$, each $\psi \in \mathfrak{M}$, and each $T \in \Gamma \cap (0, \infty)$,

$$\sup\{\psi(x(T)) : x \in Y(0,T) \text{ and } x(0) = \xi_1\} + \delta$$

$$\leq \sup\{\psi(x(T)) : x \in Y(0,T) \text{ and } x(0) = \xi_2\}.$$

(Note that (C2) holds if $\mathfrak{M} = \{\phi\}$.)

The following result is proved in Sect. 5.19.

Theorem 5.28 *Assume that the function π is continuous at any point of Ω. Let $\epsilon \in (0,1)$, $M > 0$. Then there exist $\delta, M_1 > 0$ and $L \in \Gamma \cap [1,\infty)$ such that for every $T \in \Gamma \cap (6L,\infty)$, every $\psi \in \mathfrak{M}$, and every $x \in Y(0,T)$ which satisfies*

$$\rho(x(0), \widehat{\theta}) \leq M,$$

$$\psi(x(T)) \geq \sup\{\psi(z(T)) : z \in Y(0,T) \text{ and } z(0) = x(0)\} - \delta,$$

the value $\pi(x(0))$ is finite and there exist

$$\lambda \in \phi(B(\widehat{\theta}, M_1) \cap \Omega) \cap [\pi(x(0)) - \epsilon, \pi(x(0)) + \epsilon]$$

and $\tau \in [0,3L] \cap \Gamma$ such that for all $t \in [\tau, T-3L] \cap \Gamma$,

$$\rho(x(t), \phi^{-1}(\lambda) \cap B(\widehat{\theta}, M_1) \cap \Omega) \leq \epsilon$$

and $|\phi(x(t)) - \lambda| \leq \epsilon$. Moreover, if (B1) and (B2) hold and $\rho(x(0), \Omega) \leq \delta$, then $\tau = 0$.

The next result follows easily from Theorem 5.28 and Propositions 5.3 and 5.23.

Theorem 5.29 *Assume that the function π is continuous at any point of Ω and (P1) hods. Let $\epsilon \in (0,1)$, $M > 0$. Then there exist $\delta, M_1 > 0$ and $L \in \Gamma \cap [1,\infty)$ such that for every $T \in \Gamma \cap (6L,\infty)$, every $\psi \in \mathfrak{M}$, and every $x \in Y(0,T)$ which satisfies*

$$\rho(x(0), \widehat{\theta}) \leq M,$$

$$\psi(x(T)) \geq \sup\{\psi(z(T)) : z \in Y(0,T) \text{ and } z(0) = x(0)\} - \delta,$$

the value $\pi(x(0))$ is finite and there exists $\tau \in [0,3L] \cap \Gamma$ such that for all $t \in [\tau, T-3L] \cap \Gamma$,

$$\rho(x(t), F(x(0)) \leq \epsilon$$

and $|\phi(x(t)) - \pi(x(0))| \leq \epsilon$. Moreover, if (B1) and (B2) hold and $\rho(x(0), \Omega) \leq \delta$, then $\tau = 0$.

5.19 Proof of Theorem 5.28

Assumption (A5) implies that there exist $M_1 > M + 1$ and an integer $L_0 \ge 0$ such that the following property holds:

(i) For each $T \in \Gamma \cap [L_0, \infty)$ and each $y \in Y(0, T)$ satisfying $\rho(y(0), \widehat{\theta}) \le M$, we have

$$\rho(y(t), B(\widehat{\theta}, M_1 - 1)) \le 1/4, \ t \in [L_0, T] \cap \Gamma,$$

and if (B1) holds, then $L_0 = 0$.
Lemma 5.21 implies that there exists $\epsilon_0 \in (0, \epsilon/8)$ such that the following property holds:

(ii) For each

$$c \in \phi(B(\widehat{\theta}, M_1) \cap \Omega),$$

there exists

$$c_0 \in \phi(B(\widehat{\theta}, M_1) \cap \Omega)$$

such that

$$|c - c_0| \le \epsilon/8,$$

and for each

$$\tilde{c} \in \phi(B(\widehat{\theta}, M_1) \cap \Omega) \cap [c - \epsilon_0, c + \epsilon_0],$$

the inequality

$$\mathrm{dist}(\phi^{-1}(\tilde{c}) \cap B(\widehat{\theta}, M_1) \cap \Omega, \phi^{-1}(c_0) \cap B(\widehat{\theta}, M_1) \cap \Omega) \le \epsilon/4$$

is true.
Lemma 5.11 and the compactness of the set $\Omega \cap B(\widehat{\theta}, M_1 + 1)$ imply that there exists $\epsilon_1 \in (0, \epsilon_0/4)$ such that the following property holds:

(iii) For each $z_1, z_2 \in X$ which satisfy

$$\rho(z_i, \Omega \cap B(\widehat{\theta}, M_1 + 1)) \le 2\epsilon_1, \ i = 1, 2$$

and

$$\rho(z_1, z_2) \le 2\epsilon_1,$$

we have

$$|\phi(z_1) - \phi(z_2)| \le \epsilon_0/8, \ |\pi(z_1) - \pi(z_2)| \le \epsilon_0/8.$$

Assumption (A6) and Theorems 5.15 and 5.24 imply that there exist $\epsilon_2 \in (0, \epsilon_1/4)$ and an integer $L_1 \geq L_0$ such that the following property holds:

(iv) For every $T \in \Gamma \cap (2L_1, \infty)$ and every $x \in Y(0, T)$ which satisfies

$$\rho(x(0), \widehat{\theta}) \leq M_1 \text{ and } \phi(x(T)) \geq \pi(x(0)) - \epsilon_2,$$

the value $\pi(x(0))$ is finite and there exists $\tau \in [0, L_1] \cap \Gamma$ such that the inequality

$$\rho(x(t), \Omega) \leq \epsilon_1/4$$

holds for all $t \in [\tau, T - L_1] \cap \Gamma$, and if (B1) and (B2) hold and $\rho(x(0), \Omega) \leq \epsilon_2$, then $\tau = 0$.

Fix

$$\gamma_0 \in (0, \epsilon_2/4).$$

Lemma 5.11 and assumption (C2) imply that there exists

$$\delta_0 \in (0, \gamma_0/4)$$

such that the following properties holds:

(v) For each $\xi_1, \xi_2 \in X$ satisfying

$$\rho(\xi_i, X_0 \cap B(\widehat{\theta}, M_1 + 2)) \leq 2\delta_0, \ i = 1, 2$$

and $\rho(\xi_1, \xi_2) \leq 2\delta_0$, we have

$$|\phi(\xi_1) - \phi(\xi_2)| \leq \gamma_0/8;$$

(vi) For each $\xi_1, \xi_2 \in \Omega \cap B(\widehat{\theta}, M_1 + 2)$ satisfying $\phi(\xi_1) \leq \phi(\xi_2) - \gamma_0/4$, each $T \in \Gamma \cap (0, \infty)$, and each $\psi \in \mathfrak{M}$,

$$\sup\{\psi(x(T)) : x \in Y(0, T), \ x(0) = \xi_1\} + 2\delta_0$$

$$\leq \sup\{\psi(x(T)) : x \in Y(0, T), \ x(0) = \xi_2\}.$$

Assumption (C1) imply that there exists

$$\delta_1 \in (0, \delta_0/16)$$

such that the following properties holds:

(vii) For each $z \in X \cap B(\widehat{\theta}, M_1+2)$ satisfying $\rho(z, \Omega) \leq \delta_1$, there exist $\xi_1, \xi_2 \in \Omega$ such that

$$\rho(z, \xi_i) \leq \delta_0/16, \ i = 1, 2$$

and such that for each $T \in (0, \infty) \cap \Gamma$ and each $\psi \in \mathfrak{M}$,

$$\sup\{\psi(x(T)) : \ x \in Y(0, T), \ x(0) = \xi_1\} - \delta_0/16$$

$$\leq \sup\{\psi(x(T)) : \ x \in Y(0, T), \ x(0) = z\}$$

$$\leq \sup\{\psi(x(T)) : \ x \in Y(0, T), \ x(0) = \xi_2\} + \delta_0/16.$$

Theorems 5.15 and 5.24 imply that there exist $\delta \in (0, \delta_1/4)$ and an integer $L_2 \geq 1$ such that the following property holds:

(viii) For every $T \in \Gamma \cap (2L_2, \infty)$ and every $x \in Y(0, T)$ which satisfies

$$\rho(x(0), \widehat{\theta}) \leq M \text{ and } \phi(x(T)) \geq \pi(x(0)) - \delta,$$

the value $\pi(x(0))$ is finite and there exists $\tau \in [0, L_2] \cap \Gamma$ such that the inequality

$$\rho(x(t), \Omega) \leq \delta_1$$

holds for all $t \in [\tau, T - L_2] \cap \Gamma$, and if (B1) and (B2) hold and $\rho(x(0), \Omega) \leq \delta_2$, then $\tau = 0$.

Theorems 3.7 and 4.7 imply that there exists an integer

$$L \geq 4L_2 + 4L_1 + 4L_0 \tag{5.216}$$

such that the following property holds:

(ix) For every $T \in (L, \infty) \cap \Gamma$ and every $x \in Y(0, T)$ satisfying $\rho(x(0), \widehat{\theta}) \leq M$, we have

$$\text{mes}(\{t \in [0, T] \cap \Gamma : \ \rho(x(t), \Omega) > \delta\}) < L.$$

Assume that

$$T \in (6L, \infty) \cap \Gamma, \ \psi \in \mathfrak{M}, \tag{5.217}$$

$x \in Y(0, T)$ satisfies

$$\rho(x(0), \widehat{\theta}) \leq M, \tag{5.218}$$

$$\psi(x(T)) \geq \sup\{\psi(z(T)) : \ z \in Y(0, T), \ z(0) = x(0)\} - \delta. \tag{5.219}$$

Property (i), (5.217), and (5.218) imply that for every $t \in [L_0, \infty) \cap \Gamma$,

$$\rho(x(t), B(\widehat{\theta}, M_1 - 1)) \le 4^{-1}$$

and

$$\rho(x(t), \widehat{\theta}) \le M_1 - 1/2. \tag{5.220}$$

Property (ix) and (5.217)–(5.220) imply that there exist

$$\tau_0 \in [L, 2L] \cap \Gamma, \ \ \tau_1 \in [T - 2L, T - L] \cap \Gamma \tag{5.221}$$

such that for $i = 0, 1$ we have

$$\rho(x(\tau_i), \widehat{\theta}) \le M_1 - 1/2, \tag{5.222}$$

$$\rho(x(\tau_i), \Omega) \le \delta, \ \ i = 0, 1. \tag{5.223}$$

There exists $y \in Y(0, T)$ such that

$$y(0) = x(0), \tag{5.224}$$

$$\phi(y(T)) \ge \phi(z(T)) - \delta/4 \text{ for every } z \in Y(0, T) \text{ satisfying } z(0) = y(0). \tag{5.225}$$

Properties (i) and (viii) and Eqs. (5.216)–(5.218), (5.224), and (5.225) imply that there exists

$$\tau_2 \in [0, L_0 + L_2] \cap \Gamma \tag{5.226}$$

such that

$$\rho(y(t), \widehat{\theta}) \le M_1 - 1/2, \ \ \rho(y(t), \Omega) \le \delta_1, \ \ t \in [\tau_2, T - L_2], \tag{5.227}$$

and if (B1) and (B2) holds and $\rho(x(0), \Omega) \le \delta$, then $\tau_2 = 0$.

By (5.216), (5.217), (5.221), (5.226), and (5.227),

$$\tau_0, \tau_1 \in [\tau_2, T - L], \tag{5.228}$$

$$\rho(y(\tau_i), \widehat{\theta}) \le M_1 - 1/2, \ \ \rho(y(\tau_i), \Omega) \le \delta_1, \ \ i = 0, 1. \tag{5.229}$$

Thus we have shown that there exist $\tau_0, \tau_1 \in [0, T] \cap \Gamma$ such that (5.222), (5.223), and (5.229) hold,

$$\tau_0 \le 2L, \ \ T - 2L \le \tau_1 \le T - L, \tag{5.230}$$

and if (B1) and (B2) hold and $\rho(x(0), \Omega) \le \delta$, then $\tau_0 = 0$.

We will show that for $i = 0, 1$,

$$|\phi(x(\tau_i)) - \phi(y(\tau_i))| \le \gamma_0. \tag{5.231}$$

Let $i \in \{0, 1\}$. Assume that (5.231) does not hold. Then

$$|\phi(x(\tau_i)) - \phi(y(\tau_i))| > \gamma_0. \tag{5.232}$$

Property (vii), (5.222), and (5.223) imply that there exist

$$\xi_1, \xi_2 \in \Omega \tag{5.233}$$

such that

$$\rho(x(\tau_i), \xi_j) \le \delta_0/16, \;\; j = 1, 2, \tag{5.234}$$

and for each $S \in (0, \infty) \cap \Gamma$ and each $\Lambda \in \mathfrak{M}$,

$$\begin{aligned} &\sup\{\Lambda(z(S)) : \; z \in Y(0, S), \; z(0) = \xi_1\} - \delta_0/16 \\ &\le \sup\{\Lambda(z(S)) : \; z \in Y(0, S), \; z(0) = x(\tau_i)\} \\ &\le \sup\{\Lambda(z(S)) : \; z \in Y(0, S), \; z(0) = \xi_2\} + \delta_0/16. \end{aligned} \tag{5.235}$$

Property (vii) and (5.229) imply that there exist

$$\eta_1, \eta_2 \in \Omega \tag{5.236}$$

such that

$$\rho(y(\tau_i), \eta_j) \le \delta_0/16, \;\; j = 1, 2 \tag{5.237}$$

and that for each $S \in (0, \infty) \cap \Gamma$ and each $\Lambda \in \mathfrak{M}$,

$$\begin{aligned} &\sup\{\Lambda(z(S)) : \; z \in Y(0, S), \; z(0) = \eta_1\} - \delta_0/16 \\ &\le \sup\{\Lambda(z(S)) : \; z \in Y(0, S), \; z(0) = y(\tau_i)\} \\ &\le \sup\{\Lambda(z(S)) : \; z \in Y(0, S), \; z(0) = \eta_2\} + \delta_0/16. \end{aligned} \tag{5.238}$$

Property (v) and Eqs. (5.222), (5.229), (5.233), (5.234), (5.236), and (5.237) imply that for $j = 1, 2$,

$$|\phi(\xi_j) - \phi(x(\tau_i))| \le \gamma_0/8, \; |\phi(y(\tau_i)) - \phi(\eta_j)| \le \gamma_0/8. \tag{5.239}$$

For each pair $p, q \in \{1, 2\}$, it follows from (5.232) and (5.239) that

$$|\phi(\xi_p)-\phi(\eta_q)| \geq |\phi(x(\tau_i))-\phi(y(\tau_i))|-|\phi(x(\tau_i))-\phi(\xi_p)|-|\phi(y(\tau_i))-\phi(\eta_q)|$$
$$\geq \gamma_0 - \gamma_0/4. \tag{5.240}$$

Equations (5.239) and (5.240) imply that

$$[\min\{\phi(\eta_1), \phi(\eta_2)\}, \max\{\phi(\eta_1), \phi(\eta_2)\}]$$
$$\cap [\min\{\phi(\xi_1), \phi(\xi_2)\}, \max\{\phi(\xi_1), \phi(\xi_2)\}] = \emptyset. \tag{5.241}$$

In view of (5.241), there are two cases:

$$\phi(\eta_1) > \phi(\xi_2); \tag{5.242}$$

$$\phi(\eta_2) < \phi(\xi_1). \tag{5.243}$$

Assume that (5.243) is true. By (5.240) and (5.243),

$$\phi(\xi_1) - \phi(\eta_2) \geq 3\gamma_0/4. \tag{5.244}$$

Property (vi) (with $\xi_1 = \eta_2$, $\xi_2 = \xi_1$) and Eqs. (5.235) (with $S = T - \tau_i$, $\Lambda = \phi$), (5.222), (5.224), (5.225), (5.229), (5.230), (5.233), (5.234), (5.236), (5.237), (5.244), and (5.238) (with $\Lambda = \phi$, $S = T - \tau_i$) imply that

$$\phi(y(T)) + \delta/4$$
$$\geq \sup\{\phi(z(T)) : \ z \in Y(0, T), \ z(0) = x(0)\}$$
$$\geq \sup\{\phi(z(T - \tau_i)) : \ z \in Y(0, T - \tau_i), \ z(0) = x(\tau_i)\}$$
$$\geq \sup\{\phi(z(T - \tau_i)) : \ z \in Y(0, T - \tau_i), \ z(0) = \xi_1\} - \delta_0/16$$
$$\geq \sup\{\phi(z(T - \tau_i)) : \ z \in Y(0, T - \tau_i), \ z(0) = \eta_2\} + 2\delta_0 - \delta_0/16,$$
$$\geq \sup\{\phi(z(T - \tau_i)) : \ z \in Y(0, T - \tau_i), \ z(0) = y(\tau_i)\} + 2\delta_0 - \delta_0/8$$
$$\geq \phi(y(T)) + 2\delta_0 - \delta_0/8.$$

This contradicts the choice of δ. The contradiction we have reached proves that (5.242) holds. By (5.240) and (5.242),

$$\phi(\eta_1) - \phi(\xi_2) > 3\gamma_0/4. \tag{5.245}$$

Property (vi) (with $\xi_1 = \xi_2, \xi_2 = \eta_1$) and Eqs. (5.235) and (5.238) (with $S = T - \tau_i$, $\Lambda = \psi$), (5.219), (5.222), (5.223), (5.229), (5.230), (5.233), (5.234), (5.236), (5.237), and (5.245) imply that

$$\psi(x(T)) + \delta$$

$$\geq \sup\{\psi(z(T)) : z \in Y(0, T), z(0) = x(0)\}$$

$$\geq \sup\{\psi(z(T - \tau_i)) : z \in Y(0, T - \tau_i), z(0) = y(\tau_i)\}$$

$$\geq \sup\{\psi(z(T - \tau_i)) : z \in Y(0, T - \tau_i), z(0) = \eta_1\} - \delta_0/16$$

$$\geq \sup\{\psi(z(T - \tau_i)) : z \in Y(0, T - \tau_i), z(0) = \xi_2\} + 2\delta_0 - \delta_0/16$$

$$\geq \sup\{\psi(z(T - \tau_i)) : z \in Y(0, T - \tau_i), z(0) = x(\tau_i)\} + 2\delta_0 - \delta_0/8$$

$$\geq \psi(x(T)) + 2\delta_0 - \delta_0/8.$$

This contradicts the choice of δ. The contradiction we have reached proves that

$$|\phi(x(\tau_i)) - \phi(y(\tau_i))| \leq \gamma_0, \ i = 0, 1. \tag{5.246}$$

In view of (5.216), (5.217), and (5.230),

$$\tau_1 - \tau_0 \geq T - 4L \geq 2L > 4L_1.$$

Proposition 5.1 and Eqs. (5.1), (5.224), (5.225), (5.230), and (5.246) imply that

$$\phi(x(\tau_1)) \geq \phi(y(\tau_1)) - \gamma_0 \geq \phi(y(T)) - \gamma_0$$

$$\geq \pi(x(0)) - \delta/4 - \gamma_0 \geq \pi(x(\tau_0)) - \gamma_0 - \delta/4. \tag{5.247}$$

Equations (5.222), (5.247), and (5.247) and property (iv) applied to the restriction of x to $[\tau_0, \tau_1] \cap \Gamma$ imply that the value $\pi(x(\tau_0))$ is finite and there exists $\tau \in [0, L_1] \cap \Gamma$ such that

$$\rho(x(t), \Omega) \leq \epsilon_1/4, \ t \in [\tau_0 + \tau, \tau_1 - L] \cap \Gamma, \tag{5.248}$$

and if (B1) and (B2) hold and $\rho(x(0), \Omega) \leq \epsilon_2$, then $\tau = 0$. Note that if $\rho(x(0), \Omega) \leq \delta$ and (B1) and (B2) hold, then $\tau_0 = 0$, $\tau = 0$. Otherwise we assume that

$$\tau = L_1. \tag{5.249}$$

By (5.222) and (5.223), there exist $\tilde{\xi}_i \in X$, $i = 0, 1$, such that for $i = 0, 1$,

$$\tilde{\xi}_i \in \Omega, \ \ \rho(x(\tau_i), \tilde{\xi}_i) \le \delta, \ \ \rho(\tilde{\xi}_i, \widehat{\theta}) \le M_1. \tag{5.250}$$

Equations (5.223), (5.233), (5.234), and (5.250) and property (ii) applied with $c = \phi(\tilde{\xi}_0)$ imply that there exists

$$\lambda \in \phi(B(\widehat{\theta}, M_1) \cap \Omega) \cap [\phi(\tilde{\xi}_0) - \epsilon/8, \phi(\tilde{\xi}_0) + \epsilon/8] \tag{5.251}$$

such that the following property holds:

(x) For each

$$\tilde{c} \in \phi(B(\widehat{\theta}, M_1) \cap \Omega) \cap [\lambda - \epsilon_0, \lambda + \epsilon_0],$$

the inequality

$$\mathrm{dist}(\phi^{-1}(\tilde{c}) \cap B(\widehat{\theta}, M_1) \cap \Omega, \phi^{-1}(\lambda) \cap B(\widehat{\theta}, M_1) \cap \Omega) \le \epsilon/4$$

is true.

Assume that

$$t \in [\tau_0 + \tau, \tau_1 - L] \cap \Gamma. \tag{5.252}$$

Property (i) and Eqs. (5.220), (5.249), and (5.252) imply that

$$\rho(x(t), \widehat{\theta}) \le M_1 - 1/2. \tag{5.253}$$

It follows from (5.248) and (5.252) that there exists

$$\xi(t) \in \Omega \tag{5.254}$$

such that

$$\rho(x(t), \xi(t)) < \epsilon_1. \tag{5.255}$$

By (5.253) and (5.255),

$$\rho(\xi(t), \widehat{\theta}) \le M_1. \tag{5.256}$$

Proposition 5.1 and (5.252) imply that

$$\pi(x(0)) \ge \pi(x(\tau_0)) \ge \pi(x(t)) \ge \pi(x(\tau_1)). \tag{5.257}$$

Property (iii) and Eqs. (5.250) and (5.254)–(5.256) imply that

$$|\pi(x(\tau_i)) - \pi(\tilde{\xi}_i)| \le \epsilon_0/8,\ i = 0, 1, \tag{5.258}$$

$$|\phi(x(\tau_i)) - \phi(\tilde{\xi}_i)| \le \epsilon_0/8,\ i = 0, 1, \tag{5.259}$$

$$|\pi(\xi(t)) - \pi(x(t))| \le \epsilon_0/8,\ |\phi(\xi(t)) - \phi(x(t))| \le \epsilon_0/8. \tag{5.260}$$

Corollary 5.9, (5.219), (5.250), and (5.257)–(5.259) imply that

$$\phi(x(\tau_0)) \le \phi(\tilde{\xi}_0) + \epsilon_0/8 = \pi(\tilde{\xi}_0) + \epsilon_0/8 \le \pi(x(\tau_0)) + \epsilon_0/4$$

$$\le \pi(x(0)) + \epsilon_0/4 \le \phi(x(\tau)) + \epsilon_0/4 + \delta \le \phi(x(\tau_1)) + \epsilon_0/4 + \delta. \tag{5.261}$$

In view of (5.1) and (5.261),

$$|\phi(x(\tau_0)) - \pi(x_0)| \le \epsilon_0/4. \tag{5.262}$$

By (5.259) and (5.262),

$$|\phi(\tilde{\xi}_0) - \pi(x_0)| \le \epsilon_0/2. \tag{5.263}$$

Equations (5.1), (5.261), and (5.262) imply that

$$|\phi(x(\tau_0)) - \phi(x(t))| \le \epsilon_0/4 + \delta. \tag{5.264}$$

It follows from (5.259), (5.260), and (5.264) that

$$|\phi(x(\tau_0)) - \phi(x(t))| \le \epsilon_0/4 + \delta. \tag{5.265}$$

It follows from (5.259), (5.260), and (5.264) that

$$|\phi(\tilde{\xi}_0) - \phi(\xi(t))|$$

$$\le |\phi(\tilde{\xi}_0) - \phi(x(\tau_0))| + |\phi(x(\tau_0)) - \phi(x(t))| + |\phi(x(t)) - \phi(\xi(t))|$$

$$\le \epsilon_0/8 + \epsilon_0/4 + \delta + \epsilon_0/8 < \epsilon_0. \tag{5.266}$$

Property (x) (applied with $\tilde{c} = \phi(\xi(t))$ and Eqs. (5.254), (5.256), and (5.266) imply that

$$\rho(\xi(t), \phi^{-1}(\lambda) \cap B(\widehat{\theta}, M_1) \cap \Omega) \le \epsilon/4$$

and together with (5.255) this implies that

$$\rho(x(t), \phi^{-1}(\lambda) \cap B(\widehat{\theta}, M_1) \cap \Omega) \leq \epsilon. \tag{5.267}$$

It follows from (5.251), (5.260), and (5.266) that

$$|\phi(x(t)) - \lambda|$$

$$\leq |\phi(x(t)) - \phi(\xi(t))| + |\phi(\xi(t)) - \phi(\tilde{\xi}_0)| + |\phi(\tilde{\xi}_0) - \lambda|$$

$$\leq \epsilon_0/8 + \epsilon_0 + \epsilon/8 < \epsilon. \tag{5.268}$$

Thus (5.267) and (5.268) hold for all $t \in [\tau_0 + \tau, \tau_1 - L_1] \cap \Gamma$. By (5.251) and (5.263),

$$|\lambda - \pi(x(0))| \leq \epsilon.$$

This completes the proof of Theorem 5.28.

5.20 Continuity of the Function π

In this section we consider a particular case when X is a vector space equipped with an inner product which induces the norm

$$\|x\| = \langle x, x\rangle^{1/2}, \ x \in X$$

and the metric

$$\rho(x, y) = \|x - y\|, \ x, y \in X.$$

Assume that a mapping $F : X \to 2^Y$ is proper ($\mathrm{Dom}(F) \neq \emptyset$). Recall (see Chap. 2) that for each pair of numbers $T_1 < T_2$, $Y(T_1, T_2)$ is the set of all functions $x \in W^{1,1}(T_1, T_2; X)$ such that

$$x'(t) \in F(x(t)), \ t \in [T_1, T_2] \text{ almost everywhere (a.e.).}$$

We also suppose that the assumptions made in Chap. 2 hold.

Assume that X_+ is a nonempty closed, convex cone in X such that $\lambda X_+ \subset X_+$ for each $\lambda \geq 0$. We assume that X is ordered by X_+: for each $x, y \in X$,

$$x \leq y \text{ if and only if } y - x \in X_+.$$

We suppose that for each $x \in X$, $\emptyset + x = \emptyset$.

Proposition 5.30 *Assume that for each $x, y \in X$ satisfying $x \le y$,*

$$F(x) \subset F(y), \ \phi(x) \le \phi(y),$$

$\xi \in \Omega$, and that for each $\epsilon > 0$ there exists $\delta > 0$ and $\xi_1 \in \Omega$ such that $\|\xi - \xi_1\| \le \epsilon$ and $B(\xi, \delta) \subset \xi_1 + X_+$. Then π is lower semicontinuous at ξ.

Proof Let $\epsilon > 0$. There exists $\epsilon_1 \in (0, \epsilon)$ such that

$$|\phi(z) - \phi(\xi)| \le \epsilon/4 \tag{5.269}$$

for each $z \in B(\xi, \epsilon_1)$. There exist

$$\delta \in (0, \epsilon_1), \ \xi_1 \in \Omega \cap B(\xi, \epsilon_1) \tag{5.270}$$

such that the following property holds:

(a) For each $x \in B(\xi, \delta)$, we have $\xi_1 \le x$.

Assume that

$$x \in B(\xi, \delta). \tag{5.271}$$

Corollary 5.9 and (5.270) imply that

$$\pi(\xi_1) = \phi(\xi_1). \tag{5.272}$$

Assume that $T > 0$ and $z \in Y(0, T)$ satisfies

$$z(0) = \xi_1. \tag{5.273}$$

By definition for a.e. $t \in [0, T]$,

$$z'(t) \in F(z(t)). \tag{5.274}$$

Define

$$z_1(t) = z(t) - \xi_1 + x, \ t \in [0, T]. \tag{5.275}$$

Property (a) and Eqs. (5.271), (5.273), and (5.275) imply that

$$z_1(0) = x, \ z_1(t) \ge z(t), \ t \in [0, T].$$

For a.e. $t \in [0, T]$,

$$z_1'(t) = z'(t) \in F(z(t)) \subset F(z_1(t)).$$

It is not difficult to see that

$$\sup\{\phi(y(T)) : \; y \in Y(0, T), \; y(0) = x\}$$

$$\geq \sup\{\phi(y(T)) : \; y \in Y(0, T), \; y(0) = \xi_1\},$$

$$\phi(z_1(T)) \geq \phi(z(T)), \; \pi(x) \geq \pi(\xi_1).$$

In view of (5.269), (5.270), and (5.272), for every $x \in B(\xi, \delta)$,

$$\pi(x) \geq \pi(\xi_1) = \phi(\xi_1) \geq \phi(\xi) - \epsilon/4 \geq \pi(\xi) - \epsilon/4.$$

Proposition 5.30 is proved.

Proposition 5.31 *Assume that for each $x, y \in X$ satisfying $x \leq y$,*

$$F(x) \subset F(y), \; \phi(x) \leq \phi(y),$$

$\xi \in \Omega$ and that for each $\epsilon > 0$ there exist $\delta > 0$ and $\xi_1 \in \Omega$ such that $\|\xi - \xi_1\| \leq \epsilon$ and $B(\xi, \delta) \subset \xi_1 - X_+$. Then π is upper semicontinuous at ξ.

Proof Let $\epsilon > 0$. There exists $\epsilon_1 \in (0, \epsilon)$ such that

$$|\phi(z) - \phi(\xi)| \leq \epsilon/4 \tag{5.276}$$

for each $z \in B(\xi, \epsilon_1)$. There exist

$$\delta \in (0, \epsilon_1), \; \xi_1 \in \Omega \cap B(\xi, \epsilon_1), \tag{5.277}$$

and for each $x \in B(\xi, \delta)$, we have

$$x \leq \xi_1. \tag{5.278}$$

Assume that

$$x \in B(\xi, \delta), \tag{5.279}$$

$T > 0$, and $z \in Y(0, T)$ satisfies

$$z(0) = x. \tag{5.280}$$

By definition for a.e. $t \in [0, T]$,

$$z'(t) \in F(z(t)).$$

Define

$$z_1(t) = z(t) + \xi_1 - x, \ t \in [0, T]. \tag{5.281}$$

Equations (5.278), (5.279), and (5.281) imply that

$$z_1(t) \geq z(t), \ t \in [0, T]. \tag{5.282}$$

By (5.282), for a.e. $t \in [0, T]$,

$$z_1'(t) = z'(t) \in F(z(t)) \subset F(z_1(t)).$$

Clearly,

$$\phi(z_1(T)) \geq \phi(z(T)),$$

$$\sup\{\phi(y(T)) : \ y \in Y(0, T), \ y(0) = \xi_1\}$$

$$\geq \sup\{\phi(y(T)) : \ y \in Y(0, T), \ y(0) = x\},$$

$$\pi(z_1) \geq \pi(x).$$

Together with Corollary 5.9, (5.276), and (5.277), this implies that

$$\pi(x) \leq \pi(\xi_1) = \phi(\xi_1) \leq \phi(\xi) + \epsilon/4 = \pi(\xi) + \epsilon/4,$$

$$\pi(x) \leq \pi(\xi) + \epsilon/4$$

for all $x \in B(\xi, \delta)$. Proposition 5.31 is proved.

5.21 Examples

In this chapter we studied the class of dynamical systems satisfying assumptions (A1)–(A5). The most restrictive assumption is (A4). Here we present two examples of infinite dimensional systems for which (A4) holds. Both of them are discussed in [18].

Example 5.32 (The One-Dimensional Heat Equation) For $f \in L^\infty(0, 1)$, define $S(t)f$ to be the unique solution $u(\cdot, t)$ of the problem

$$u_t = u_{xx}, \ 0 < x < 1, \ t > 0,$$

$$u = 0 \text{ at } x = 0, 1, \ u(x, 0) = f(x).$$

Set $x(0) = f$, $x(t) = S(t)f$, $t > 0$. It was shown in [18] that for this dynamical system (A4) holds.

Example 5.33 (A Semilinear Wave Equation) Let $\Omega \subset R^n$, $n \geq 3$, be a bounded and open set with boundary $\partial\Omega$. Consider the semilinear wave equation

$$u_{tt} + \beta u_t - \Delta u + f(u) = 0 \text{ in } \Omega \tag{5.283}$$

with boundary condition

$$u|_{\partial\Omega} = 0 \tag{5.284}$$

and initial conditions

$$u(x,0) = u_0(x), \;\; u_t(x,0) = u_1(x), \tag{5.285}$$

where $\beta > 0$ is a constant and a continuous function $f : R^1 \to R^1$ satisfies the growth condition

$$|f(u)| \leq c_0(|u|^{n/n-2} + 1),$$

where $c_0 > 0$ is a constant and a condition

$$\liminf_{|u|\to\infty} f(u)/u \geq -\lambda_1,$$

where λ_1 is the first eigenvalue of the operator $-\Delta$ with the boundary conditions (5.284). Let

$$X = H_0^1(\Omega) \times L^2(\Omega).$$

Then given $(u_0, u_1) \in X$, there exists a weak solution $x(\cdot) = (u, u_t)$ on $[0, \infty)$ satisfying (5.283)–(5.285). It was shown in [18] that for this dynamical system (A4) holds.

Chapter 6
Discrete-Time Nonautonomous Problems on Half-Axis

In this chapter we study the turnpike phenomenon for discrete-time nonautonomous problems on subintervals of half-axis in metric spaces, which are not necessarily compact. For these optimal control problems the turnpike is not a singleton. We show that the turnpike phenomenon is stable under small perturbations of an objective function. Using the Baire category approach we show that for some classes of problems a typical (generic) problem has a turnpike property.

6.1 Preliminaries and Boundedness Results

Let (E, ρ_E) and (F, ρ_F) be metric spaces. The space $E \times F$ is equipped with the metric

$$\rho_{E\times F}((x_1, u_1), (x_2, u_2)) = \rho_E(x_1, x_2) + \rho_F(u_1, u_2),\ (x_i, u_i) \in E \times F,\ i = 1, 2.$$

We suppose that $\mathcal{A}$ is a nonempty subset of $\{0, 1, \dots, \} \times E, \mathcal{U} : \mathcal{A} \to 2^F$ is a point to set mapping with a graph

$$\mathcal{M} = \{(t, x, u) :\ (t, x) \in \mathcal{A},\ u \in \mathcal{U}(t, x)\}, \tag{6.1}$$

$G : \mathcal{M} \to E$ and $f : \mathcal{M} \to R^1$.

Let $0 \le T_1 < T_2$ be integers. We denote by $X(T_1, T_2)$ the set of all pairs of sequences $(\{x_t\}_{t=T_1}^{T_2}, \{u_t\}_{t=T_1}^{T_2-1})$ such that for each integer $t \in \{T_1, \dots, T_2\}$,

$$(t, x_t) \in \mathcal{A}, \tag{6.2}$$

for each integer $t \in \{T_1, \dots, T_2 - 1\}$,

A. J. Zaslavski, *Turnpike Phenomenon in Metric Spaces*, Springer Optimization and Its Applications 201, https://doi.org/10.1007/978-3-031-27208-0_6

$$u_t \in \mathcal{U}(t, x_t), \tag{6.3}$$

$$x_{t+1} = G(t, x_t, u_t) \tag{6.4}$$

and which are called trajectory–control pairs.

Let $T_1 \geq 0$ be an integer. Denote by $X(T_1, \infty)$ the set of all pairs of sequences $\{x_t\}_{t=T_1}^{\infty} \subset E$, $\{u_t\}_{t=T_1}^{\infty} \subset F$ such that for each integer $T_2 > T_1$, $(\{x_t\}_{t=T_1}^{T_2}, \{u_t\}_{t=T_1}^{T_2-1}) \in X(T_1, T_2)$. Elements of $X(T_1, \infty)$ are called trajectory–control pairs. Let $0 \leq T_1 < T_2$ be integers. A sequence $\{x_t\}_{t=T_1}^{T_2} \subset E$ ($\{x_t\}_{t=T_1}^{\infty} \subset E$, respectively) is called a trajectory if there exists a sequence $\{u_t\}_{t=T_1}^{T_2-1} \subset F$ ($\{u_t\}_{t=T_1}^{\infty} \subset F$, respectively) referred to as a control such that $(\{x_t\}_{t=T_1}^{T_2}, \{u_t\}_{t=T_1}^{T_2-1}) \in X(T_1, T_2)$ ($(\{x_t\}_{t=T_1}^{\infty}, \{u_t\}_{t=T_1}^{\infty}) \in X(T_1, \infty)$, respectively).

Let $\theta_0 \in E$, $\theta_1 \in F$, $a_0 > 0$ and let $\psi : [0, \infty) \to [0, \infty)$ be an increasing function such that

$$\psi(t) \to \infty \text{ as } t \to \infty. \tag{6.5}$$

We suppose that the function f satisfies

$$f(t, x, u) \geq \psi(\rho_E(x, \theta_0)) - a_0 \text{ for each } (t, x, u) \in \mathcal{M}. \tag{6.6}$$

For each pair of integers $T_2 > T_1 \geq 0$ and each pair of points $y, z \in E$ satisfying $(T_1, y), (T_2, z) \in \mathcal{A}$ we consider the following problems:

$$\sum_{t=T_1}^{T_2-1} f(t, x_t, u_t) \to \min,\ (\{x_t\}_{t=T_1}^{T_2}, \{u_t\}_{t=T_1}^{T_2-1}) \in X(T_1, T_2),\ x_{T_1} = y,\ x_{T_2} = z, \tag{P_1}$$

$$\sum_{t=T_1}^{T_2-1} f(t, x_t, u_t) \to \min,\ (\{x_t\}_{t=T_1}^{T_2}, \{u_t\}_{t=T_1}^{T_2-1}) \in X(T_1, T_2),\ x_{T_1} = y, \tag{P_2}$$

$$\sum_{t=T_1}^{T_2-1} f(t, x_t, u_t) \to \min,\ (\{x_t\}_{t=T_1}^{T_2}, \{u_t\}_{t=T_1}^{T_2-1}) \in X(T_1, T_2). \tag{P_3}$$

For each pair of integers $T_2 > T_1 \geq 0$ and each pair of points $y, z \in E$ satisfying $(T_1, y), (T_2, z) \in \mathcal{A}$ we define

$$U^f(T_1, T_2, y, z) = \inf\{\sum_{t=T_1}^{T_2-1} f(t, x_t, u_t) :\ (\{x_t\}_{t=T_1}^{T_2}, \{u_t\}_{t=T_1}^{T_2-1}) \in X(T_1, T_2),$$

$$x_{T_1} = y,\ x_{T_2} = z\}, \tag{6.7}$$

$$U^f(T_1, T_2, y) = \inf\{U^f(T_1, T_2, y, h) : h \in E, (T_2, h) \in \mathcal{A}\}, \tag{6.8}$$

$$\widehat{U}^f(T_1, T_2, z) = \inf\{U^f(T_1, T_2, h, z) : h \in E, (T_1, h) \in \mathcal{A}\}, \tag{6.9}$$

$$U^f(T_1, T_2) = \inf\{U^f(T_1, T_2, h, \xi) : h, \xi \in E, (T_1, h), (T_2, \xi) \in \mathcal{A}\}. \tag{6.10}$$

We suppose that $b_f > 0$ is an integer, $(\{x_t^f\}_{t=0}^\infty, \{u_t^f\}_{t=0}^\infty) \in X(0, \infty)$,

$$\{x_t^f : t = 0, 1, \dots\} \text{ is bounded }, \tag{6.11}$$

$$\Delta_f := \sup\{|f(t, x_t^f, u_t^f)| : t = 0, 1, \dots\} < \infty \tag{6.12}$$

and that the following assumptions hold.

(A1) For each $S_1 > 0$ there exist $S_2 > 0$ and an integer $c > 0$ such that

$$\sum_{t=T_1}^{T_2-1} f(t, x_t^f, u_t^f) \le \sum_{t=T_1}^{T_2-1} f(t, x_t, u_t) + S_2$$

for each pair of integers $T_1 \ge 0$, $T_2 \ge T_1 + c$ and each $(\{x_t\}_{t=T_1}^{T_2}, \{u_t\}_{t=T_1}^{T_2-1}) \in X(T_1, T_2)$ satisfying $\rho_E(\theta_0, x_j) \le S_1$, $j = T_1, T_2$.

(A2) For each $\epsilon > 0$ there exists $\delta > 0$ such that for each $(T_i, z_i) \in \mathcal{A}$, $i = 1, 2$ satisfying $\rho_E(z_i, x_{T_i}^f) \le \delta$, $i = 1, 2$ and $T_2 \ge b_f$ there exist integers $\tau_1, \tau_2 \in (0, b_f]$,

$$(\{x_t^{(1)}\}_{t=T_1}^{T_1+\tau_1}, \{u_t^{(1)}\}_{t=T_1}^{T_1+\tau_1-1}) \in X(T_1, T_1 + \tau_1),$$

$$(\{x_t^{(2)}\}_{t=T_2-\tau_2}^{T_2}, \{u_t^{(2)}\}_{t=T_2-\tau_2}^{T_2-1}) \in X(T_2 - \tau_2, T_2)$$

such that

$$x_{T_1}^{(1)} = z_1, \; x_{T_1+\tau_1}^{(1)} = x_{T_1+\tau_1}^f,$$

$$\sum_{t=T_1}^{T_1+\tau_1-1} f(t, x_t^{(1)}, u_t^{(1)}) \le \sum_{t=T_1}^{T_1+\tau_1-1} f(t, x_t^f, u_t^f) + \epsilon,$$

$$x_{T_2}^{(2)} = z_2, \; x_{T_2-\tau_2}^{(2)} = x_{T_2-\tau_2}^f,$$

$$\sum_{t=T_2-\tau_2}^{T_2-1} f(t, x_t^{(2)}, u_t^{(2)}) \le \sum_{t=T_2-\tau_2}^{T_2-1} f(t, x_t^f, u_t^f) + \epsilon.$$

Section 6.6 contains examples of optimal control problems satisfying assumptions (A1) and (A2). Many examples can also be found in [119–121, 131, 137, 138, 147].

The following result is proved in Section 3.8 of [148].

Theorem 6.1

1. *There exists* $S > 0$ *such that for each pair of integers* $T_2 > T_1 \geq 0$ *and each* $(\{x_t\}_{t=T_1}^{T_2}, \{u_t\}_{t=T_1}^{T_2-1}) \in X(T_1, T_2)$,

$$\sum_{t=T_1}^{T_2-1} f(t, x_t, u_t) + S \geq \sum_{t=T_1}^{T_2-1} f(t, x_t^f, u_t^f).$$

2. *For each* $(\{x_t\}_{t=0}^{\infty}, \{u_t\}_{t=0}^{\infty}) \in X(0, \infty)$ *either*

$$\sum_{t=0}^{T-1} f(t, x_t, u_t) - \sum_{t=0}^{T-1} f(t, x_t^f, u_t^f) \to \infty \text{ as } T \to \infty$$

or

$$\sup\{|\sum_{t=0}^{T-1} f(t, x_t, u_t) - \sum_{t=0}^{T-1} f(t, x_t^f, u_t^f)| : T \in \{1, 2, \dots\}\} < \infty. \tag{6.13}$$

Moreover, if (6.13) holds, then

$$\sup\{\rho_E(x_t, \theta_0) : t = 0, 1, \dots\} < \infty.$$

We say that $(\{x_t\}_{t=0}^{\infty}, \{u_t\}_{t=0}^{\infty}) \in X(0, \infty)$ is (f)-goof if (6.13) holds [31, 117, 129, 133, 142, 147]. The next boundedness result is proved in Section 3.8 [148]. It has a prototype in [117].

Theorem 6.2 *Let* $c > 0$ *be an integer and* $M_0 > 0$*. Then there exists* $M_1 > 0$ *such that for each pair of integers* $T_1 \geq 0$, $T_2 \geq T_1 + c$ *and each* $(\{x_t\}_{t=T_1}^{T_2}, \{u_t\}_{t=T_1}^{T_2-1}) \in X(T_1, T_2)$ *satisfying*

$$\sum_{t=T_1}^{T_2-1} f(t, x_t, u_t) \leq \sum_{t=T_1}^{T_2-1} f(t, x_t^f, u_t^f) + M_0$$

the inequality $\rho_E(\theta_0, x_t) \leq M_1$ *holds for all* $t = T_1, \dots, T_2 - 1$.

Let $L > 0$ be an integer. Denote by $\mathcal{A}_L$ the set of all $(S, z) \in \mathcal{A}$ for which there exists an integer $\tau \in (0, L]$ and $(\{x_t\}_{t=S}^{S+\tau}, \{u_t\}_{t=S}^{S+\tau-1}) \in X(S, S + \tau)$ such that

$$x_S = z,\ x_{S+\tau} = x^f_{S+\tau},\ \sum_{t=S}^{S+\tau-1} f(t, x_t, u_t) \le L.$$

Denote by $\widehat{\mathcal{A}}_L$ the set of all $(S, z) \in \mathcal{A}$ such that $S \ge L$ and there exists an integer $\tau \in (0, L]$ and $(\{x_t\}_{t=S-\tau}^{S}, \{u_t\}_{t=S-\tau}^{S-1}) \in X(S-\tau, S)$ satisfying

$$x_{S-\tau} = x^f_{S-\tau},\ x_S = z,\ \sum_{t=S-\tau}^{S-1} f(t, x_t, u_t) \le L.$$

The following Theorems 6.3–6.5 are also boundedness results. They are proved in Section 3.8 of [148].

Theorem 6.3 *Let $L > 0$ be an integer and $M_0 > 0$. Then there exists $M_1 > 0$ such that for each integer $T_1 \ge 0$, each integer $T_2 \ge T_1 + 2L$ and each $(\{x_t\}_{t=T_1}^{T_2}, \{u_t\}_{t=T_1}^{T_2-1}) \in X(T_1, T_2)$ satisfying*

$$(T_1, x_{T_1}) \in \mathcal{A}_L,\ (T_2, x_{T_2}) \in \widehat{\mathcal{A}}_L,$$

$$\sum_{t=T_1}^{T_2-1} f(t, x_t, u_t) \le U^f(T_1, T_2, x_{T_1}, x_{T_2}) + M_0$$

the inequality $\rho_E(\theta_0, x_t) \le M_1$ holds for all $t = T_1, \dots, T_2 - 1$.

Theorem 6.4 *Let $L > 0$ be an integer and $M_0 > 0$. Then there exists $M_1 > 0$ such that for each integer $T_1 \ge 0$, each integer $T_2 \ge T_1 + L$ and each $(\{x_t\}_{t=T_1}^{T_2}, \{u_t\}_{t=T_1}^{T_2-1}) \in X(T_1, T_2)$ satisfying*

$$(T_1, x_{T_1}) \in \mathcal{A}_L,\ \sum_{t=T_1}^{T_2-1} f(t, x_t, u_t) \le U^f(T_1, T_2, x_{T_1}) + M_0$$

the inequality $\rho_E(\theta_0, x_t) \le M_1$ holds for all $t = T_1, \dots, T_2 - 1$.

Theorem 6.5 *Let $L > 0$ be an integer and $M_0 > 0$. Then there exists $M_1 > 0$ such that for each integer $T_1 \ge 0$, each integer $T_2 \ge T_1 + L$ and each $(\{x_t\}_{t=T_1}^{T_2}, \{u_t\}_{t=T_1}^{T_2-1}) \in X(T_1, T_2)$ satisfying*

$$(T_2, x_{T_2}) \in \widehat{\mathcal{A}}_L,\ \sum_{t=T_1}^{T_2-1} f(t, x_t, u_t) \le \widehat{U}^f(T_1, T_2, x_{T_2}) + M_0$$

the inequality $\rho_E(\theta_0, x_t) \le M_1$ holds for all $t = T_1, \dots, T_2 - 1$.

6.2 Turnpike Properties

We say that f possesses the turnpike property (or TP for short) if for each $\epsilon > 0$ and each $M > 0$ there exist $\delta > 0$ and an integer $L > 0$ such that for each integer $T_1 \geq 0$, each integer $T_2 \geq T_1 + 2L$ and each $(\{x_t\}_{t=T_1}^{T_2}, \{u_t\}_{t=T_1}^{T_2-1}) \in X(T_1, T_2)$ which satisfies

$$\sum_{t=T_1}^{T_2-1} f(t, x_t, u_t) \leq \min\{U^f(T_1, T_2) + M, U^f(T_1, T_2, x_{T_1}, x_{T_2}) + \delta\}$$

there exist integers $\tau_1 \in [T_1, T_1 + L]$, $\tau_2 \in [T_2 - L, T_2]$ such that

$$\rho_E(x_t, x_t^f) \leq \epsilon, \ t = \tau_1 \dots, \tau_2.$$

Moreover, if $\rho_E(x_{T_2}, x_{T_2}^f) \leq \delta$, then $\tau_2 = T_2$ and if $\rho_E(x_{T_1}, x_{T_1}^f) \leq \delta$, $T_1 \geq L$, then $\tau_1 = T_1$.

We say that f possesses the strong turnpike property (or STP for short) if for each $\epsilon > 0$ and each $M > 0$ there exist $\delta > 0$ and an integer $L > 0$ such that for each integer $T_1 \geq 0$, each integer $T_2 \geq T_1 + 2L$ and each $(\{x_t\}_{t=T_1}^{T_2}, \{u_t\}_{t=T_1}^{T_2-1}) \in X(T_1, T_2)$ which satisfies

$$\sum_{t=T_1}^{T_2-1} f(t, x_t, u_t) \leq \min\{U^f(T_1, T_2) + M, U^f(T_1, T_2, x_{T_1}, x_{T_2}) + \delta\}$$

there exist integers $\tau_1 \in [T_1, T_1 + L]$, $\tau_2 \in [T_2 - L, T_2]$ such that

$$\rho_E(x_t, x_t^f) \leq \epsilon, \ t = \tau_1 \dots, \tau_2.$$

Moreover, if $\rho_E(x_{T_2}, x_{T_2}^f) \leq \delta$, then $\tau_2 = T_2$ and if $\rho_E(x_{T_1}, x_{T_1}^f) \leq \delta$, then $\tau_1 = T_1$.

Theorem 6.1 implies the following two results.

Theorem 6.6 *Assume that f has TP and that $\epsilon, M > 0$. Then there exist $\delta > 0$ and an integer $L > 0$ such that for each integer $T_1 \geq 0$, each integer $T_2 \geq T_1 + 2L$ and each $(\{x_t\}_{t=T_1}^{T_2}, \{u_t\}_{t=T_1}^{T_2-1}) \in X(T_1, T_2)$ which satisfies*

$$\sum_{t=T_1}^{T_2-1} f(t, x_t, u_t) \leq \min\{\sum_{t=T_1}^{T_2-1} f(t, x_t^f, u_t^f) + M, U^f(T_1, T_2, x_{T_1}, x_{T_2}) + \delta\}$$

there exist integers $\tau_1 \in [T_1, T_1 + L]$, $\tau_2 \in [T_2 - L, T_2]$ such that

$$\rho_E(x_t, x_t^f) \leq \epsilon, \ t = \tau_1, \dots, \tau_2.$$

Moreover, if $\rho_E(x_{T_2}, x^f_{T_2}) \le \delta$, *then* $\tau_2 = T_2$ *and if* $\rho_E(x_{T_1}, x^f_{T_1}) \le \delta$, $T_1 \ge L$, *then* $\tau_1 = T_1$.

Theorem 6.7 *Assume that* f *has STP and that* $\epsilon, M > 0$. *Then there exist* $\delta > 0$ *and an integer* $L > 0$ *such that for each integer* $T_1 \ge 0$, *each integer* $T_2 \ge T_1 + 2L$ *and each* $(\{x_t\}_{t=T_1}^{T_2}, \{u_t\}_{t=T_1}^{T_2-1}) \in X(T_1, T_2)$ *which satisfies*

$$\sum_{t=T_1}^{T_2-1} f(t, x_t, u_t) \le \min\{\sum_{t=T_1}^{T_2-1} f(t, x_t^f, u_t^f) + M, U^f(T_1, T_2, x_{T_1}, x_{T_2}) + \delta\}$$

there exist integers $\tau_1 \in [T_1, T_1 + L]$, $\tau_2 \in [T_2 - L, T_2]$ *such that*

$$\rho_E(x_t, x_t^f) \le \epsilon, \; t = \tau_1, \dots, \tau_2.$$

Moreover, if $\rho_E(x_{T_2}, x^f_{T_2}) \le \delta$, *then* $\tau_2 = T_2$ *and if* $\rho_E(x_{T_1}, x^f_{T_1}) \le \delta$, *then* $\tau_1 = T_1$.

It is not difficult to show (see Section 3.3 of [148]) that the following result is true.

Proposition 6.8 *Let* $L > 0$ *be an integer,* $T_1 \ge 0$, $T_2 \ge T_1 + 2L$ *be integers,* $(T_1, z_1) \in \mathcal{A}_L$, $(T_2, z_2) \in \widehat{\mathcal{A}}_L$. *Then*

$$U^f(T_1, T_2, z_1, z_2) \le \sum_{t=T_1}^{T_2-1} f(t, x_t^f, u_t^f) + 2L(1 + a_0).$$

Proposition 6.8 and Theorems 6.6 and 6.7 imply the following two results.

Theorem 6.9 *Assume that* f *has TP,* $L_0 > 0$ *is an integer and that* $\epsilon > 0$. *Then there exist* $\delta > 0$ *and an integer* $L > L_0$ *such that for each integer* $T_1 \ge 0$, *each integer* $T_2 \ge T_1 + 2L$ *and each* $(\{x_t\}_{t=T_1}^{T_2}, \{u_t\}_{t=T_1}^{T_2-1}) \in X(T_1, T_2)$ *which satisfies*

$$(T_1, x_{T_1}) \in \mathcal{A}_{L_0}, \; (T_2, x_{T_2}) \in \widehat{\mathcal{A}}_{L_0},$$

$$\sum_{t=T_1}^{T_2-1} f(t, x_t, u_t) \le U^f(T_1, T_2, x_{T_1}, x_{T_2}) + \delta$$

there exist integers $\tau_1 \in [T_1, T_1 + L]$, $\tau_2 \in [T_2 - L, T_2]$ *such that*

$$\rho_E(x_t, x_t^f) \le \epsilon, \; t = \tau_1, \dots, \tau_2.$$

Moreover, if $\rho_E(x_{T_2}, x^f_{T_2}) \le \delta$ *then* $\tau_2 = T_2$ *and if* $\rho_E(x_{T_1}, x^f_{T_1}) \le \delta$, $T_1 \ge L$, *then* $\tau_1 = T_1$.

Theorem 6.10 *Assume that f has STP, $L_0 > 0$ is an integer and that $\epsilon > 0$. Then there exist $\delta > 0$ and an integer $L > L_0$ such that for each integer $T_1 \geq 0$, each integer $T_2 \geq T_1 + 2L$ and each $(\{x_t\}_{t=T_1}^{T_2}, \{u_t\}_{t=T_1}^{T_2-1}) \in X(T_1, T_2)$ which satisfies*

$$(T_1, x_{T_1}) \in \mathcal{A}_{L_0}, \ (T_2, x_{T_2}) \in \widehat{\mathcal{A}}_{L_0},$$

$$\sum_{t=T_1}^{T_2-1} f(t, x_t, u_t) \leq U^f(T_1, T_2, x_{T_1}, x_{T_2}) + \delta$$

there exist integers $\tau_1 \in [T_1, T_1 + L]$, $\tau_2 \in [T_2 - L, T_2]$ such that

$$\rho_E(x_t, x_t^f) \leq \epsilon, \ t = \tau_1, \dots, \tau_2.$$

Moreover, if $\rho_E(x_{T_2}, x_{T_2}^f) \leq \delta$ then $\tau_2 = T_2$ and if $\rho_E(x_{T_1}, x_{T_1}^f) \leq \delta$, then $\tau_1 = T_1$.

It is not difficult to show (see Section 3.3 of [148]) that the following result is true.

Proposition 6.11 *Let $L > 0$ be an integer, $(T_1, z) \in \mathcal{A}_L$, $T_2 \geq T_1 + L$ be an integer. Then*

$$U^f(T_1, T_2, z) \leq \sum_{t=T_1}^{T_2-1} f(t, x_t^f, u_t^f) + L(1 + a_0).$$

Proposition 6.11 and Theorems 6.6, 6.7 imply the following two results.

Theorem 6.12 *Assume that f has TP, $L_0 > 0$ is an integer and that $\epsilon > 0$. Then there exist $\delta > 0$ and an integer $L > L_0$ such that for each integer $T_1 \geq 0$, each integer $T_2 \geq T_1 + 2L$ and each $(\{x_t\}_{t=T_1}^{T_2}, \{u_t\}_{t=T_1}^{T_2-1}) \in X(T_1, T_2)$ which satisfies*

$$(T_1, x_{T_1}) \in \mathcal{A}_{L_0}, \ \sum_{t=T_1}^{T_2-1} f(t, x_t, u_t) \leq U^f(T_1, T_2, x_{T_1}) + \delta$$

there exist integers $\tau_1 \in [T_1, T_1 + L]$, $\tau_2 \in [T_2 - L, T_2]$ such that

$$\rho_E(x_t, x_t^f) \leq \epsilon, \ t = \tau_1, \dots, \tau_2.$$

Moreover, if $\rho_E(x_{T_2}, x_{T_2}^f) \leq \delta$ then $\tau_2 = T_2$ and if $\rho_E(x_{T_1}, x_{T_1}^f) \leq \delta$, $T_1 \geq L$, then $\tau_1 = T_1$.

Theorem 6.13 *Assume that f has STP, $L_0 > 0$ is an integer and that $\epsilon > 0$. Then there exist $\delta > 0$ and an integer $L > L_0$ such that for each integer $T_1 \geq 0$, each integer $T_2 \geq T_1 + 2L$ and each $(\{x_t\}_{t=T_1}^{T_2}, \{u_t\}_{t=T_1}^{T_2-1}) \in X(T_1, T_2)$ which satisfies*

$$(T_1, x_{T_1}) \in \mathcal{A}_L, \ \sum_{t=T_1}^{T_2-1} f(t, x_t, u_t) \leq U^f(T_1, T_2, x_{T_1}) + \delta$$

there exist integers $\tau_1 \in [T_1, T_1 + L]$, $\tau_2 \in [T_2 - L, T_2]$ *such that*

$$\rho_E(x_t, x_t^f) \leq \epsilon, \ t = \tau_1, \ldots, \tau_2.$$

Moreover, if $\rho_E(x_{T_2}, x_{T_2}^f) \leq \delta$ *then* $\tau_2 = T_2$ *and if* $\rho_E(x_{T_1}, x_{T_1}^f) \leq \delta$, *then* $\tau_1 = T_1$.

The next theorem is proved in Section 3.10 of [148].

Theorem 6.14 *f has TP if and only if the following properties hold:*

(P1) for each (f)-good pair of sequences $(\{x_t\}_{t=0}^{\infty}, \{u_t\}_{t=0}^{\infty}) \in X(0, \infty)$,

$$\lim_{t\to\infty} \rho_E(x_t, x_t^f) = 0;$$

(P2) for each $\epsilon > 0$ *and each* $M > 0$ *there exist* $\delta > 0$ *and an integer* $L > 0$ *such that for each integer* $T \geq 0$ *and each* $(\{x_t\}_{t=T}^{T+L}, \{u_t\}_{t=T}^{T+L-1}) \in X(T, T + L)$ *which satisfies*

$$\sum_{t=T}^{T+L-1} f(t, x_t, u_t) \leq \min\{U^f(T, T + L, x_T, x_{T+L}) + \delta,$$

$$\sum_{t=T}^{T+L-1} f(t, x_t^f, u_t^f) + M\}$$

there exists an integer $s \in [T, T + L]$ *such that* $\rho_E(x_s, x_s^f) \leq \epsilon$.

We say that f possesses the weak turnpike property (or WTP for short) if for each $\epsilon > 0$ and each $M > 0$ there exist natural numbers Q, l such that for each integer $T_1 \geq 0$, each integer $T_2 \geq T_1 + lQ$ and each $(\{x_t\}_{t=T_1}^{T_2}, \{u_t\}_{t=T_1}^{T_2-1}) \in X(T_1, T_2)$ which satisfies

$$\sum_{t=T_1}^{T_2-1} f(t, x_t, u_t) \leq \sum_{t=T_1}^{T_2-1} f(t, x_t^f, u_t^f) + M$$

there exist finite sequences of integers $\{a_i\}_{i=1}^q$, $\{b_i\}_{i=1}^q \subset \{T_1, \ldots, T_2\}$ such that an integer $q \leq Q$,

$$0 \leq b_i - a_i \leq l, \ i = 1, \ldots, q,$$

$$b_i \leq a_{i+1} \text{ for all integers } i \text{ satisfying } 1 \leq i < q,$$

$$\rho_E(x_t, x_t^f) \leq \epsilon \text{ for all integers } t \in [T_1, T_2] \setminus \cup_{i=1}^q [a_i, b_i].$$

WTP was studied in [108, 109] for discrete-time unconstrained problems, in [110] for variational problems and in [127] for discrete-time constrained problems.

The next result is proved in Section 3.12 of [148].

Theorem 6.15 *f has WTP if and only if f has (P1) and (P2).*

The proof of Theorem 6.14 is based on the following result (see Lemma 3.35 of [148]).

Lemma 6.16 *Assume that (P1) holds and that $\epsilon > 0$. Then there exists $\delta > 0$ and an integer $L > 0$ such that for each integer $T_1 \geq L$, each integer $T_2 \geq T_1 + 2b_f$ and each $(\{x_t\}_{t=T_1}^{T_2}, \{u_t\}_{t=T_1}^{T_2-1}) \in X(T_1, T_2)$ which satisfies*

$$\rho_E(x_{T_i}, x_{T_i}^f) \leq \delta,\ i = 1, 2,$$

$$\sum_{t=T_1}^{T_2-1} f(t, x_t, u_t) \leq U^f(T_1, T_2, x_{T_1}, x_{T_2}) + \delta$$

the inequality $\rho_E(x_t, x_t^f) \leq \epsilon$ holds for all $t = T_1, \ldots, T_2$.

6.3 Perturbed Problems

In this section we suppose that the following assumption holds.

(A3) For each $\epsilon > 0$ there exists $\delta > 0$ such that for each $(T_i, z_i) \in \mathcal{A}$, $i = 1, 2$ satisfying $\rho_E(z_i, x_{T_i}^f) \leq \delta$, $i = 1, 2$ and $T_2 \geq b_f$ there exist integers $\tau_1, \tau_2 \in (0, b_f]$,

$$(\{x_t^{(1)}\}_{t=T_1}^{T_1+\tau_1}, \{u_t^{(1)}\}_{t=T_1}^{T_1+\tau_1-1}) \in X(T_1, T_1 + \tau_1),$$

$$(\{x_t^{(2)}\}_{t=T_2-\tau_2}^{T_2}, \{u_t^{(2)}\}_{t=T_2-\tau_2}^{T_2-1}) \in X(T_2 - \tau_2, T_2)$$

such that

$$x_{T_1}^{(1)} = z_1,\ x_{T_1+\tau_1}^{(1)} = x_{T_1+\tau_1}^f,$$

$$\sum_{t=T_1}^{T_1+\tau_1-1} f(t, x_t^{(1)}, u_t^{(1)}) \leq \sum_{t=T_1}^{T_1+\tau_1-1} f(t, x_t^f, u_t^f) + \epsilon,$$

$$\rho(x_t^{(1)}, x_t^f) \leq \epsilon,\ t = T_1, \ldots, T_1 + \tau_1,$$

$$x^{(2)}_{T_2} = z_2,\ x^{(2)}_{T_2-\tau_2} = x^f_{T_2-\tau_2},$$

$$\sum_{t=T_2-\tau_2}^{T_2-1} f(t, x^{(2)}_t, u^{(2)}_t) \le \sum_{t=T_2-\tau_2}^{T_2-1} f(t, x^f_t, u^f_t) + \epsilon,$$

$$\rho(x^{(2)}_t, x^f_t) \le \epsilon,\ t = T_2 - \tau_2, \dots, T_2.$$

Clearly, (A3) implies (A2). Assume that $\phi : E \times E \to [0, 1]$ is a continuous function satisfying $\phi(x, x) = 0$ for all $x \in E$ and such that the following properties hold:

(i) for each $\epsilon > 0$ there exists $\delta > 0$ such that if $x, y \in E$ and $\phi(x, y) \le \delta$, then $\rho_E(x, y) \le \epsilon$.
(ii) for each $\epsilon > 0$ there exists $\delta > 0$ such that if $x, y \in E$ and $\rho_E(x, y) \le \delta$, then $\phi(x, y) \le \epsilon$.

For each $r \in (0, 1)$ set

$$f_r(t, x, u) = f(t, x, u) + r\phi(x, x^f_t),\ (t, x, u) \in \mathcal{M}.$$

Clearly, for any $r \in (0, 1)$, (A1), (A3) hold for f_r with $(x^{f_r}_t, u^{f_r}_t) = (x^f_t, u^f_t)$, $t = 0, 1, \dots$. The next result is proved in Section 3.17 of [148].

Theorem 6.17 *Let $r \in (0, 1)$. Then f_r has TP, (P1), and (P2).*

6.4 Examples

Example 6.18 We consider an example which is a particular case of the problem introduced in Sect. 6.1.

Let $a_1 > 0$, $\psi_1 : [0, \infty) \to [0, \infty)$ be an increasing function such that $\psi_1(t) \to \infty$ as $t \to \infty$, $\mu : \{0, 1, \dots\} \to R^1$, $\pi : \{0, 1, \dots\} \times E \to R^1$, $L : \mathcal{M} \to [0, \infty)$,

$$\pi(t, x) \ge -a_1 \text{ for all } (t, x) \in \{0, 1, \dots\} \times E,$$

π be bounded on bounded subsets of $\{0, 1, \dots\} \times E$,

$$\mu(t) \ge -a_1 \text{ for all } t \in \{0, 1, \dots\},$$

and for all $(t, x, u) \in \mathcal{M}$,

$$L(t, x, u) - \pi(t, G(t, x, u)) \ge -a_1 + \psi_1(\rho_E(\theta_0, x)), \tag{6.14}$$

$$L(t, x, u) = 0 \text{ if and only if } x = x^f_t,\ u = u^f_t, \tag{6.15}$$

$$f(t,x,u)=\mu(t)+L(t,x,u)+\pi(t,x)-\pi(t+1,G(t,x,u)). \tag{6.16}$$

By the relation above, for all $(t,x,u)\in\mathcal{M}$.

$$f(t,x,u)\geq -3a_1+\psi_1(\rho_E(\theta_0,x))$$

and (6.6) holds with $a_0=3a_1$, $\psi_0=\psi_1$.

Let $T_2>T_1\geq 0$ be integers and $(\{x_t\}_{t=T_1}^{T_2},\{u_t\}_{t=T_1}^{T_2-1})\in X(T_1,T_2)$. It is not difficult to see that

$$\sum_{t=T_1}^{T_2-1} f(t,x_t,u_t)=\sum_{t=T_1}^{T_2-1}\mu(t)+\sum_{t=T_1}^{T_2-1}L(t,x_t,u_t)+\sum_{t=T_1}^{T_2-1}(\pi(t,x_t)-\pi(t+1,x_{t+1}))$$

$$=\sum_{t=T_1}^{T_2-1}\mu(t)+\sum_{t=T_1}^{T_2-1}L(t,x_t,u_t)+\pi(T_1,x_{T_1})-\pi(T_2,x_{T_2}),$$

$$\sum_{t=T_1}^{T_2-1} f(t,x_t^f,u_t^f)=\sum_{t=T_1}^{T_2-1}\mu(t)+\pi(T_1,x_{T_1}^f)-\pi(T_2,x_{T_2}^f)$$

and

$$\sum_{t=T_1}^{T_2-1} f(t,x_t,u_t)-\sum_{t=T_1}^{T_2-1} f(t,x_t^f,u_t^f)$$

$$\geq \pi(T_1,x_{T_1})-\pi(T_2,x_{T_2})-\pi(T_1,x_{T_1}^f)+\pi(T_2,x_{T_2}^f).$$

Since π is bounded on bounded subsets of $\{0,1,\dots,\}\times E$ we conclude that (A1) holds.

We suppose that the following assumptions hold:

(B1) for each $\epsilon>0$ there exists $\delta>0$ such that for each integer $T\geq 0$, each pair $(T,z_1),(T+1,z_2)\in\mathcal{A}$ satisfying $\rho_E(z_1,x_T^f)$, $\rho_E(z_2,x_{T+1}^f)\leq\delta$, there exists $u\in\mathcal{U}(T,z_1)$ such that

$$\rho_F(u,u_T^f)\leq\epsilon,\ z_2=G(T,z_1,u);$$

(B2) for each $\epsilon>0$ there exists $\delta>0$ such that for each $(T,z,\xi)\in\mathcal{M}$ satisfying $\rho_E(z,x_T^f)\leq\delta$, $\rho_F(\xi,u_T^f)\leq\delta$, we have $f(T,z,\xi)\leq f(T,x_T^f,u_T^f)+\epsilon$.

It is clear that (B1) and (B2) imply (A2) and (A3).

Example 6.19 We consider an example which is a particular case of the problem introduced in Sect. 6.1. Let $a_1>0$, $\psi_1:[0,\infty)\to[0,\infty)$ be an increasing function such that $\psi_1(t) \to \infty$ as $t \to \infty$, $\mu : \{0, 1, \dots\} \to R^1$, $\pi : \{0, 1, \dots\} \times E \to R^1$, $L : \mathcal{M} \to [0, \infty)$,

$$\pi(t, x) \geq -a_1 \text{ for all } (t, x) \in \{0, 1, \dots\} \times E,$$

π be bounded on bounded subsets of $\{0, 1, \dots\} \times E$,

$$\mu(t) \geq -a_1 \text{ for all } t \in \{0, 1, \dots\},$$

and for all $(t, x, u) \in \mathcal{M}$, (6.14)–(6.16) hold. Then as it was shown in Example 6.18, (A1) and (6.6) hold with $a_0 = 3a_1$, $\psi_0 = \psi_1$.

We suppose that there exists a natural number d_f such that the following assumptions hold:

(B3) for each $\epsilon > 0$ there exists $\delta > 0$ such that for each integer $T \geq 0$, each pair (T, z_1), $(T + d_f, z_2) \in \mathcal{A}$ satisfying $\rho_E(z_1, x_T^f)$, $\rho_E(z_2, x_{T+d_f}^f) \leq \delta$, there exists $(\{x_t\}_{t=T}^{T+d_f}, \{u_t\}_{t=T}^{T+d_f-1}) \in X(T, T + d_f)$ such that

$$x_T = z_1,\ x_{T+d_f} = z_2,\ \rho(u_t, u_t^f) \leq \epsilon,\ t = T, \dots, T + d_f - 1;$$

(B4) for each $\epsilon > 0$ there exists $\delta > 0$ such that for each $(T, z) \in \mathcal{A}$ satisfying $\rho_E(z, x_T^f) \leq \delta$ and each $u \in \mathcal{U}(T, z)$ satisfying, $\rho_F(u, u_T^f) \leq \delta$, we have

$$\rho_E(x_{T+1}^f, G(T, z, u)) \leq \epsilon,\ f(T, z, u) \leq f(T, x_T^f, u_T^f) + \epsilon.$$

(B3) and (B4) imply (A2) and (A3).

Assume now that for any nonempty bounded set $\Omega \subset E$ the function π is bounded on $\{0, 1, \dots, \} \times \Omega$, the function μ is bounded and that the following property holds:

for each $M, \epsilon > 0$ there exists $\delta > 0$ such that for each $(t, x, u) \in \mathcal{M}$ which satisfies

$$\rho_E(\theta_0, x) \leq M \text{ and } L(t, x, u) \leq \delta$$

we have $\rho_E(x, x_t^f) \leq \epsilon$.

It was shown in Section 3.6 of [148] that f has TP.

6.5 The Space of Objective Functions and the Stability Result

Denote by $\mathfrak{M}_\psi$ the set of all functions $g : \mathcal{M} \to R^1$, which satisfy

$$g(t, x, u) \geq \psi(\rho_E(x, \theta_0)) - a_0 \text{ for each } (t, x, u) \in \mathcal{M}. \tag{6.17}$$

For each $g \in \mathfrak{M}_\psi$ and each pair of integers $T_2 > T_1 \geq 0$ and each pair of points $y, z \in E$ satisfying $(T_1, y), (T_2, z) \in \mathcal{A}$ we define

$$U^g(T_1, T_2, y, z) = \inf\{\sum_{t=T_1}^{T_2-1} g(t, x_t, u_t) : (\{x_t\}_{t=T_1}^{T_2}, \{u_t\}_{t=T_1}^{T_2-1}) \in X(T_1, T_2),$$

$$x_{T_1} = y, \ x_{T_2} = z\},$$

$$U^g(T_1, T_2, y) = \inf\{\sum_{t=T_1}^{T_2-1} g(t, x_t, u_t) :$$

$$(\{x_t\}_{t=T_1}^{T_2}, \{u_t\}_{t=T_1}^{T_2-1}) \in X(T_1, T_2), \ x_{T_1} = y\},$$

$$U^g(T_1, T_2, z) = \inf\{\sum_{t=T_1}^{T_2-1} g(t, x_t, u_t) :$$

$$(\{x_t\}_{t=T_1}^{T_2}, \{u_t\}_{t=T_1}^{T_2-1}) \in X(T_1, T_2), \ x_{T_2} = z\},$$

$$U^f(T_1, T_2) = \inf\{\sum_{t=T_1}^{T_2-1} g(t, x_t, u_t) : (\{x_t\}_{t=T_1}^{T_2}, \{u_t\}_{t=T_1}^{T_2-1}) \in X(T_1, T_2)\}. \quad (6.18)$$

We assume that $f : \mathcal{M} \to R^1$ satisfies all the assumptions introduced in Sect. 6.1 (see (A1) and (A2)). We equip the space $\mathfrak{M}_\psi$ with the uniformity determined by the base

$$E(N, \epsilon) = \{(g_1, g_2) \in \mathfrak{M}_\psi \times \mathfrak{M}_\psi : |g_1(t, x, u) - g_2(t, x.u)| \leq \epsilon$$

$$\text{for each } (t, x, u) \in \mathcal{M} \text{ satisfying } \rho_E(x, \theta_0), \ \rho_F(u, \theta_1) \leq N\}$$

$$\cap \{(g_1, g_2) \in \mathfrak{M}_\psi \times \mathfrak{M}_\psi : |g_1(t, x, u) - g_2(t, x, u)| \leq \epsilon$$

$$\text{for each } (t, x, u) \in \mathcal{M} \text{ satisfying } \min\{g_1(t, x, u), \ g_2(t, x, u)\} \leq N\}, \quad (6.19)$$

where $N, \epsilon > 0$. Clearly, the uniform space $\mathfrak{M}_\psi$ is Hausdorff and has a countable base. Therefore $\mathfrak{M}_\psi$ is metrizable (by a metric d_ψ). It is not difficult to see that the uniform space $\mathfrak{M}_\psi$ is complete. It is equipped with a topology induced by the uniformity which is called sometimes as a weak topology.

The following proposition is easily deduced from growth condition (6.17).

Proposition 6.20 *For each integer $M_0 > 0$ there exists $M_1 > 0$ such that for each integer $T_1 \geq 0$, each integer $T_2 \in [T_1, T_1 + M_0]$, each $g \in \mathfrak{M}_\psi$ and each $(\{x_t\}_{t=T_1}^{T_2}, \{u_t\}_{t=T_1}^{T_2-1}) \in X(T_1, T_2)$ which satisfies*

$$\sum_{t=T_1}^{T_2-1} g(t, x_t, u_t) \leq M_0$$

the inequality $\rho_E(x_t, \theta_0) \leq M_1$ holds for every $t \in \{T_1, \ldots, T_2 - 1\}$.

Proposition 6.21 *Let $g \in \mathfrak{M}_\psi$ and $D, c, \epsilon > 0$. Then there exists a neighborhood V of g in $\mathfrak{M}_\psi$ such that for each pair of integers $T_1 \geq 0$, $T_2 \in (T_1, T_1 + c]$, each $h \in V$ and each $(\{x_t\}_{t=T_1}^{T_2}, \{u_t\}_{t=T_1}^{T_2-1}) \in X(T_1, T_2)$ which satisfies*

$$\min\{\sum_{t=T_1}^{T_2-1} g(t, x_t, u_t), \sum_{t=T_1}^{T_2-1} h(t, x_t, u_t)\} \leq D \tag{6.20}$$

the inequality

$$|\sum_{t=T_1}^{T_2-1} g(t, x_t, u_t) - \sum_{t=T_1}^{T_2-1} h(t, x_t, u_t)| \leq \epsilon$$

holds.

Proof Fix

$$N > D + a_0 c, \ \delta \in (0, \min\{1, \epsilon/c\}). \tag{6.21}$$

Define

$$V = \{h \in \mathfrak{M}_\psi : (g, h) \in E(N, \delta)\}. \tag{6.22}$$

Assume that $h \in V$, $T_1 \geq 0$ and $T_2 \in (T_1, T_1 + c]$ are integers and that $(\{x_t\}_{t=T_1}^{T_2}, \{u_t\}_{t=T_1}^{T_2-1}) \in X(T_1, T_2)$ satisfies (6.20). Let

$$j \in \{T_1, \ldots, T_2 - 1\}.$$

By (6.17) and (6.20),

$$\min\{g(j, x_j, u_j), h(j, x_j, u_j)\} \leq D + a_0 c.$$

Together with (6.19), (6.21), (6.22) and the inclusion $h \in V$ this implies that

$$|g(j, x_j, u_j) - h(j, x_j, u_j)| \leq \delta.$$

Combined with (6.21) this implies that

$$|\sum_{t=T_1}^{T_2-1} g(t, x_t, u_t) - \sum_{t=T_1}^{T_2-1} h(t, x_t, u_t)| \le \delta c < \epsilon.$$

Proposition 6.21 is proved.

In Sect. 6.7 we prove the following result which shows the stability of the turnpike phenomenon under small perturbations of the objective function.

Theorem 6.22 *Assume that $f \in \mathfrak{M}_\psi$ has TP, L is a natural number and $\epsilon \in (0, 1)$. Then there exists an integer $L_1 > L$, $\delta \in (0, \epsilon)$ and a neighborhood V of f in $\mathfrak{M}_\psi$ such that for each $g \in V$ and each pair of integers $T_1 \ge 0$, $T_2 \ge T_1 + 4L_1$, $U^g(T_1, T_2, z)$ is finite for each $z \in E$ satisfying $(T_1, z) \in \mathcal{A}_L$, $U^g(T_1, T_2, z_1, z_2)$ is finite for each $z_1, z_2 \in E$ satisfying $(T_1, z_1) \in \mathcal{A}_L$ and $(T_2, z_2) \in \widehat{\mathcal{A}}_L$ and that for each $(\{x_t\}_{t=T_1}^{T_2}, \{u_t\}_{t=T_1}^{T_2-1}) \in X(T_1, T_2)$ which satisfies at least one of the following conditions:*

(i)

$$\sum_{t=T_1}^{T_2-1} g(t, x_t, u_t) \le \min\{U^g(T_1, T_2) + L,\ U^g(T_1, T_2, x_{T_1}, x_{T_2}) + \delta\};$$

(ii)

$$\sum_{t=T_1}^{T_2-1} g(t, x_t, u_t) \le U^g(T_1, T_2, x_{T_1}),\ (T_1, x_{T_1}) \in \mathcal{A}_L;$$

(iii)

$$\sum_{t=T_1}^{T_2-1} g(t, x_t, u_t) \le U^g(T_1, T_2, x_{T_1}, x_{T_2}) + \delta,$$

$$(T_1, x_{T_1}) \in \mathcal{A}_L,\ (T_2, x_{T_2}) \in \widehat{\mathcal{A}}_L$$

there exist integers $\tau_1 \in [T_1, T_1 + L_1]$, $\tau_2 \in [T_2 - L_1, T_2]$ such that

$$\rho_E(x_t, x_t^f) \le \epsilon,\ t = \tau_1 \dots, \tau_2.$$

Moreover, if $\rho_E(x_{T_2}, x_{T_2}^f) \le \delta$, then $\tau_2 = T_2$ and if $T_1 \ge L_1$ and $\rho_E(x_{T_1}, x_{T_1}^f) \le \delta$, then $\tau_1 = T_1$.

It is clear that $U^g(T_1, T_2, z)$ is finite for each $(T_1, z) \in \mathcal{A}_L$ if $T_2 - T_1 \geq L$ and $U^g(T_1, T_2, z_1, z_2)$ is finite for each $(T_1, z_1) \in \mathcal{A}_L$ and each $(T_2, z_2) \in \widehat{\mathcal{A}}_L$ if $T_2 - T_1 \geq 2L$.

6.6 Auxiliary Results

Lemma 6.23 *Assume that $f \in \mathfrak{M}_\psi$ has TP, $M > 0$ and $\epsilon \in (0, 1)$. Then there exists an integer $L_0 \geq 4$ such that for each integer $L_1 > L_0$ there exists a neighborhood V of f in $\mathfrak{M}_\psi$ such that for each $g \in V$ and each pair of integers $T_1 \geq 0$, $T_2 \in (T_1, T_1 + L_1]$ and for each $(\{x_t\}_{t=T_1}^{T_2}, \{u_t\}_{t=T_1}^{T_2-1}) \in X(T_1, T_2)$ which satisfies*

$$\sum_{t=T_1}^{T_2-1} g(t, x_t, u_t) \leq \sum_{t=T_1}^{T_2-1} f(t, x_t^f, u_t^f) + M \tag{6.23}$$

the inequality

$$Card(\{t \in \{T_1, \dots, T_2\} : \rho_E(x_t, x_t^f) > \epsilon\}) \leq L_0/2$$

holds.

Proof Theorems 6.14 and 6.15 imply that there exists an integer $L_0 \geq 4$ such that the following property holds:

(a) for each pair of integers $T_1 \geq 0$, $T_2 \geq T_1 + L_0/2$ and for each $(\{x_t\}_{t=T_1}^{T_2}, \{u_t\}_{t=T_1}^{T_2-1}) \in X(T_1, T_2)$ which satisfies

$$\sum_{t=T_1}^{T_2-1} f(t, x_t, u_t) \leq \sum_{t=T_1}^{T_2-1} f(t, x_t^f, u_t^f) + M + 4$$

the inequality

$$\mathrm{Card}(\{t \in \{T_1, \dots, T_2\} : \rho_E(x_t, x_t^f) > \epsilon\}) \leq L_0/4$$

holds.

Let $L_1 > L_0$ be an integer.

Proposition 6.21 implies that there exists a neighborhood V of f in $\mathfrak{M}_\psi$ such that the following property holds:

(b) for each pair of integers $T_1 \geq 0$, $T_2 \in (T_1, T_1 + L_1]$, each $g \in V$ and each $(\{x_t\}_{t=T_1}^{T_2}, \{u_t\}_{t=T_1}^{T_2-1}) \in X(T_1, T_2)$ which satisfies

$$\min\{\sum_{t=T_1}^{T_2-1} f(t, x_t, u_t), \sum_{t=T_1}^{T_2-1} g(t, x_t, u_t)\} \le M + L_1 \Delta_f$$

the inequality

$$|\sum_{t=T_1}^{T_2-1} f(t, x_t, u_t) - \sum_{t=T_1}^{T_2-1} g(t, x_t, u_t)| \le 1 \tag{6.24}$$

holds.

Assume that $T_1 \ge 0$, $T_2 \in (T_1, T_1 + L_1]$ are integers, $g \in V$ and that $(\{x_t\}_{t=T_1}^{T_2}, \{u_t\}_{t=T_1}^{T_2-1}) \in X(T_1, T_2)$ satisfies (6.23). By (6.12) and (6.23),

$$\sum_{t=T_1}^{T_2-1} g(t, x_t, u_t) \le M + \Delta_f L_1.$$

Together with property (b) this implies (6.24). By (6.23) and (6.24),

$$\sum_{t=T_1}^{T_2-1} f(t, x_t, u_t) \le \sum_{t=T_1}^{T_2-1} g(t, x_t, u_t) + 1 \le \sum_{t=T_1}^{T_2-1} f(t, x_t^f, u_t^f) + M + 1.$$

Combined with property (a) this implies that

$$\mathrm{Card}(\{t \in \{T_1, \dots, T_2\} : \rho_E(x_t, x_t^f) > \epsilon\}) \le L_0/2.$$

Lemma 6.23 is proved.

Lemma 6.24 *Assume that $f \in \mathfrak{M}_\psi$ has TP, $\epsilon > 0$, $M \ge 8$ and L is a natural number. Then there exist integers $L_0 \ge L + 4$,*

$$L_1 > 8L_0 + 2L_0 L(a_0 + 1)$$

and a neighborhood V of f in $\mathfrak{M}_\psi$ such that for each $g \in V$, each pair of integers $T_1 \ge 0$, $T_2 \ge T_1 + L_1$ and each $(\{x_t\}_{t=T_1}^{T_2}, \{u_t\}_{t=T_1}^{T_2-1}) \in X(T_1, T_2)$ which satisfies at least one of the following conditions:

(i)

$$\sum_{t=T_1}^{T_2-1} g(t, x_t, u_t) \le U^g(T_1, T_2) + M;$$

(ii)

$$\sum_{t=T_1}^{T_2-1} g(t, x_t, u_t) \le U^g(T_1, T_2, x_{T_1}) + M, \ (T_1, x_{T_1}) \in \mathcal{A}_L;$$

(iii)

$$\sum_{t=T_1}^{T_2-1} g(t, x_t, u_t) \le U^g(T_1, T_2, x_{T_1}, x_{T_2}) + M,$$

$$(T_1, x_{T_1}) \in \mathcal{A}_L, \ (T_2, x_{T_2}) \in \widehat{\mathcal{A}}_L;$$

(iv)

$$\sum_{t=T_1}^{T_2-1} g(t, x_t, u_t) \le \widehat{U}^g(T_1, T_2, x_{T_2}) + M, \ (T_2, x_{T_2}) \in \widehat{\mathcal{A}}_L$$

the inequality

$$Card(\{t \in \{T_1, \dots, T_2\} : \ \rho_E(x_t, x_t^f) \le \epsilon\}) \ge L_0/2$$

holds.

Proof Lemma 6.23 implies that there exists an integer $L_0 \ge 4 + L$ and a neighborhood V_0 of f in $\mathfrak{M}_\psi$ such that the following property holds:

(a) for each $g \in V_0$ and each pair of integers $T_1 \ge 0$, $T_2 \in (T_1, T_1 + 8(L_0 + 1)]$ and each $(\{x_t\}_{t=T_1}^{T_2}, \{u_t\}_{t=T_1}^{T_2-1}) \in X(T_1, T_2)$ which satisfies

$$\sum_{t=T_1}^{T_2-1} g(t, x_t, u_t) \le \sum_{t=T_1}^{T_2-1} f(t, x_t^f, u_t^f) + M$$

the inequality

$$\text{Card}(\{t \in \{T_1, \dots, T_2\} : \ \rho_E(x_t, x_t^f) > \epsilon\}) \le L_0/2$$

is true.

Proposition 6.21 implies that there exists a neighborhood $V \subset V_0$ of f in $\mathfrak{M}_\psi$ such that the following property holds:

(b) for each pair of integers $T_1 \ge 0$, $T_2 \in (T_1, T_1 + 8L_0]$, each $g \in V$ and each $(\{x_t\}_{t=T_1}^{T_2}, \{u_t\}_{t=T_1}^{T_2-1}) \in X(T_1, T_2)$ which satisfies

$$\min\{\sum_{t=T_1}^{T_2-1} f(t, x_t, u_t), \sum_{t=T_1}^{T_2-1} g(t, x_t, u_t)\} \leq 2L_0(\Delta_f + 1) + L(1 + a_0)$$

the inequality

$$|\sum_{t=T_1}^{T_2-1} f(t, x_t, u_t) - \sum_{t=T_1}^{T_2-1} g(t, x_t, u_t)| \leq 8^{-1}$$

is true.

Choose an integer

$$L_1 > L_0(8 + 2L(a_0 + 1)). \tag{6.25}$$

Assume that

$$g \in V, T_1 \geq 0, \ T_2 \geq T_1 + L_1 \tag{6.26}$$

are integers and that $(\{x_t\}_{t=T_1}^{T_2}, \{u_t\}_{t=T_1}^{T_2-1}) \in X(T_1, T_2)$ satisfies at least one of conditions (i)–(iv). We construct $(\{\widehat{x}_t\}_{t=T_1}^{T_2}, \{\widehat{u}_t\}_{t=T_1}^{T_2-1}) \in X(T_1, T_2)$. If condition (i) holds we set

$$\widehat{x}_t = x_t^f, \ t = T_1, \dots, T_2, \ \widehat{u}_t = u_t^f, \ t = T_1, \dots, T_2 - 1.$$

If condition (ii) holds, then there exist $(\{\widehat{x}_t\}_{t=T_1}^{T_2}, \{\widehat{u}_t\}_{t=T_1}^{T_2-1}) \in X(T_1, T_2)$ and an integer $\tau_1 \in (0, L]$ such that

$$\widehat{x}_{T_1} = x_{T_1}^f, \ \sum_{t=T_1}^{T_1+\tau_1-1} f(t, \widehat{x}_t, \widehat{u}_t) \leq L,$$

$$\widehat{x}_t = x_t^f, \ t = T + \tau_1, \dots, T_2,$$

$$\widehat{u}_t = u_t^f, \ t = T + \tau_1, \dots, T_2 - 1.$$

If condition (iii) holds, then there exist $(\{\widehat{x}_t\}_{t=T_1}^{T_2}, \{\widehat{u}_t\}_{t=T_1}^{T_2-1}) \in X(T_1, T_2)$ and integers $\tau_1, \tau_2 \in (0, L]$ such that

$$\widehat{x}_{T_1} = x_{T_1}, \ \widehat{x}_{T_2} = x_{T_2},$$

$$\widehat{x}_t = x_t^f, \ t = T + \tau_1, \dots, T_2 - \tau_2, \ \widehat{u}_t = u_t^f, \ t = T + \tau_1, \dots, T_2 - \tau_2 - 1,$$

$$\sum_{t=T_1}^{T_1+\tau_1-1} f(t, \widehat{x}_t, \widehat{u}_t) \leq L, \ \sum_{t=T_2-\tau_2}^{T_2-1} f(t, \widehat{x}_t, \widehat{u}_t) \leq L.$$

If condition (iv) holds, then there exist $(\{\widehat{x}_t\}_{t=T_1}^{T_2}, \{\widehat{u}_t\}_{t=T_1}^{T_2-1}) \in X(T_1, T_2)$ and an integer $\tau_1 \in (0, L]$ such that

$$\widehat{x}_{T_2} = x_{T_2},$$

$$\widehat{x}_t = x_t^f,\ t = T_1, \dots, T_2 - \tau_1,\ \widehat{u}_t = u_t^f,\ t = T_1, \dots, T_2 - \tau_1 - 1,$$

$$\sum_{t=T_2-\tau_1}^{T_2-1} f(t, \widehat{x}_t, \widehat{u}_t) \le L.$$

There exists a natural number q such that

$$qL_0 \le T_2 - T_1 < (q+1)L_0. \tag{6.27}$$

By (6.25)–(6.27),

$$q > (T_2 - T_1)L_0^{-1} - 1 \ge L_1 L_0^{-1} - 1 \ge 8. \tag{6.28}$$

It follows from property (b), (6.28), the relation $L_0 \ge L+4$ and the construction of $(\{\widehat{x}_t\}_{t=T_1}^{T_2}, \{\widehat{u}_t\}_{t=T_1}^{T_2-1})$ that for all $i = 1, \dots, q-2$,

$$\left| \sum_{t=T_1+iL_0}^{T_1+(i+1)L_0-1} g(t, \widehat{x}_t, \widehat{u}_t) - \sum_{t=T_1+iL_0}^{T_1+(i+1)L_0-1} f(t, x_t^f, \widehat{u}_t^f) \right| \le 8^{-1}. \tag{6.29}$$

By (6.17), (6.27), the relation $L_0 \ge L+4$ and the construction of the trajectory $(\{\widehat{x}_t\}_{t=T_1}^{T_2}, \{\widehat{u}_t\}_{t=T_1}^{T_2-1})$,

$$\sum_{t=T_1+(q-1)L_0}^{T_2-1} f(t, \widehat{x}_t, \widehat{u}_t) \le \sum_{t=T_1+(q-1)L_0}^{T_2-1} f(t, x_t^f, \widehat{u}_t^f) + L(a_0+1). \tag{6.30}$$

Property (b) and Eqs. (6.12), (6.26), (6.27), and (6.30) imply that

$$\sum_{t=T_1+(q-1)L_0}^{T_2-1} g(t, \widehat{x}_t, \widehat{u}_t) \le \sum_{t=T_1+(q-1)L_0}^{T_2-1} f(t, x_t^f, u_t^f) + L(a_0+1) + 1. \tag{6.31}$$

By (6.6), the relation $L_0 \ge L+4$ and the construction of $(\{\widehat{x}_t\}_{t=T_1}^{T_2}, \{\widehat{u}_t\}_{t=T_1}^{T_2-1})$,

$$\sum_{t=T_1}^{T_1+L_0-1} f(t, \widehat{x}_t, \widehat{u}_t)$$

$$\leq \sum_{t=T_1}^{T_1+L_0-1} f(t, x_t^f, \widehat{u}_t^f) + L(a_0 + 1). \tag{6.32}$$

Property (b) and Eqs. (6.12), (6.26), and (6.32) imply that

$$\sum_{t=T_1}^{T_1+L_0-1} g(t, \widehat{x}_t, \widehat{u}_t) \leq \sum_{t=T_1}^{T_1+L_0-1} f(t, \widehat{x}_t, \widehat{u}_t) + 1$$

$$\leq \sum_{t=T_1}^{T_1+L_0-1} f(t, x_t^f, \widehat{u}_t^f) + L(a_0 + 1) + 1. \tag{6.33}$$

It follows from (6.27), (6.29), and (6.31) that in the all cases

$$\sum_{t=T_1}^{T_2-1} g(t, \widehat{x}_t, \widehat{u}_t) \leq \sum_{t=T_1}^{T_2-1} f(t, x_t^f, \widehat{u}_t^f) + q + 1 + 2L(a_0 + 1). \tag{6.34}$$

In order to complete the proof of our lemma we need only to show that

$$\mathrm{Card}(\{t \in \{T_1, \dots, T_2\} : \rho_E(x_t, x_t^f) \leq \epsilon\}) \geq L_0/2.$$

Assume the contrary. Then

$$\mathrm{Card}(\{t \in \{T_1, \dots, T_2\} : \rho_E(x_t, x_t^f) \leq \epsilon\}) < L_0/2. \tag{6.35}$$

Property (a), (6.26), (6.27), and (6.35) imply that for all $i = 0, \dots, q-2$,

$$\sum_{t=T_1+iL_0}^{T_1+(i+1)L_0-1} g(t, x_t, u_t) > \sum_{t=T_1+iL_0}^{T_1+(i+1)L_0-1} f(t, x_t^f, \widehat{u}_t^f) + M,$$

$$\sum_{t=T_1+(q-1)L_0}^{T_2-1} g(t, x_t, u_t) > \sum_{t=T_1+(q-1)L_0}^{T_2-1} f(t, x_t^f, u_t^f) + M.$$

By the equations above

$$\sum_{t=T_1}^{T_2-1} g(t, x_t, u_t) \geq \sum_{t=T_1}^{T_2-1} f(t, x_t^f, u_t^f) + L(a_0 + 1) + Mq. \tag{6.36}$$

It follows from (6.34), (6.36), the inequality $M \geq 8$, conditions (i)–(iv) and the construction of $(\{\widehat{x}_t\}_{t=T_1}^{T_2}, \{\widehat{u}_t\}_{t=T_1}^{T_2-1})$ that

$$\sum_{t=T_1}^{T_2-1} f(t, x_t^f, u_t^f) \le \sum_{t=T_1}^{T_2-1} g(t, x_t, u_t) - Mq \le \sum_{t=T_1}^{T_2-1} g(t, \widehat{x}_t, \widehat{u}_t) - qM + M$$

$$\le -(q-1)M + \sum_{t=T_1}^{T_2-1} f(t, x_t^f, u_t^f) + q + 1 + 2L(1 + a_0)$$

and

$$q \le 3 + 2L(1 + a_0). \tag{6.37}$$

In view of (6.26), (6.27) and (6.37),

$$L_1 \le L_0(q+1) \le L_0(4 + 2L(1 + a_0)).$$

This contradicts (6.25). The contradiction we have reached completes the proof of Lemma 6.24.

6.7 Proof of Theorem 6.22

We may assume without loss of generality that

$$L > 4 + 4b_f + (b_f + 1)(\Delta_f + 1) \tag{6.38}$$

and that the following property holds:

(a) if $(T, z) \in \mathcal{A}$ and $\rho_E(z, x_t^f) \le \epsilon$, then $z \in \mathcal{A}_L$ and if in addition $T \ge L$, the $z \in \widehat{\mathcal{A}}_L$.

Lemma 6.16 and (A2) imply that there exists $\delta \in (0, \epsilon)$ and an integer $\bar{L}_0 > 4b_f$ such that the following properties hold:

(b) for each integer $T_1 \ge \bar{L}_0$, each integer $T_2 \ge T_1 + 2b_f$ and each $(\{x_t\}_{t=T_1}^{T_2}, \{u_t\}_{t=T_1}^{T_2-1}) \in X(T_1, T_2)$ which satisfies

$$\rho_E(x_{T_i}, x_{T_i}^f) \le 4\delta,\ i = 1, 2,$$

$$\sum_{t=T_1}^{T_2-1} f(t, x_t, u_t) \le U^f(T_1, T_2, x_{T_1}, x_{T_2}) + 4\delta$$

the inequality $\rho_E(x_t, x_t^f) \le \epsilon$ holds for all $t = T_1, \dots, T_2$;

(c) for each $(T_i, z_i) \in \mathcal{A}$, $i = 1, 2$ satisfying $\rho_E(z_i, x^f_{T_i}) \le \delta$, $i = 1, 2$ and $T_2 \ge b_f$ there exist

$$(\{x^{(1)}_t\}_{t=T_1}^{T_1+b_f}, \{u^{(1)}_t\}_{t=T_1}^{T_1+b_f-1}) \in X(T_1, T_1 + b_f),$$

$$(\{x^{(2)}_t\}_{t=T_2-b_f}^{T_2}, \{u^{(2)}_t\}_{t=T_2-b_f}^{T_2-1}) \in X(T_2 - b_f, T_2)$$

such that

$$x^{(1)}_{T_1} = z_1,\ x^{(1)}_{T_1+b_f} = x^f_{T_1+b_f},$$

$$\sum_{t=T_1}^{T_1+b_f-1} f(t, x^{(1)}_t, u^{(1)}_t) \le \sum_{t=T_1}^{T_1+b_f-1} f(t, x^f_t, u^f_t) + 1,$$

$$x^{(2)}_{T_2} = z_2,\ x^{(2)}_{T_2-b_f} = x^f_{T_2-b_f},$$

$$\sum_{t=T_2-b_f}^{T_2-1} f(t, x^{(2)}_t, u^{(2)}_t) \le \sum_{t=T_2-b_f}^{T_2-1} f(t, x^f_t, u^f_t) + 1.$$

Lemma 6.24 implies that there exists integers

$$L_0 > L + 2\bar{L}_0 + 8,\ L_1 > 8L_0 + 2L_0L(a_0 + 1) \tag{6.39}$$

a neighborhood V_0 of f in $\mathfrak{M}_\psi$ such that the following property holds:

(d) for each $g \in V_0$, each pair of integers $T_1 \ge 0$, $T_2 \ge T_1 + L_1$ and each $(\{x_t\}_{t=T_1}^{T_2}, \{u_t\}_{t=T_1}^{T_2-1}) \in X(T_1, T_2)$ which satisfies at least one of the following conditions:

$$\sum_{t=T_1}^{T_2-1} g(t, x_t, u_t) \le U^g(T_1, T_2) + L;$$

$$\sum_{t=T_1}^{T_2-1} g(t, x_t, u_t) \le U^g(T_1, T_2, x_{T_1}) + L,\ (T_1, x_{T_1}) \in \mathcal{A}_L;$$

$$\sum_{t=T_1}^{T_2-1} g(t, x_t, u_t) \le U^g(T_1, T_2, x_{T_1}, x_{T_2}) + L,$$

$$(T_1, x_{T_1}) \in \mathcal{A}_L,\ (T_2, x_{T_2}) \in \widehat{\mathcal{A}}_L;$$

$$\sum_{t=T_1}^{T_2-1} g(t, x_t, u_t) \le \widehat{U}^g(T_1, T_2, x_{T_2}) + L, \ (T_2, x_{T_2}) \in \widehat{\mathcal{A}}_L$$

the inequality

$$\mathrm{Card}(\{t \in \{T_1, \dots, T_2\} : \ \rho_E(x_t, x_t^f) \le \delta\}) \ge L_0/2$$

is true.

Proposition 6.21 implies that there exists a neighborhood $V \subset V_0$ of f in $\mathfrak{M}_\psi$ such that the following property holds:

(e) for each pair of integers $S_1 \ge 0$, $S_2 \in (S_1, S_1 + L_1]$, each $g \in V$ and each $(\{x_t\}_{t=S_1}^{S_2}, \{u_t\}_{t=S_1}^{S_2-1}) \in X(S_1, S_2)$ which satisfies

$$\min\{\sum_{t=S_1}^{S_2-1} f(t, x_t, u_t), \sum_{t=S_1}^{S_2-1} g(t, x_t, u_t)\} \le (L_1 + 2)\Delta_f + 8 + a_0$$

the inequality

$$|\sum_{t=S_1}^{S_2-1} f(t, x_t, u_t) - \sum_{t=S_1}^{S_2-1} g(t, x_t, u_t)| \le \delta$$

is true.

Assume that $g \in V$ and pair of integers

$$T_1 \ge 0, \ T_2 \ge T_1 + 4L_1 \tag{6.40}$$

and that $(\{x_t\}_{t=T_1}^{T_2}, \{u_t\}_{t=T_1}^{T_2-1}) \in X(T_1, T_2)$ satisfies at least one of the following conditions:

(i)

$$\sum_{t=T_1}^{T_2-1} g(t, x_t, u_t) \le \min\{U^g(T_1, T_2) + L, \ U^g(T_1, T_2, x_{T_1}, x_{T_2}) + \delta\};$$

(ii)

$$\sum_{t=T_1}^{T_2-1} g(t, x_t, u_t) \le U^g(T_1, T_2, x_{T_1}) + \delta, \ (T_1, x_{T_1}) \in \mathcal{A}_L;$$

(iii)

$$\sum_{t=T_1}^{T_2-1} g(t, x_t, u_t) \le U^g(T_1, T_2, x_{T_1}, x_{T_2}) + \delta,$$

$$(T_1, x_{T_1}) \in \mathcal{A}_L, \ (T_2, x_{T_2}) \in \widehat{\mathcal{A}}_L.$$

Property (d), conditions (i)–(iii), and (6.40) imply that

$$\mathrm{Card}(\{t \in \{T_1, \dots, T_2\} : \ \rho_E(x_t, x_t^f) \le \delta\}) \ge L_0/2. \tag{6.41}$$

Set

$$S_1 = \min\{t \in \{\max\{T_1, \bar{L}_0\}, \dots, T_2\} : \ \rho_E(x_t, x_t^f) \le \delta\}. \tag{6.42}$$

(Note that if $T_1 \ge \bar{L}_0$ and $\rho_E(x_{T_1}, x_{T_1}^f) \le \delta$, then $S_1 = T_1$.) Clearly, S_1 is well-defined. If $S_1 \ge T_1 + L_1$, then it follows from properties (a), (d) and conditions (i)–(iii) and (6.40) that

$$\mathrm{Card}(\{t \in \{T_1, \dots, S_1\} : \ \rho_E(x_t, x_t^f) \le \delta\}) \ge L_0/2.$$

This contradicts (6.42) (see (6.39)). The contradiction we have reached proves that

$$S_1 < T_1 + L_1. \tag{6.43}$$

Set

$$S_2 = \max\{t \in \{S_1, \dots, T_2\} : \ \rho_E(x_t, x_t^f) \le \delta\}. \tag{6.44}$$

If $S_2 \le T_2 - L_1$, then it follows from properties (a), (d) and conditions (i)–(iii) and Eqs. (6.39), (6.40) and (6.44) that

$$\mathrm{Card}(\{t \in \{S_2, \dots, T_2\} : \ \rho_E(x_t, x_t^f) \le \delta\}) \le L_0/2.$$

This contradicts (6.44). Thus

$$S_2 > T_2 - L_1. \tag{6.45}$$

In order to complete the proof of Theorem 6.22 it is sufficient to show that

$$\rho_E(x_t, x_t^f) \le \epsilon, \ t \in \{S_1, \dots, S_2\}.$$

Assume the contrary. Then there exists an integer

$$\tau_0 \in \{S_1, \dots, S_2\} \tag{6.46}$$

such that

$$\rho_E(x_{\tau_0}, x^f_{\tau_0}) > \epsilon. \tag{6.47}$$

In view of (6.40) and (6.43)–(6.45),

$$S_2 - S_1 \geq T_2 - T_1 - 2L_1 \geq 2L_1. \tag{6.48}$$

By (6.48), there exists an integer S_3 such that

$$\tau_0 \in [S_3, S_3 + \lfloor L_0/4 \rfloor + 1] \subset [S_1, S_2]. \tag{6.49}$$

Set

$$S_4 = S_3 + \lfloor L_0/4 \rfloor + 1. \tag{6.50}$$

Set

$$S_5 = \max\{t \in \{S_1, \dots, S_3\} : \rho_E(x_t, x^f_t) \leq \delta\}. \tag{6.51}$$

$$S_6 = \min\{t \in \{S_4, \dots, S_2\} : \rho_E(x_t, x^f_t) \leq \delta\}. \tag{6.52}$$

It follows from (6.49)–(6.52) that

$$S_6 - S_5 \geq S_4 - S_3 \geq \lfloor L_0/4 \rfloor + 1. \tag{6.53}$$

Assume that

$$S_6 - S_5 \geq L_1.$$

Properties (a) and (d), conditions (i)–(iii) and Eqs. (6.40), (6.50) and (6.51) imply that

$$\mathrm{Card}(\{t \in \{S_5, \dots, S_6\} : \rho_E(x_t, x^f_t) \leq \delta\}) \geq L_0/2.$$

On the other hand, by (6.50)–(6.52),

$$\mathrm{Card}(\{t \in \{S_5, \dots, S_6\} : \rho_E(x_t, x^f_t) \leq \delta\}) \leq S_4 - S_3 + 2 \leq \lfloor L_0/4 \rfloor + 3.$$

The contradiction we have reached proves that

$$S_6 - S_5 \leq L_1.$$

Together with (6.53) this implies that

$$\lfloor L_0/4 \rfloor + 1 \le S_6 - S_5 \le L_1. \tag{6.54}$$

Property (c) and Eqs. (6.39), (6.51), (6.52), and (6.54) imply that there exist

$$(\{y_t\}_{t=S_5}^{S_6}, \{v_t\}_{t=S_5}^{S_6-1}) \in X(S_5, S_6)$$

such that

$$y_{S_5} = x_{S_5}, \ y_{S_6} = y_{S_6},$$

$$y_t = x_t^f, \ t \in \{S_5 + b_f, \dots, S_6 - b_f\}, \ v_t = u_t^f, \ t \in \{S_5 + b_f, \dots, S_6 - b_f - 1\},$$

$$\sum_{t=S_5}^{S_5+b_f-1} f(t, y_t, v_t) \le \sum_{t=S_5}^{S_5+b_f-1} f(t, x_t^f, u_t^f) + 1,$$

$$\sum_{t=S_6-b_f}^{S_6-1} f(t, y_t, v_t) \le \sum_{t=S_6-b_f}^{S_6} f(t, x_t^f, u_t^f) + 1. \tag{6.55}$$

By (6.12), (6.54), and (6.55),

$$\sum_{t=S_5}^{S_6-1} f(t, y_t, v_t) \le \sum_{t=S_5}^{S_6-1} f(t, x_t^f, u_t^f) + 2 \le 2 + \Delta_f(L_1 + 2). \tag{6.56}$$

In view of conditions (i)–(iii),

$$\sum_{t=S_5}^{S_6-1} g(t, x_t, u_t) \le U^g(S_5, S_6, x_{S_5}, x_{S_6}) + \delta. \tag{6.57}$$

Property (e) and Eqs. (6.40) and (6.54)–(6.56) imply that

$$U^g(S_5, S_6, x_{S_5}, x_{S_6})$$

$$\le \inf\{\sum_{t=S_5}^{S_6-1} g(t, \xi_t, \eta_t) : (\{\xi_t\}_{t=S_5}^{S_6}, \{\eta_t\}_{t=S_5}^{S_6-1}) \in X(S_5, S_6),$$

$$\xi_{S_i} = x_{S_i}, \ i = 5, 6, \ \sum_{t=S_5}^{S_6-1} f(t, \xi_t, \eta_t) \le \Delta_f(L_1 + 2) + 3\}$$

$$\leq \delta + \inf\{\sum_{t=S_5}^{S_6-1} f(t, \xi_t, \eta_t) :$$

$$(\{\xi_t\}_{t=S_5}^{S_6}, \{\eta_t\}_{t=S_5}^{S_6-1}) \in X(S_5, S_6),\ \xi_{S_i} = x_{S_i},\ i = 5, 6,$$

$$\sum_{t=S_5}^{S_6-1} f(t, \xi_t, \eta_t) \leq \Delta_f(L_1 + 2) + 3\} \leq \delta + U^f(S_5, S_6, x_{S_5}, x_{S_6}). \tag{6.58}$$

By (6.57) and (6.58),

$$\sum_{t=S_5}^{S_6-1} g(t, x_t, u_t) \leq \Delta_f(L_1 + 2) + 4. \tag{6.59}$$

Property (e) and (6.54) and (6.59) imply that

$$|\sum_{t=S_5}^{S_6-1} f(t, x_t, u_t) - \sum_{t=S_5}^{S_6-1} g(t, x_t, u_t)| \leq \delta.$$

Together with (6.57) and (6.58) this implies that

$$\sum_{t=S_5}^{S_6-1} f(t, x_t, u_t) \leq \sum_{t=S_5}^{S_6-1} g(t, x_t, u_t) + \delta$$

$$\leq U^g(S_5, S_6, x_{S_5}, x_{S_6}) + 2\delta \leq U^f(S_5, S_6, x_{S_5}, x_{S_6}) + 3\delta.$$

By the equation above, property (b) and Eqs. (6.39), (6.42), (6.51), (6.52) and (6.54),

$$\rho_E(x_t, x_t^f) \leq \epsilon,\ t \in \{S_5, \ldots, S_6\}.$$

Together with (6.49) and (6.50) this implies that

$$\rho_E(x_{\tau_0}, x_{\tau_0}^f) \leq \epsilon.$$

This contradicts (6.47). The contradiction we have reached competes the proof of Theorem 6.22.

6.8 Problems with Discount

We use the notation, definitions and assumptions introduced in Sects. 6.1–6.3 and 6.5.

Let $g \in \mathfrak{M}_\psi$ and $\alpha = \{\alpha_t\}_{t=0}^\infty \subset (0, 1]$. For each pair of integers $T_2 > T_1 \geq 0$ and each pair of points $y, z \in E$ satisfying $(T_1, y), (T_2, z) \in \mathcal{A}$ we define

$$U^{\alpha g}(T_1, T_2, y, z) = \inf\{\sum_{t=T_1}^{T_2-1} \alpha_t g(t, x_t, u_t) :$$

$$(\{x_t\}_{t=T_1}^{T_2}, \{u_t\}_{t=T_1}^{T_2-1}) \in X(T_1, T_2),$$

$$x_{T_1} = y,\ x_{T_2} = z\},$$

$$U^{\alpha g}(T_1, T_2, y) = \inf\{\sum_{t=T_1}^{T_2-1} \alpha_t g(t, x_t, u_t) :$$

$$(\{x_t\}_{t=T_1}^{T_2}, \{u_t\}_{t=T_1}^{T_2-1}) \in X(T_1, T_2),\ x_{T_1} = y\},$$

$$\widehat{U}^{\alpha g}(T_1, T_2, y) = \inf\{\sum_{t=T_1}^{T_2-1} \alpha_t g(t, x_t, u_t) :$$

$$(\{x_t\}_{t=T_1}^{T_2}, \{u_t\}_{t=T_1}^{T_2-1}) \in X(T_1, T_2),\ x_{T_2} = y\},$$

$$U^{\alpha g}(T_1, T_2) = \inf\{\sum_{t=T_1}^{T_2-1} \alpha_t g(t, x_t, u_t) :\ (\{x_t\}_{t=T_1}^{T_2}, \{u_t\}_{t=T_1}^{T_2-1}) \in X(T_1, T_2)\}. \tag{6.60}$$

We assume that all the assumptions of f posed in Sects. 6.1–6.3 and 6.5 hold and prove the following stability result for problems with discount.

Theorem 6.25 *Assume that $f \in \mathfrak{M}_\psi$ has TP, L is a natural number and $\epsilon \in (0, 1)$. Then there exists an integer $L_1 > L$, $\delta \in (0, \epsilon)$ $\lambda \in (0, 1)$ and a neighborhood V of f in $\mathfrak{M}_\psi$ such that for each $g \in V$ and each pair of integers $T_1 \geq 0$, $T_2 \geq T_1 + 4L_1$, each $\alpha = \{\alpha_t\}_{t=0}^\infty \subset (0, 1]$ which satisfies*

$$\alpha_{t_1}\alpha_{t_2}^{-1} \geq \lambda \text{ for each } t_1, t_2 \in \{T_1, \dots, T_2\} \text{ satisfying } |t_2 - t_1| \leq L_1$$

and that for each $(\{x_t\}_{t=T_1}^{T_2}, \{u_t\}_{t=T_1}^{T_2-1}) \in X(T_1, T_2)$ which satisfies at least one of the following conditions:

(i)

$$\sum_{t=T_1}^{T_2-1} \alpha_t g(t, x_t, u_t) \le U^{\alpha g}(T_1, T_2) + \delta \min\{\alpha_t : t = T_1, \ldots, T_2 - 1\};$$

(ii)

$$(T_1, x_{T_1}) \in \mathcal{A}_L, \sum_{t=T_1}^{T_2-1} \alpha_t g(t, x_t, u_t)$$

$$\le U^{\alpha g}(T_1, T_2, x_{T_1}) + \delta \min\{\alpha_t : t = T_1, \ldots, T_2 - 1\};$$

(iii)

$$(T_1, x_{T_1}) \in \mathcal{A}_L, (T_2, x_{T_2}) \in \widehat{\mathcal{A}}_L$$

and

$$\sum_{t=T_1}^{T_2-1} \alpha_t g(t, x_t, u_t) \le U^{\alpha g}(T_1, T_2, x_{T_1}, x_{T_2}) + \delta \min\{\alpha_t : t = T_1, \ldots, T_2 - 1\}$$

there exist integers $\tau_1 \in [T_1, T_1 + L_1]$, $\tau_2 \in [T_2 - L_1, T_2]$ *such that*

$$\rho_E(x_t, x_t^f) \le \epsilon, \ t = \tau_1 \ldots, \tau_2.$$

Moreover, if $\rho_E(x_{T_2}, x_{T_2}^f) \le \delta$, *then* $\tau_2 = T_2$ *and if* $T_1 \ge L_1$ *and* $\rho_E(x_{T_1}, x_{T_1}^f) \le \delta$, *then* $\tau_1 = T_1$.

6.9 Auxiliary Results for Theorem 6.25

We assume in this section that f has TP.

Lemma 6.26 *Assume that* $M \ge 2$ *and* $\epsilon \in (0, 1)$. *Then there exists an integer* $L_0 \ge 4$ *such that for each integer* $L_1 > L_0$ *there exist* $\lambda \in (0, 1)$ *and a neighborhood* V *of* f *in* $\mathfrak{M}_\psi$ *such that for each* $g \in V$ *and each pair of integers* $T_1 \ge 0$, $T_2 \in (T_1, T_1 + L_1]$, *each* $\alpha = \{\alpha_t\}_{t=0}^{\infty} \subset (0, 1]$ *which satisfies*

$$\alpha_{t_1} \alpha_{t_2}^{-1} \ge \lambda \text{ for each } t_1, t_2 \in \{T_1, \ldots, T_2 - 1\}$$

and for each $(\{x_t\}_{t=T_1}^{T_2}, \{u_t\}_{t=T_1}^{T_2-1}) \in X(T_1, T_2)$ *which satisfies*

$$\sum_{t=T_1}^{T_2-1} \alpha_t g(t, x_t, u_t) \leq \sum_{t=T_1}^{T_2-1} \alpha_t f(t, x_t^f, u_t^f) + 4^{-1} M \min\{\alpha_t : t = T_1, \dots, T_2 - 1\} \tag{6.61}$$

the inequality

$$Card(\{t \in \{T_1, \dots, T_2\} : \rho_E(x_t, x_t^f) > \epsilon\}) \leq L_0/2 \tag{6.62}$$

holds.

Proof Lemma 6.23 implies that there exists an integer $L_0 \geq 4$ such that for each integer $L_1 > L_0$ the following property holds:

(a) there exists a neighborhood V of f in $\mathfrak{M}_\psi$ such that for each $g \in V$ and each pair of integers $T_1 \geq 0$, $T_2 \in (T_1, T_1 + L_1]$ and each $(\{x_t\}_{t=T_1}^{T_2}, \{u_t\}_{t=T_1}^{T_2-1}) \in X(T_1, T_2)$ which satisfies

$$\sum_{t=T_1}^{T_2-1} g(t, x_t, u_t) \leq \sum_{t=T_1}^{T_2-1} f(t, x_t^f, u_t^f) + M$$

equation (6.62) is true.

Let $L_1 > L_0$ be an integer and let V be a neighborhood of f in $\mathfrak{M}_\psi$ such that property (a) holds. Fix

$$\lambda \in (2^{-1}, 1)$$

such that

$$4(1 - \lambda)(\Delta_f + 1)(2L_1 + 2L_1 a_0 + 2) < 1. \tag{6.63}$$

Assume that

$$g \in V, \ T_1 \geq 0, \ T_2 \in (T_1, T_1 + L_1] \tag{6.64}$$

are integers, $\alpha = \{\alpha_t\}_{t=0}^{\infty} \subset (0, 1]$ satisfies

$$\alpha_{t_1} \alpha_{t_2}^{-1} \geq \lambda \text{ for each } t_1, t_2 \in \{T_1, \dots, T_2 - 1\} \text{ satisfying } |t_1 - t_2| \leq L_1 \tag{6.65}$$

and $(\{x_t\}_{t=T_1}^{T_2}, \{u_t\}_{t=T_1}^{T_2-1}) \in X(T_1, T_2)$ satisfies (6.61). We show that (6.62) holds. Assume the contrary. Then by property (a) and (6.64),

$$\sum_{t=T_1}^{T_2-1} g(t, x_t, u_t) > \sum_{t=T_1}^{T_2-1} f(t, x_t^f, u_t^f) + M. \tag{6.66}$$

It follows from (6.17), (6.64), and (6.66) that

$$\sum_{t=T_1}^{T_2-1} \alpha_t g(t, x_t, u_t) = -a_0 \sum_{t=T_1}^{T_2-1} \alpha_t + \sum_{t=T_1}^{T_2-1} \alpha_t (g(t, x_t, u_t) + a_0)$$

$$\geq -a_0 \sum_{t=T_1}^{T_2-1} \alpha_t$$

$$+ \min\{\alpha_t : t = T_1, \ldots, T_2 - 1\}\{\sum_{t=T_1}^{T_2-1} (g(t, x_t, u_t) + a_0)$$

$$\geq -a_0 \sum_{t=T_1}^{T_2-1} \alpha_t + \min\{\alpha_t : t = T_1, \ldots, T_2 - 1\}(\sum_{t=T_1}^{T_2-1} (f(t, x_t^f, u_t^f) + a_0) + M). \tag{6.67}$$

Equations (6.6), (6.12) and (6.63)–(6.65) imply that

$$\max\{\alpha_t : t = T_1, \ldots, T_2 - 1\} \sum_{t=T_1}^{T_2-1} (f(t, x_t^f, u_t^f) + a_0)$$

$$- \min\{\alpha_t : t = T_1, \ldots, T_2 - 1\} \sum_{t=T_1}^{T_2-1} (f(t, x_t^f, u_t^f) + a_0)$$

$$\leq \max\{\alpha_t : t = T_1, \ldots, T_2 - 1\}(1 - \lambda) \sum_{t=T_1}^{T_2-1} (f(t, x_t^f, u_t^f) + a_0)$$

$$\leq \max\{\alpha_t : t = T_1, \ldots, T_2 - 1\}(1 - \lambda)(\Delta_f + a_0)L_1$$

$$\leq \max\{\alpha_t : t = T_1, \ldots, T_2 - 1\}. \tag{6.68}$$

By (6.6), (6.67), and (6.68),

$$\sum_{t=T_1}^{T_2-1} \alpha_t g(t, x_t, u_t) \geq -a_0 \sum_{t=T_1}^{T_2-1} \alpha_t$$

$$+ \max\{\alpha_t : t = T_1, \ldots, T_2 - 1\} \sum_{t=T_1}^{T_2-1} (f(t, x_t^f, u_t^f) + a_0)$$

$$- \max\{\alpha_t : t = T_1, \dots, T_2 - 1\} + M \min\{\alpha_t : t = T_1, \dots, T_2 - 1\}$$

$$\ge \sum_{t=T_1}^{T_2-1} \alpha_t f(t, x_t^f, u_t^f) - \max\{\alpha_t : t = T_1, \dots, T_2 - 1\}$$

$$+ M \min\{\alpha_t : t = T_1, \dots, T_2 - 1\}$$

$$\ge \sum_{t=T_1}^{T_2-1} \alpha_t f(t, x_t^f, u_t^f) + 2^{-1} M \min\{\alpha_t : t = T_1, \dots, T_2 - 1\}.$$

This contradicts (6.61). The contradiction we have reached proves (6.62) and Lemma 6.26.

Lemma 6.27 *Assume that* $\epsilon \in (0, 1)$ *and* L *is a natural number. Then there exist integers* $L_0 > L + 4$,

$$L_1 > 8L_0 + 2L_0 L(a_0 + 1) + L_0(2L_0 + 4),$$

$\lambda \in (0, 1)$ *and a neighborhood* V *of* f *in* $\mathfrak{M}_\psi$ *such that for each* $g \in V$, *each pair of integers* $T_1 \ge 0$, $T_2 \ge T_1 + L_1$, *each* $\alpha = \{\alpha_t\}_{t=0}^{\infty} \subset (0, 1]$ *which satisfies*

$$\alpha_{t_1} \alpha_{t_2}^{-1} \ge \lambda \text{ for each } t_1, t_2 \in \{T_1, \dots, T_2 - 1\} \text{ satisfying } |t_1 - t_2| \le L_1 \qquad (6.69)$$

the following assertion holds.

Assume that $(\{x_t\}_{t=T_1}^{T_2}, \{u_t\}_{t=T_1}^{T_2-1}) \in X(T_1, T_2)$ *satisfies at least one of the following conditions:*

(i)

$$\sum_{t=T_1}^{T_2-1} \alpha_t g(t, x_t, u_t) \le U^{\alpha g}(T_1, T_2) + 4 \min\{\alpha_t : t = T_1, \dots, T_2 - 1\};$$

(ii)

$$\sum_{t=T_1}^{T_2-1} \alpha_t g(t, x_t, u_t) \le U^{\alpha g}(T_1, T_2, x_{T_1}) + 4 \min\{\alpha_t : t = T_1, \dots, T_2 - 1\},$$

$$(T_1, x_{T_1}) \in \mathcal{A}_L;$$

(iii)

$$\sum_{t=T_1}^{T_2-1} \alpha_t g(t, x_t, u_t) \le U^{\alpha g}(T_1, T_2, x_{T_1}, x_{T_2}) + 4 \min\{\alpha_t : t = T_1, \dots, T_2 - 1\},$$

$$(T_1, x_{T_1}) \in \mathcal{A}_L, \ (T_2, x_{T_2}) \in \widehat{\mathcal{A}}_L;$$

(iv)

$$\sum_{t=T_1}^{T_2-1} \alpha_t g(t, x_t, u_t) \le \widehat{U}^{\alpha g}(T_1, T_2, x_{T_2}) + 4\min\{\alpha_t : t = T_1, \dots, T_2 - 1\},$$

$$(T_2, x_{T_2}) \in \widehat{\mathcal{A}}_L$$

Then

$$Card(\{t \in \{T_1, \dots, T_2\} : \rho_E(x_t, x_t^f) \le \epsilon\}) \ge L_0/2. \tag{6.70}$$

Proof Lemma 6.26 implies that there exists an integer $L_0 > 4 + L$, $\lambda_0 \in (0, 1)$ and a neighborhood V_0 of f in $\mathfrak{M}_\psi$ such that the following property holds:

(a) for each $g \in V_0$, each pair of integers $T_1 \ge 0$, $T_2 \in (T_1, T_1 + 2L_0]$, each $\alpha = \{\alpha_t\}_{t=0}^{\infty} \subset (0, 1]$ which satisfies

$$\alpha_{t_1}\alpha_{t_2}^{-1} \ge \lambda_0 \text{ for each } t_1, t_2 \in \{T_1, \dots, T_2 - 1\}$$

and for each $(\{x_t\}_{t=T_1}^{T_2}, \{u_t\}_{t=T_1}^{T_2-1}) \in X(T_1, T_2)$ which satisfies

$$\sum_{t=T_1}^{T_2-1} \alpha_t g(t, x_t, u_t) \le \sum_{t=T_1}^{T_2-1} \alpha_t f(t, x_t^f, u_t^f) + 8\min\{\alpha_t : t = T_1, \dots, T_2 - 1\}$$

the inequality

$$\text{Card}(\{t \in \{T_1, \dots, T_2\} : \rho_E(x_t, x_t^f) > \epsilon\}) \le L_0/2$$

is true.

Choose an integer

$$q_0 > 8L + 16(L_0 + 1)(a_0 + 1) \tag{6.71}$$

and

$$\lambda \in (\lambda_0, 1)$$

such that

$$\lambda^{q_0+1} > 2^{-1}, \ (1 - \lambda)\lambda^{-1}(L_0 + 1)(\Delta_f + a_0 + 1) < 8^{-1}. \tag{6.72}$$

Proposition 6.21 implies that there exists a neighborhood $V \subset V_0$ of f in $\mathfrak{M}_\psi$ such that the following property holds:

(b) for each pair of integers $T_1 \geq 0$, $T_2 \in (T_1, T_1 + 8L_0]$, each $g \in V$ and each $(\{x_t\}_{t=T_1}^{T_2}, \{u_t\}_{t=T_1}^{T_2-1}) \in X(T_1, T_2)$ which satisfies

$$\min\{\sum_{t=T_1}^{T_2-1} f(t, x_t, u_t), \sum_{t=T_1}^{T_2-1} g(t, x_t, u_t)\} \leq 2L_0(\Delta_f + 1) + L(1 + 2a_0)$$

the inequality

$$|\sum_{t=T_1}^{T_2-1} f(t, x_t, u_t) - \sum_{t=T_1}^{T_2-1} g(t, x_t, u_t)| \leq 8^{-1}$$

is true.

Choose an integer

$$L_1 > L_0(8 + 2L(a_0 + 1)) + L_0(2a_0 + 2q_0 + 4). \tag{6.73}$$

Assume that

$$T_1 \geq 0,\ T_2 \in (T_1, T_1 + L_1] \tag{6.74}$$

are integers, $\alpha = \{\alpha_t\}_{t=0}^{\infty} \subset (0, 1]$ satisfies (6.69),

$$g \in V \tag{6.75}$$

and that $(\{x_t\}_{t=T_1}^{T_2}, \{u_t\}_{t=T_1}^{T_2-1}) \in X(T_1, T_2)$ satisfies at least one of the conditions (i)–(iv). In order to complete the proof of the lemma it is sufficient to show that (6.70) holds.

We construct $(\{\widehat{x}_t\}_{t=T_1}^{T_2}, \{\widehat{u}_t\}_{t=T_1}^{T_2-1}) \in X(T_1, T_2)$. If condition (i) holds, then we set

$$\widehat{x}_t = x_t^f,\ t = T_1, \dots, T_2,\ \widehat{u}_t = u_t^f,\ t = T_1, \dots, T_2 - 1.$$

If condition (ii) holds, then there exist $(\{\widehat{x}_t\}_{t=T_1}^{T_2}, \{\widehat{u}_t\}_{t=T_1}^{T_2-1}) \in X(T_1, T_2)$ and an integer $\tau_1 \in [0, L]$ such that

$$\widehat{x}_{T_1} = x_{T_1},\ \sum_{t=T_1}^{T_1+\tau_1-1} f(t, \widehat{x}_t, \widehat{u}_t) \leq L,$$

$$\widehat{x}_t = x_t^f,\ t = T + \tau_1, \dots, T_2,$$

$$\widehat{u}_t = u_t^f,\ t = T + \tau_1, \dots, T_2 - 1.$$

If condition (iii) holds, then there exist $(\{\widehat{x}_t\}_{t=T_1}^{T_2}, \{\widehat{u}_t\}_{t=T_1}^{T_2-1}) \in X(T_1, T_2)$ and integers $\tau_1, \tau_2 \in [0, L]$ such that

$$\widehat{x}_{T_1} = x_{T_1},\ \widehat{x}_{T_2} = x_{T_2},$$

$$\widehat{x}_t = x_t^f,\ t = T + \tau_1, \dots, T_2 - \tau_2,\ \widehat{u}_t = u_t^f,\ t = T + \tau_1, \dots, T_2 - \tau_2 - 1,$$

$$\sum_{t=T_1}^{T_1+\tau_1-1} f(t, \widehat{x}_t, \widehat{u}_t) \le L,\ \sum_{t=T_2-\tau_2}^{T_2-1} f(t, \widehat{x}_t, \widehat{u}_t) \le L.$$

If condition (iv) holds, then there exist $(\{\widehat{x}_t\}_{t=T_1}^{T_2}, \{\widehat{u}_t\}_{t=T_1}^{T_2-1}) \in X(T_1, T_2)$ and an integer $\tau_1 \in [0, L]$ such that

$$\widehat{x}_{T_2} = x_{T_2},$$

$$\widehat{x}_t = x_t^f,\ t = T_1, \dots, T_2 - \tau_1,\ \widehat{u}_t = u_t^f,\ t = T_1, \dots, T_2 - \tau_1 - 1,$$

$$\sum_{t=T_2-\tau_1}^{T_2-1} f(t, \widehat{x}_t, \widehat{u}_t) \le L.$$

There exists a natural number q such that

$$qL_0 \le T_2 - T_1 < (q + 1)L_0. \tag{6.76}$$

By (6.73), (6.74), and (6.76),

$$q \ge (T_2 - T_1)L_0^{-1} - 1 \ge L_1 L_0^{-1} - 1 \ge 8. \tag{6.77}$$

It follows from property (b), (6.12), (6.75), the relation $L_0 > L + 4$ and the construction of $(\{\widehat{x}_t\}_{t=T_1}^{T_2}, \{\widehat{u}_t\}_{t=T_1}^{T_2-1})$ that for all $i = 1, \dots, q - 2$,

$$\left| \sum_{t=T_1+iL_0}^{T_1+(i+1)L_0-1} g(t, \widehat{x}_t, \widehat{u}_t) - \sum_{t=T_1+iL_0}^{T_1+(i+1)L_0-1} f(t, x_t^f, \widehat{u}_t^f) \right| \le 8^{-1}. \tag{6.78}$$

By (6.6), (6.76) and the construction of $(\{\widehat{x}_t\}_{t=T_1}^{T_2}, \{\widehat{u}_t\}_{t=T_1}^{T_2-1})$,

$$\sum_{t=T_1+(q-1)L_0}^{T_2-1} f(t, \widehat{x}_t, \widehat{u}_t) \le \sum_{t=T_1+(q-1)L_0}^{T_2-1} f(t, x_t^f, u_t^f) + L(a_0 + 1). \tag{6.79}$$

Property (b) and Eqs. (6.12), (6.75), (6.76), and (6.79) imply that

$$\sum_{t=T_1+(q-1)L_0}^{T_2-1} g(t,\widehat{x}_t,\widehat{u}_t) \le \sum_{t=T_1+(q-1)L_0}^{T_2-1} f(t,\widehat{x}_t,\widehat{u}_t)+1$$

$$\le \sum_{t=T_1+(q-1)L_0}^{T_2-1} f(t,x_t^f,u_t^f)+L(1+a_0)+1. \tag{6.80}$$

By (6.6), (6.76), the relation $L_0 > L+4$ and the construction of the trajectory–control pair $(\{\widehat{x}_t\}_{t=T_1}^{T_2}, \{\widehat{u}_t\}_{t=T_1}^{T_2-1})$,

$$\sum_{t=T_1}^{T_1+L_0-1} f(t,\widehat{x}_t,\widehat{u}_t) \le \sum_{t=T_1}^{T_1+L_0-1} f(t,x_t^f,u_t^f)+L(a_0+1). \tag{6.81}$$

Property (b) and Eqs. (6.12), (6.75), and (6.81) imply that

$$\sum_{t=T_1}^{T_1+L_0-1} g(t,\widehat{x}_t,\widehat{u}_t) \le \sum_{t=T_1}^{T_1+L_0-1} f(t,\widehat{x}_t,\widehat{u}_t)+1$$

$$\le \sum_{t=T_1}^{T_1+L_0-1} f(t,x_t^f,u_t^f)+L(a_0+1)+1. \tag{6.82}$$

Let

$$\{S_1,S_2\} \in \{(T_1+iL_0,\, T_1+(i+1)L_0 : i=1,\dots,q-2\}.$$

Equations (6.12), (6.17), (6.69), and (6.73) imply that

$$\sum_{t=S_1}^{S_2-1} \alpha_t g(t,\widehat{x}_t,\widehat{u}_t) = \sum_{t=S_1}^{S_2-1} \alpha_t (g(t,\widehat{x}_t,\widehat{u}_t)+a_0) - a_0 \sum_{t=S_1}^{S_2-1} \alpha_t$$

$$\le \max\{\alpha_t : t \in \{S_1,\dots,S_2-1\}\} \sum_{t=S_1}^{S_2-1} (g(t,\widehat{x}_t,\widehat{u}_t)+a_0) - a_0 \sum_{t=S_1}^{S_2-1} \alpha_t$$

$$\le \max\{\alpha_t : t \in \{S_1,\dots,S_2-1\}\} \sum_{t=S_1}^{S_2-1} (f(t,x_t^f,u_t^f)+a_0)$$

$$+ 8^{-1} \max\{\alpha_t : t \in \{S_1, \dots, S_2 - 1\}\} - a_0 \sum_{t=S_1}^{S_2-1} \alpha_t$$

$$\leq \sum_{t=S_1}^{S_2-1} \alpha(f(t, x_t^f, u_t^f) + a_0)$$

$$+ (1 - \lambda) \max\{\alpha_t : t \in \{S_1, \dots, S_2 - 1\}\} \sum_{t=S_1}^{S_2-1} (f(t, x_t^f, u_t^f) + a_0)$$

$$+ 8^{-1} \max\{\alpha_t : t \in \{S_1, \dots, S_2 - 1\}\} - a_0 \sum_{t=S_1}^{S_2-1} \alpha_t$$

$$\leq \sum_{t=S_1}^{S_2-1} \alpha_t f(t, x_t^f, u_t^f) + (1 - \lambda) \max\{\alpha_t : t \in \{S_1, \dots, S_2 - 1\}\} L_0(\Delta_f + a_0)$$

$$+ 8^{-1} \max\{\alpha_t : t \in \{S_1, \dots, S_2 - 1\}\}$$

$$\leq \sum_{t=S_1}^{S_2-1} \alpha_t f(t, x_t^f, u_t^f) + \lambda^{-1}(1-\lambda) \min\{\alpha_t : t \in \{S_1, \dots, S_2-1\}\} L_0(\Delta_f + a_0)$$

$$+ 8^{-1}\lambda^{-1} \min\{\alpha_t : t \in \{S_1, \dots, S_2 - 1\}\}$$

$$\leq \sum_{t=S_1}^{S_2-1} \alpha_t f(t, x_t^f, u_t^f) + 2^{-1} \min\{\alpha_t : t \in \{S_1, \dots, S_2 - 1\}\}. \tag{6.83}$$

By (6.12), (6.17), (6.69), (6.72), (6.73), (6.76), (6.80), and (6.82), for every

$$\{S_1, S_2\} \in \{(T_1, T_1 + L_0), \ (T_1 + (q - 1)L_0, T_2)\}$$

we have

$$\sum_{t=S_1}^{S_2-1} \alpha_t g(t, \widehat{x}_t, \widehat{u}_t) \leq \sum_{t=S_1}^{S_2-1} \alpha_t (g(t, \widehat{x}_t, \widehat{u}_t) + a_0) - a_0 \sum_{t=S_1}^{S_2-1} \alpha_t$$

$$\leq \max\{\alpha_t : t \in \{S_1, \dots, S_2 - 1\}\} \sum_{t=S_1}^{S_2-1} (g(t, \widehat{x}_t, \widehat{u}_t) + a_0) - a_0 \sum_{t=S_1}^{S_2-1} \alpha_t$$

$$\leq \max\{\alpha_t : t \in \{S_1, \ldots, S_2 - 1\}\}(\sum_{t=S_1}^{S_2-1} (f(t, x_t^f, u_t^f) + a_0) + L(a_0 + 1) + 1)$$

$$- a_0 \sum_{t=S_1}^{S_2-1} \alpha_t$$

$$\leq \max\{\alpha_t : t \in \{S_1, \ldots, S_2 - 1\}\}(L(a_0 + 1) + 1) - a_0 \sum_{t=S_1}^{S_2-1} \alpha_t$$

$$+ \sum_{t=S_1}^{S_2-1} \alpha_t(f(t, x_t^f, u_t^f) + a_0)$$

$$+ (1 - \lambda) \max\{\alpha_t : t \in \{S_1, \ldots, S_2 - 1\}\} \sum_{t=S_1}^{S_2-1} \alpha(f(t, x_t^f, u_t^f) + a_0)$$

$$\leq \sum_{t=S_1}^{S_2-1} \alpha_t f(t, x_t^f, u_t^f) + \min\{\alpha_t : t \in \{S_1, \ldots, S_2 - 1\}\}\lambda^{-1}(L(a_0 + 1) + 1)$$

$$+ \min\{\alpha_t : t \in \{S_1, \ldots, S_2 - 1\}\}\lambda^{-1}(1 - \lambda)(2L_0\Delta_f + a_0)$$

$$\leq \sum_{t=S_1}^{S_2-1} \alpha_t f(t, x_t^f, u_t^f) + 2^{-1} \min\{\alpha_t : t \in \{S_1, \ldots, S_2 - 1\}\}$$

$$+ 2 \min\{\alpha_t : t \in \{S_1, \ldots, S_2 - 1\}\}(L(a_0 + 1) + 1). \quad (6.84)$$

In order to complete the proof of the lemma it is sufficient to show that (6.70) holds. Assume the contrary. Then

$$\mathrm{Card}(\{t \in \{T_1, \ldots, T_2\} : \rho_E(x_t, x_t^f) \leq \epsilon\}) < L_0/2. \quad (6.85)$$

Property (a) and Eqs. (6.69), (6.75), (6.76) and (6.85) imply that for all $i = 0, \ldots, q - 2$,

$$\sum_{t=T_1+iL_0}^{T_1+(i+1)L_0-1} \alpha_t g(t, x_t, tu_t)$$

$$> \sum_{t=T_1+iL_0}^{T_1+(i+1)L_0-1} \alpha_t f(t, x_t^f, u_t^f) + 8 \min\{\alpha_t : t \in \{T_1 + iL_0, \dots, T_i + (i+1)L_0 - 1\}\} \tag{6.86}$$

and

$$\sum_{t=T_1+(q-1)L_0}^{T_2-1} \alpha_t g(t, x_t, tu_t)$$

$$> \sum_{t=T_1+(q-1)L_0}^{T_2-1} \alpha_t f(t, x_t^f, u_t^f) + 8 \min\{\alpha_t : t \in \{T_1 + (q-1)L_0, \dots, T_2 - 1\}\}. \tag{6.87}$$

It follows from (6.76), (6.86), and (6.87) that

$$\sum_{t=T_1}^{T_2-1} \alpha_t g(t, x_t, u_t) > \sum_{t=T_1}^{T_2-1} \alpha_t f(t, x_t^f, u_t^f)$$

$$+ 8(\sum_{i=0}^{q-2} \min\{\alpha_t : t \in \{T_1 + iL_0, \dots, T_i + (i+1)L_0 - 1\}\}$$

$$+ \min\{\alpha_t : t \in \{T_1 + (q-1)L_0, \dots, T_2 - 1\}\}). \tag{6.88}$$

In view of (6.76), (6.83), and (6.84),

$$\sum_{t=T_1}^{T_2-1} \alpha_t g(t, \widehat{x}_t, \widehat{u}_t) \le \sum_{t=T_1}^{T_2-1} \alpha_t f(t, x_t^f, u_t^f)$$

$$+ 2^{-1} \sum_{i=0}^{q-2} \min\{\alpha_t : t \in \{T_1 + iL_0, \dots, T_i + (i+1)L_0 - 1\}\}$$

$$+ 2^{-1} \min\{\alpha_t : t \in \{T_1 + (q-1)L_0, \dots, T_2 - 1\}\}$$

$$+ 2(L(a_0 + 1) + 1) \min\{\alpha_t : t \in \{T_1, \dots, T_i + L_0 - 1\}\}$$

$$+ \min\{\alpha_t : t \in \{T_1 + (q-1)L_0, \dots, T_2 - 1\}\}. \tag{6.89}$$

By (6.88), (6.89), conditions (i)–(iv) and the construction of the trajectory–control pair $(\{\widehat{x}_t\}_{t=T_1}^{T_2}, \{\widehat{u}_t\}_{t=T_1}^{T_2-1})$,

$$\sum_{t=T_1}^{T_2-1} \alpha_t f(t, x_t^f, u_t^f) + 8(\sum_{i=0}^{q-2} \min\{\alpha_t : t \in \{T_1 + iL_0, \dots, T_1 + (i+1)L_0 - 1\}\}$$

$$+ \min\{\alpha_t : t \in \{T_1 + (q-1)L_0, \dots, T_2 - 1\}\})$$

$$< \sum_{t=T_1}^{T_2-1} \alpha_t g(t, x_t, u_t) \le \sum_{t=T_1}^{T_2-1} \alpha_t g(t, \widehat{x}_t, \widehat{u}_t)$$

$$+ 4 \min\{\alpha_t : t \in \{T_1, \dots, T_2 - 1\}\}$$

$$\le \sum_{t=T_1}^{T_2-1} \alpha_t f(t, x_t^f, u_t^f)$$

$$+ 2^{-1}(\sum_{i=0}^{q-2} \min\{\alpha_t : t \in \{T_1 + iL_0, \dots, T_1 + (i+1)L_0 - 1\}\}$$

$$+ \min\{\alpha_t : t \in \{T_1 + (q-1)L_0, \dots, T_2 - 1\}\})$$

$$+ 4 \min\{\alpha_t : t \in \{T_1, \dots, T_2 - 1\}\}$$

$$+ 2(L(a_0 + 1) + 1)(\min\{\alpha_t : t \in \{T_1, \dots, T_1 + L_0 - 1\}\}$$

$$+ \min\{\alpha_t : t \in \{T_1 + (q-1)L_0, \dots, T_2 - 1\}\}). \tag{6.90}$$

In view of (6.90),

$$(L(a_0 + 1) + 1)(\min\{\alpha_t : t \in \{T_1, \dots, T_i + L_0 - 1\}\}$$

$$+ \min\{\alpha_t : t \in \{T_1 + (q-1)L_0, \dots, T_2 - 1\}\})$$

$$> 2(\sum_{i=0}^{q-2} \min\{\alpha_t : t \in \{T_1 + iL_0, \dots, T_1 + (i+1)L_0 - 1\}\}$$

$$+ \min\{\alpha_t : t \in \{T_1 + (q-1)L_0, \dots, T_2 - 1\}\}). \tag{6.91}$$

There exists

$$j \in \{0, \dots, q-2\}$$

such that

$$\min\{\alpha_t : t \in \{T_1 + iL_0, \dots, T_i + (i+1)L_0 - 1\}\}$$

$$\leq \min\{\alpha_t : t \in \{T_1 + jL_0, \dots, T_i + (j+1)L_0 - 1\}\} \tag{6.92}$$

for every $i \in \{0, \dots, q-2\}$. In view of (6.73) and (6.77),

$$q - 2 \geq L_1 L_0^{-1} - 3 \geq 2q_0 + 2. \tag{6.93}$$

By (6.93), there exists an integer $j_0 \geq 0$ such that

$$j_0 + q_0 \leq q - 2, \;\; j_0 \leq j \leq j_0 + q_0. \tag{6.94}$$

Equations (6.69), (6.72), (6.73), (6.91), (6.92), and (6.94) imply that

$$(2L(a_0+1)+1)\min\{\alpha_t : t \in \{T_1 + jL_0, \dots, T_1 + (j+1)L_0 - 1\}\}\lambda^{-1}$$

$$\geq \sum_{i=0}^{q-2} \min\{\alpha_t : t \in \{T_1 + iL_0, \dots, T_1 + (i+1)L_0 - 1\}\}$$

$$\geq \sum_{i=j_0}^{j_0+q_0-1} \min\{\alpha_t : t \in \{T_1 + iL_0, \dots, T_1 + (i+1)L_0 - 1\}\}$$

$$\geq \min\{\alpha_t : t \in \{T_1 + jL_0, \dots, T_1 + (j+1)L_0 - 1\}\}\lambda^{q_0} q_0$$

and

$$2L(a_0+1)+1 \geq q_0 \lambda^{q_0+1} \geq 2^{-1} q_0.$$

This contradicts (6.71). The contradiction we have reached proves Lemma 6.27.

6.10 Proof of Theorem 6.25

In view of (A2), we may assume without loss of generality that

$$L > 4 + 4b_f + (b_f + 1)(\Delta_f + 1)$$

and that the following property holds:

if $(T, z) \in \mathcal{A}$ and $\rho_E(z, x_T^f) \leq \epsilon$, then $z \in \mathcal{A}_L$ and if in addition $T \geq L$, then $z \in \widehat{\mathcal{A}}_L$.

Lemma 6.16 and (A2) imply that there exists $\delta \in (0, \epsilon)$ and an integer $\bar{L}_0 > 0$ such that the following properties hold:

(a) for each integer $T_1 \geq \bar{L}_0$, each integer $T_2 \geq T_1 + 2b_f$ and each $(\{x_t\}_{t=T_1}^{T_2}, \{u_t\}_{t=T_1}^{T_2-1}) \in X(T_1, T_2)$ which satisfies

$$\rho_E(x_{T_i}, x_{T_i}^f) \leq 4\delta,\ i = 1, 2,$$

$$\sum_{t=T_1}^{T_2-1} f(t, x_t, u_t) \leq U^f(T_1, T_2, x_{T_1}, x_{T_2}) + 8\delta$$

the inequality $\rho_E(x_t, x_t^f) \leq \epsilon$ holds for all $t = T_1, \dots, T_2$;

(b) for each $(T_i, z_i) \in \mathcal{A}$, $i = 1, 2$ satisfying $\rho_E(z_i, x_{T_i}^f) \leq \delta$, $i = 1, 2$ and $T_2 > b_f$ there exist

$$(\{x_t^{(1)}\}_{t=T_1}^{T_1+b_f}, \{u_t^{(1)}\}_{t=T_1}^{T_1+b_f-1}) \in X(T_1, T_1 + b_f),$$

$$(\{x_t^{(2)}\}_{t=T_2-b_f}^{T_2}, \{u_t^{(2)}\}_{t=T_2-b_f}^{T_2-1}) \in X(T_2 - b_f, T_2)$$

such that

$$x_{T_1}^{(1)} = z_1,\ x_{T_1+b_f}^{(1)} = x_{T_1+b_f}^f,$$

$$\sum_{t=T_1}^{T_1+b_f-1} f(t, x_t^{(1)}, u_t^{(1)}) \leq \sum_{t=T_1}^{T_1+b_f-1} f(t, x_t^f, u_t^f) + 1,$$

$$x_{T_2}^{(2)} = z_2,\ x_{T_2-b_f}^{(2)} = x_{T_2-b_f}^f,$$

$$\sum_{t=T_2-b_f}^{T_2-1} f(t, x_t^{(2)}, u_t^{(2)}) \leq \sum_{t=T_2-b_f}^{T_2-1} f(t, x_t^f, u_t^f) + 1.$$

Lemma 6.27 implies that there exist integers

$$L_0 > L + 8 + 2\bar{L}_0,\ L_1 > 8L_0 + 2L_0L(a_0 + 1) + L_0(2L_0 + 4) \tag{6.95}$$

a neighborhood V_0 of f in $\mathfrak{M}_\psi$ and a number λ satisfying

$$1 > \lambda > 2^{-1},\ \lambda^{-1}(1 - \lambda)a_0L_1 < \delta,$$

$$(\lambda^{-1} - 1)(2 + (L_1 + 2)(\Delta_f + a_0 + 2)) < \delta \tag{6.96}$$

such that the following property holds:

(c) for each $g \in V_0$, each pair of integers $T_1 \geq 0$, $T_2 > T_1 + L_1$, each $\alpha = \{\alpha_t\}_{t=0}^{\infty} \subset (0, 1]$ which satisfies

$$\alpha_{t_1}\alpha_{t_2}^{-1} \geq \lambda \text{ for each } t_1, t_2 \in \{T_1, \dots, T_2 - 1\} \text{ satisfying } |t_1 - t_2| \leq L_1$$

and each $(\{x_t\}_{t=T_1}^{T_2}, \{u_t\}_{t=T_1}^{T_2-1}) \in X(T_1, T_2)$ which satisfies at least one of the following conditions:

$$\sum_{t=T_1}^{T_2-1} \alpha_t g(t, x_t, u_t) \leq U^{\alpha g}(T_1, T_2) + 4\min\{\alpha_t : t = T_1, \dots, T_2 - 1\};$$

$$\sum_{t=T_1}^{T_2-1} \alpha_t g(t, x_t, u_t) \leq U^{\alpha g}(T_1, T_2, x_{T_1})$$

$$+ 4\min\{\alpha_t : t = T_1, \dots, T_2 - 1\}, \ (T_1, x_{T_1}) \in \mathcal{A}_L;$$

$$\sum_{t=T_1}^{T_2-1} \alpha_t g(t, x_t, u_t) \leq U^{\alpha g}(T_1, T_2, x_{T_1}, x_{T_2}) + 4\min\{\alpha_t : t = T_1, \dots, T_2 - 1\},$$

$$(T_1, x_{T_1}) \in \mathcal{A}_L, \ (T_2, x_{T_2}) \in \widehat{\mathcal{A}}_L;$$

$$\sum_{t=T_1}^{T_2-1} \alpha_t g(t, x_t, u_t) \leq \widehat{U}^{\alpha g}(T_1, T_2, x_{T_2})$$

$$+ 4\min\{\alpha_t : t = T_1, \dots, T_2 - 1\}, \ (T_2, x_{T_2}) \in \widehat{\mathcal{A}}_L$$

the inequality

$$\mathrm{Card}(\{t \in \{T_1, \dots, T_2\} : \rho_E(x_t, x_t^f) \leq \delta\}) \geq L_0/2$$

is true.

Proposition 6.21 implies that there exists a neighborhood $V \subset V_0$ of f in $\mathfrak{M}_\psi$ such that the following property holds:

(d) for each pair of integers $S_1 \geq 0$, $S_2 \in (S_1, S_1 + L_1]$, each $g \in V$ and each $(\{x_t\}_{t=T_1}^{T_2}, \{u_t\}_{t=T_1}^{T_2-1}) \in X(T_1, T_2)$ which satisfies

$$\min\{\sum_{t=T_1}^{T_2-1} f(t, x_t, u_t), \sum_{t=T_1}^{T_2-1} g(t, x_t, u_t)\} \leq (L_1 + 2)\Delta_f + 8 + a_0$$

the inequality

$$|\sum_{t=T_1}^{T_2-1} f(t, x_t, u_t) - \sum_{t=T_1}^{T_2-1} g(t, x_t, u_t)| \le \delta$$

is true.

Assume that

$$g \in V, \ T_1 \ge 0, \ T_2 \ge T_1 + 4L_1 \tag{6.97}$$

are integers, $\alpha = \{\alpha_t\}_{t=0}^{\infty} \subset (0, 1]$ satisfies

$$\alpha_{t_1}\alpha_{t_2}^{-1} \ge \lambda \text{ for each } t_1, t_2 \in \{T_1, \dots, T_2 - 1\} \text{ satisfying } |t_2 - t_1| \le L_1 \tag{6.98}$$

and that $(\{x_t\}_{t=T_1}^{T_2}, \{u_t\}_{t=T_1}^{T_2-1}) \in X(T_1, T_2)$ satisfies at least one of the following conditions:

(i)

$$\sum_{t=T_1}^{T_2-1} \alpha_t g(t, x_t, u_t) \le U^{\alpha g}(T_1, T_2) + \delta \min\{\alpha_t : t = T_1, \dots, T_2 - 1\};$$

(ii)

$$(T_1, x_{T_1}) \in \mathcal{A}_L, \ \sum_{t=T_1}^{T_2-1} \alpha_t g(t, x_t, u_t)$$

$$\le U^{\alpha g}(T_1, T_2, x_{T_1}) + \delta \min\{\alpha_t : t = T_1, \dots, T_2 - 1\};$$

(iii)

$$(T_1, x_{T_1}) \in \mathcal{A}_L, \ (T_2, x_{T_2}) \in \widehat{\mathcal{A}}_L$$

and

$$\sum_{t=T_1}^{T_2-1} \alpha_t g(t, x_t, u_t) \le U^{\alpha g}(T_1, T_2, x_{T_1}, x_{T_2}) + \delta \min\{\alpha_t : t = T_1, \dots, T_2 - 1\}.$$

Property (c) and Eqs. (6.97) and (6.98) imply that

$$\mathrm{Card}(\{t \in \{T_1, \dots, T_2\} : \rho_E(x_t, x_t^f) \le \delta\}) \ge L_0/2. \tag{6.99}$$

Set

$$S_1 = \min\{t \in \{\max\{T_1, \bar{L}_0\}, \dots, T_2\} : \ \rho_E(x_t, x_t^f) \le \delta\}. \tag{6.100}$$

(Note that if $T_1 \ge \bar{L}_0$ and $\rho_E(x_{T_1}, x_{T_1}^f) \le \delta$, then $S_1 = T_1$.) Clearly, in view of (6.95), (6.97) and (6.100), S_1 is well-defined. If $S_1 > T_1 + L_1$, then it follows from properties (a), (c) and conditions (i)–(iii) and Eqs. (6.97), (6.98), (6.100) that

$$\mathrm{Card}(\{t \in \{T_1, \dots, S_1\} : \ \rho_E(x_t, x_t^f) \le \delta\}) \ge L_0/2.$$

This contradicts (6.100). The contradiction we have reached proves that

$$S_1 \le T_1 + L_1. \tag{6.101}$$

Set

$$S_2 = \max\{t \in \{S_1, \dots, T_2\} : \ \rho_E(x_t, x_t^f) \le \delta\}. \tag{6.102}$$

If $S_2 \le T_2 - L_1$, then it follows from properties (b), (c), conditions (i)–(iii) and Eqs. (6.97), (6.98) and (6.102) that

$$\mathrm{Card}(\{t \in \{S_2, \dots, T_2\} : \ \rho_E(x_t, x_t^f) \le \delta\}) \ge L_0/2.$$

This contradicts (6.102). Thus

$$S_2 \ge T_2 - L_1. \tag{6.103}$$

In order to complete the proof of the theorem it is sufficient to show that

$$\rho_E(x_t, x_t^f) \le \epsilon, \ t \in \{S_1, \dots, S_2\}.$$

Assume the contrary. Then there exists an integer

$$\tau_0 \in \{S_1, \dots, S_2\}$$

such that

$$\rho_E(x_{\tau_0}, x_{\tau_0}^f) > \epsilon. \tag{6.104}$$

In view of (6.97) and (6.103),

$$S_2 - S_1 \ge T_2 - T_1 - 2L_1 \ge 2L_1. \tag{6.105}$$

By (6.95), there exists an integer S_3 such that

$$\tau_0 \in [S_3, S_3+\lfloor L_0/4\rfloor+1] \subset [S_1, S_2] \text{ and set } S_4 = S_3+\lfloor L_0/4\rfloor+1. \quad (6.106)$$

Set

$$S_5 = \max\{t \in \{S_1, \dots, S_3\} : \ \rho_E(x_t, x_t^f) \le \delta\},$$

$$S_6 = \min\{t \in \{S_4, \dots, S_2\} : \ \rho_E(x_t, x_t^f) \le \delta\}. \quad (6.107)$$

It follows from (6.106) and (6.107) that

$$S_6 - S_5 \ge S_4 - S_3 \ge \lfloor L_0/4\rfloor + 1. \quad (6.108)$$

Assume that

$$S_6 - S_5 \ge L_1.$$

Properties (b) and (c), conditions (i)–(iii) and Eqs. (6.97), (6.107) and (6.108) imply that

$$\mathrm{Card}(\{t \in \{S_5, \dots, S_6\} : \ \rho_E(x_t, x_t^f) \le \delta\}) \ge L_0/2.$$

On the other hand, by (6.106)–(6.108),

$$\mathrm{Card}(\{t \in \{S_5, \dots, S_6\} : \ \rho_E(x_t, x_t^f) \le \delta\}) \le S_4 - S_3 + 2 \le \lfloor L_0/4\rfloor + 3.$$

The contradiction we have reached proves that

$$S_6 - S_5 \le L_1.$$

Together with (6.108) this implies that

$$\lfloor L_0/4\rfloor + 1 \le S_6 - S_5 \le L_1. \quad (6.109)$$

Property (b) and Eqs. (6.95) and (6.107)–(6.109) imply that there exist

$$(\{y_t\}_{t=S_5}^{S_6}, \{v_t\}_{t=S_5}^{S_6-1}) \in X(S_5, S_6)$$

such that

$$y_{S_5} = x_{S_5}, \ y_{S_6} = y_{S_6},$$

$$y_t = x_t^f,\ t \in \{S_5+b_f, \dots, S_6-b_f\},\ v_t = u_t^f,\ t \in \{S_5+b_f, \dots, S_6-b_f-1\}, \tag{6.110}$$

$$\sum_{t=S_5}^{S_5+b_f-1} f(t, y_t, v_t) \le \sum_{t=S_5}^{S_5+b_f-1} f(t, x_t^f, u_t^f) + 1,$$

$$\sum_{t=S_6-b_f}^{S_6-1} f(t, y_t, v_t) \le \sum_{t=S_6-b_f}^{S_6-1} f(t, x_t^f, u_t^f) + 1. \tag{6.111}$$

By (6.12) and (6.109)–(6.111),

$$\sum_{t=S_5}^{S_6-1} f(t, y_t, v_t) \le 2 + \Delta_f(L_1 + 2). \tag{6.112}$$

Property (d) and Eqs. (6.97), (6.109), and (6.112) imply that

$$\sum_{t=S_5}^{S_6-1} g(t, y_t, v_t) \le 3 + \Delta_f(L_1 + 2). \tag{6.113}$$

By (6.17), (6.98), and (6.109),

$$\sum_{t=S_5}^{S_6-1} \alpha_t g(t, x_t, u_t) = \sum_{t=S_5}^{S_6-1} \alpha_t (g(t, x_t, u_t) + a_0) - a_0 \sum_{t=S_5}^{S_6-1} \alpha_t$$

$$\ge \min\{\alpha_t : t \in \{S_5, \dots, S_6 - 1\}\} \sum_{t=S_5}^{S_6-1} (g(t, x_t, u_t) + a_0) - a_0 \sum_{t=S_5}^{S_6-1} \alpha_t$$

$$\ge \min\{\alpha_t : t \in \{S_5, \dots, S_6 - 1\}\} \sum_{t=S_5}^{S_6-1} g(t, x_t, u_t) - a_0(1 - \lambda) \sum_{t=S_5}^{S_6-1} \alpha_t. \tag{6.114}$$

In view of conditions (i)–(iii) and (6.114),

$$\min\{\alpha_t : t \in \{S_5, \dots, S_6 - 1\}\} \sum_{t=S_5}^{S_6-1} g(t, x_t, u_t)$$

$$\le \sum_{t=S_5}^{S_6-1} \alpha_t g(t, x_t, u_t) + (1 - \lambda) a_0 \sum_{t=S_5}^{S_6-1} \alpha_t$$

$$\leq U^{\alpha g}(S_5, S_6, x_{S_5}, x_{S_6}) + \delta \min\{\alpha_t : t \in \{S_5, \dots, S_6 - 1\}\}$$

$$+ (1 - \lambda) a_0 \sum_{t=S_5}^{S_6-1} \alpha_t. \tag{6.115}$$

Property (d) and Eqs. (6.17), (6.96)–(6.98), (6.109), (6.110) and (6.112) imply that

$$U^{\alpha g}(S_5, S_6, x_{S_5}, x_{S_6}) \leq \inf\{\sum_{t=S_5}^{S_6-1} \alpha_t g(t, \xi_t, \eta_t) :$$

$$(\{\xi_t\}_{t=S_5}^{S_6}, \{\eta_t\}_{t=S_5}^{S_6-1}) \in X(S_5, S_6),$$

$$\sum_{t=S_5}^{S_6-1} f(t, \xi_t, \eta_t) \leq \sum_{t=S_5}^{S_6-1} f(t, y_t, v_t), \; \xi_{S_i} = x_{S_i}, \; i = 5, 6\}.$$

$$\leq \inf\{\lambda^{-1} \min\{\alpha_t : t \in \{S_5, \dots, S_6 - 1\}\} \sum_{t=S_5}^{S_6-1} (g(t, \xi_t, \eta_t) + a_0) - a_0 \sum_{t=S_5}^{S_6-1} \alpha_t :$$

$$(\{\xi_t\}_{t=S_5}^{S_6}, \{\eta_t\}_{t=S_5}^{S_6-1}) \in X(S_5, S_6),$$

$$\sum_{t=S_5}^{S_6-1} f(t, \xi_t, \eta_t) \leq \sum_{t=S_5}^{S_6-1} f(t, y_t, v_t), \; \xi_{S_i} = x_{S_i}, \; i = 5, 6\}$$

$$\leq \lambda^{-1} \min\{\alpha_t : t \in \{S_5, \dots, S_6 - 1\}\} \inf\{\sum_{t=S_5}^{S_6-1} g(t, \xi_t, \eta_t) :$$

$$(\{\xi_t\}_{t=S_5}^{S_6}, \{\eta_t\}_{t=S_5}^{S_6-1}) \in X(S_5, S_6),$$

$$\sum_{t=S_5}^{S_6-1} f(t, \xi_t, \eta_t) \leq \sum_{t=S_5}^{S_6-1} f(t, y_t, v_t), \; \xi_{S_i} = x_{S_i}, \; i = 5, 6\}$$

$$+ a_0(\lambda^{-1} - 1) \min\{\alpha_t : t \in \{S_5, \dots, S_6 - 1\}\}(S_6 - S_5)$$

$$\leq \lambda^{-1} \min\{\alpha_t : t \in \{S_5, \dots, S_6 - 1\}\} \inf\{\sum_{t=S_5}^{S_6-1} f(t, \xi_t, \eta_t) + \delta :$$

$$(\{\xi_t\}_{t=S_5}^{S_6}, \{\eta_t\}_{t=S_5}^{S_6-1}) \in X(S_5, S_6),$$

$$\sum_{t=S_5}^{S_6-1} f(t, \xi_t, \eta_t) \leq \sum_{t=S_5}^{S_6-1} f(t, y_t, v_t), \ \xi_{S_i} = x_{S_i}, \ i = 5, 6\}$$

$$+ a_0(\lambda^{-1} - 1) \min\{\alpha_t : t \in \{S_5, \dots, S_6 - 1\}\} L_1$$

$$\leq \lambda^{-1} \min\{\alpha_t : t \in \{S_5, \dots, S_6 - 1\}\} U^f(S_5, S_6, x_{S_5}, x_{S_6})$$

$$+ \delta\lambda^{-1} \min\{\alpha_t : t \in \{S_5, \dots, S_6 - 1\}\} + \delta \min\{\alpha_t : t \in \{S_5, \dots, S_6 - 1\}\}. \quad (6.116)$$

In view of (6.110) and (6.112),

$$U^f(S_5, S_6, x_{S_5}, x_{S_6}) \leq 2 + (L_f + 2)\Delta_f. \quad (6.117)$$

By (6.115) and (6.116),

$$(\sum_{t=S_5}^{S_6-1} g(t, x_t, u_t)) \min\{\alpha_t : t \in \{S_5, \dots, S_6 - 1\}\}$$

$$\leq \delta \min\{\alpha_t : t \in \{S_5, \dots, S_6 - 1\}\} + (1 - \lambda) a_0 \sum_{t=S_5}^{S_6-1} \alpha_t$$

$$+ \lambda^{-1} \min\{\alpha_t : t \in \{S_5, \dots, S_6 - 1\}\} U^f(S_5, S_6, x_{S_5}, x_{S_6})$$

$$+ \delta\lambda^{-1} \min\{\alpha_t : t \in \{S_5, \dots, S_6 - 1\}\} + \delta \min\{\alpha_t : t \in \{S_5, \dots, S_6 - 1\}\}. \quad (6.118)$$

It follows from (6.96), (6.98), (6.109), (6.117), and (6.118) that

$$\sum_{t=S_5}^{S_6-1} g(t, x_t, u_t) \leq \delta + (1 - \lambda) a_0 \lambda^{-1} L_1 + \lambda^{-1} U^f(S_5, S_6, x_{S_5}, x_{S_6}) + \delta\lambda^{-1} + \delta$$

$$\leq U^f(S_5, S_6, x_{S_5}, x_{S_6}) + 4\delta + (\lambda^{-1} - 1)(L_1 + 2)\Delta_f$$

$$\leq U^f(S_5, S_6, x_{S_5}, x_{S_6}) + 5\delta. \quad (6.119)$$

Property (d) and Eqs. (6.97), (6.109), (6.117), and (6.119) imply that

$$|\sum_{t=S_5}^{S_6-1} g(t, x_t, u_t) - \sum_{t=S_5}^{S_6-1} f(t, x_t, u_t)| \le \delta.$$

Combined with (6.119) this implies that

$$\sum_{t=S_5}^{S_6-1} f(t, x_t, u_t) \le U^f(S_5, S_6, x_{S_5}, x_{S_6}) + 6\delta.$$

Together with property (a) and Eqs. (6.107)–(6.110) this implies that

$$\rho_E(x_t, x_t^f) \le \epsilon, \ t \in \{S_5, \dots, S_6\}.$$

Combined with (6.106) this implies that

$$\rho_E(x_\tau, x_\tau^f) \le \epsilon.$$

This contradicts (6.104). The contradiction we have reached completes the proof of Theorem 6.25.

6.11 Genericity Results for Discrete-Time Problems

Assume that a set $\mathfrak{M} \subset \mathfrak{M}_\psi$ is equipped with a metric d_s which induces a uniformity which is stronger than the relative uniformity determined by the base $E(N, \epsilon)$, $N, \epsilon > 0$ (see (6.19)) and is metrizable by the metric d_ψ. Recall that the topology induced by the metric d_ψ in $\mathfrak{M}_\psi$ is called weak. The space $\mathfrak{M} \subset \mathfrak{M}_\psi$ is equipped with the relative weak topology (which is called the weak topology) and with the topology induced by the metric d_s which is called the strong topology.

Assume that $\mathfrak{M}_{reg}$ is the set of all $f \in \mathfrak{M}$ such that there exists an integer $b_f > 0$, $(\{x_t^f\}_{t=0}^\infty, \{u_t^f\}_{t=0}^\infty) \in X(0, \infty)$, $\Delta_f \ge 0$ such that (A1), (A3) and Eqs. (6.11) and (6.12) hold.

Assume that $\phi : E \times E \to [0, 1]$ is a continuous function satisfying $\phi(x, x) = 0$ for all $x \in E$ and such that the following properties hold:

(i) for each $\epsilon > 0$ there exists $\delta > 0$ such that if $x, y \in E$ and $\phi(x, y) \le \delta$, then $\rho_E(x, y) \le \epsilon$.
(ii) for each $\epsilon > 0$ there exists $\delta > 0$ such that if $x, y \in E$ and $\rho_E(x, y) \le \delta$, then $\phi(x, y) \le \epsilon$.

Let $f \in \mathfrak{M}_{reg}$ and let $r \in (0, 1)$. Define

$$f_r(t, x, u) = f(t, x, u) + r\phi(x, x_t^f), \ (t, x, u) \in \mathcal{M}.$$

It was shown in Sect. 6.3 that (A1) and (A3) hold for f_r with $(x_t^{f_r}, u_t^{f_r}) = (x_t^f, u_t^f)$, $t = 0, 1, \dots, b_{f_r} = b_f$ and $\Delta_{f_r} = \Delta_f$. Assume that for every $f \in \mathfrak{M}_{reg}$ and every $r \in (0, 1)$, $f_r \in \mathfrak{M}$ and $f_r \to f$ as $r \to 0^+$ in the strong topology. Theorem 6.17 implies that for every $f \in \mathfrak{M}_{reg}$ and every $r \in (0, 1)$, f_r has properties TP, WTP, P1, P2. Denote by $\bar{\mathfrak{M}}_{reg}$ the closure of $\mathfrak{M}_{reg}$ in the metric space $(\mathfrak{M}, d_s)$.

We prove the following result.

Theorem 6.28 *Let $M > 0$ and $\epsilon \in (0, 1)$. Then there exists an open (in the weak topology) and everywhere dense (in the strong topology) set $\mathcal{F} \subset \bar{\mathfrak{M}}_{reg}$ such that*

$$\{f_r : f \in \mathfrak{M}_{reg}, r \in (0, 1)\} \subset \mathcal{F}$$

and that for each $g \in \mathcal{F}$ there exists an open set in the weak topology $V \subset \mathfrak{M}_\psi$ containing g, $\delta \in (0, 1)$, $L > 0$ and $(\{x_t^*\}_{t=0}^\infty, \{u_t^*\}_{t=0}^\infty) \in X(0, \infty)$ and the following assertion holds.

Assume that $h \in V$, $T_1 \geq 0$, $T_2 \geq T_1 + 2L$ are integers and a trajectory–control pair $(\{x_t\}_{t=T_1}^{T_2}, \{u_t\}_{t=T_1}^{T_2-1}) \in X(T_1, T_2)$ satisfies

$$\sum_{t=T_1}^{T_2-1} h(t, x_t, u_t) \leq \min\{U^h(T_1, T_2) + M, U^h(T_1, T_2, x_{T_1}, x_{T_2}) + \delta\}.$$

Then there exist integers $\tau_1 \in [T_1, T_1 + L]$, $\tau_2 \in [T_2 - L, T_2]$ such that

$$\rho_E(x_t, x_t^f) \leq \epsilon, \ t = \tau_1 \dots, \tau_2.$$

Moreover, if $\rho_E(x_{T_2}, x_{T_2}^f) \leq \delta$, then $\tau_2 = T_2$ and if $\rho_E(x_{T_1}, x_{T_1}^f) \leq \delta$, $T_1 \geq L$, then $\tau_1 = T_1$.

Proof Clearly, the set

$$\{f_r : f \in \mathfrak{M}_{reg}, r \in (0, 1)\}$$

is everywhere dense in $\bar{\mathfrak{M}}_{reg}$ equipped with the relative strong topology. Let $f \in \mathfrak{M}_{reg}$ and $r \in (0, 1)$. Then f_r has TP. Theorem 6.22 implies that there exists an open neighborhood $V(f, r)$ of f_r in $\mathfrak{M}_\psi$ in the weak topology and $L(f, r), \delta(f, r) > 0$ such that the following property holds:

for each $h \in V(f, r)$, each pair of integers $T_1 \geq 0$, $T_2 \geq T_1 + 2L(f, r)$ and each $(\{x_t\}_{t=T_1}^{T_2}, \{u_t\}_{t=T_1}^{T_2-1}) \in X(T_1, T_2)$ which satisfies

$$\sum_{t=T_1}^{T_2-1} h(t, x_t, u_t) \leq \min\{U^h(T_1, T_2) + M, U^h(T_1, T_2, x_{T_1}, x_{T_2}) + \delta(f, r)\}$$

there exist integers $\tau_1 \in [T_1, T_1 + L(f, r)]$, $\tau_2 \in [T_2 - L(f, r), T_2]$ such that

$$\rho_E(x_t, x_t^f) \le \epsilon,\ t = \tau_1 \dots, \tau_2;$$

moreover, if $\rho_E(x_{T_2}, x_{T_2}^f) \le \delta(f, r)$, then $\tau_2 = T_2$ and if $\rho_E(x_{T_1}, x_{T_1}^f) \le \delta(f, r)$, $T_1 \ge L(f, r)$, then $\tau_1 = T_1$.

Define

$$\mathcal{F} = \cup\{V(f, r) :\ f \in \mathfrak{M}_{reg},\ r \in (0, 1)\} \cap \bar{\mathfrak{M}}_{reg}.$$

It is not difficult to see that the set $\mathcal{F}$ is open in the weak topology, everywhere dense in the strong topology and that our assertion holds.

Now we assume that the metric space $(\mathfrak{M}, d_s)$ is complete and prove the following generic result.

Theorem 6.29 *Then there exists an everywhere dense (in the strong topology) set $\mathcal{F} \subset \bar{\mathfrak{M}}_{reg}$ which is a countable intersection of open (in the weak topology) subsets of $\bar{\mathfrak{M}}_{reg}$ such that*

$$\{f_r :\ f \in \mathfrak{M}_{reg},\ r \in (0, 1)\} \subset \mathcal{F}$$

and that for each $f \in \mathcal{F}$ there exist $(\{x_t^*\}_{t=0}^\infty, \{u_t^*\}_{t=0}^\infty) \in X(0, \infty)$ such that the following assertion holds.

For each $M, \epsilon > 0$ there exists a neighborhood V of f in $\mathfrak{M}_\psi$ in the weak topology and $\delta, L > 0$ such that for each $g \in V$, each pair of integers $T_1 \ge 0$, $T_2 \ge T_1 + 2L$ and each $(\{x_t\}_{t=T_1}^{T_2}, \{u_t\}_{t=T_1}^{T_2-1}) \in X(T_1, T_2)$ satisfying

$$\sum_{t=T_1}^{T_2-1} g(t, x_t, u_t) \le \min\{U^g(T_1, T_2) + M, U^g(T_1, T_2, x_{T_1}, x_{T_2}) + \delta\}$$

there exist integers $\tau_1 \in [T_1, T_1 + L]$, $\tau_2 \in [T_2 - L, T_2]$ such that

$$\rho_E(x_t, x_t^f) \le \epsilon,\ t = \tau_1 \dots, \tau_2.$$

6.12 Proof of Theorem 6.29

The set

$$\{f_r :\ f \in \mathfrak{M}_{reg},\ r \in (0, 1)\}$$

is everywhere dense in $\bar{\mathfrak{M}}_{reg}$ with the strong topology. Let $f \in \mathfrak{M}_{reg}$, $r \in (0, 1)\}$ and $n \ge 1$ be an integer. By (A3), there exists

$$\Delta(f, r, n) \in (0, 2^{-n})$$

such that the following property holds:

(a) for each $(T_i, z_i) \in \mathcal{A}$, $i = 1, 2$ satisfying $\rho_E(z_i, x^f_{T_i}) \le \Delta(f, r, n)$, $i = 1, 2$ and $T_2 \ge b_f$ there exist

$$(\{x^{(1)}_t\}_{t=T_1}^{T_1+b_f}, \{u^{(1)}_t\}_{t=T_1}^{T_1+b_f-1}) \in X(T_1, T_1 + b_f),$$

$$(\{x^{(2)}_t\}_{t=T_2-b_f}^{T_2}, \{u^{(2)}_t\}_{t=T_2-b_f}^{T_2-1}) \in X(T_2 - b_f, T_2)$$

such that

$$x^{(1)}_{T_1} = z_1, \; x^{(1)}_{T_1+b_f} = x^f_{T_1+b_f},$$

$$x^{(2)}_{T_2} = z_2, \; x^{(2)}_{T_2-b_f} = x^f_{T_2-b_f},$$

$$\sum_{t=T_1}^{T_1+b_f-1} f_r(t, x^{(1)}_t, u^{(1)}_t) \le \sum_{t=T_1}^{T_1+b_f-1} f_r(t, x^f_t, u^f_t) + 2^{-n},$$

$$\sum_{t=T_2-b_f}^{T_2-1} f_r(t, x^{(2)}_t, u^{(2)}_t) \le \sum_{t=T_2-b_f}^{T_2-1} f_r(t, x^f_t, u^f_t) + 2^{-n},$$

$$\rho(x^{(1)}_t, x^f_t) \le 2^{-n}, \; t = T_1, \dots, T_1 + b_f,$$

$$\rho(x^{(2)}_t, x^f_t) \le 2^{-n}, \; t = T_2 - b_f, \dots, T_2.$$

Theorem 6.22 and Proposition 6.21 imply that there exists an open neighborhood $\mathcal{U}(f, r, n)$ of f_r in $\mathfrak{M}_\psi$ in the weak topology and an integer $L(f, r, n) > n$, $\delta(f, r) \in (0, 2^{-n})$ such that the following properties hold:

(b) For each $h \in \mathcal{U}(f, r, n)$, each pair of integers $T_1 \ge 0$, $T_2 \ge T_1 + 2L(f, r, n)$ and each $(\{x_t\}_{t=T_1}^{T_2}, \{u_t\}_{t=T_1}^{T_2-1}) \in X(T_1, T_2)$ which satisfies

$$\sum_{t=T_1}^{T_2-1} h(t, x_t, u_t) \le \min\{U^h(T_1, T_2) + n, U^h(T_1, T_2, x_{T_1}, x_{T_2}) + \delta(f, r, n)\}$$

there exist integers $\tau_1 \in [T_1, T_1 + L(f, r)]$, $\tau_2 \in [T_2 - L(f, r), T_2]$ such that

$$\rho_E(x_t, x^f_t) \le 2^{-n} \Delta(f, r, n), \; t = \tau_1 \dots, \tau_2;$$

moreover, if $\rho_E(x_{T_2}, x^f_{T_2}) \le \delta(f, r, n)$, then $\tau_2 = T_2$ and if $\rho_E(x_{T_1}, x^f_{T_1}) \le \delta(f, r, n)$, $T_1 \ge L(f, r, n)$, then $\tau_1 = T_1$.

(c) For each $h \in \mathcal{U}(f, r, n)$, each integers $T \geq 0$ and each trajectory–control pair $(\{x_t\}_{t=T}^{T+b_f}, \{u_t\}_{t=T}^{T+b_f-1}) \in X(T, T+b_f)$ which satisfies

$$\sum_{t=T}^{T+b_f-1} f_r(t, x_t, u_t) \leq \sum_{t=T}^{T_+b_f-1} f(t, x_t^f, u_t^f) + 4$$

the inequality

$$|\sum_{t=T}^{T+b_f-1} h(t, x_t, u_t) - \sum_{t=T}^{T+b_f-1} f_r(t, x_t, u_t)| \leq 2^{-n}\Delta(f, r, n)$$

is true.

Define

$$\mathcal{F} = (\cap_{n=1}^{\infty} \cup \{\mathcal{U}(f, r, n) : f \in \mathfrak{M}_{reg}, r \in (0, 1)\}) \cap \bar{\mathfrak{M}}_{reg}. \quad (6.120)$$

It is not difficult to see that the set $\mathcal{F}$ is a countable intersection of open in the weak topology, everywhere dense in the strong topology subsets of $\bar{\mathfrak{M}}_{reg}$. Assume that

$$f \in \mathcal{F}. \quad (6.121)$$

By (6.120) and (6.121), for each integer $n \geq 1$ there exists

$$f^{(n)} \in \mathfrak{M}_{reg}, \ r^{(n)} \in (0, 1)$$

such that

$$f \in \mathcal{U}(f^{(n)}, r^{(n)}, n). \quad (6.122)$$

Property (b) and (6.122) imply that the following property holds:

(d) for each integer $n \geq 1$, each pair of integers $T_1 \geq 0$, $T_2 \geq T_1 + 2L(f^{(n)}, r^{(n)}, n)$ and each $(\{x_t\}_{t=T_1}^{T_2}, \{u_t\}_{t=T_1}^{T_2-1}) \in X(T_1, T_2)$ which satisfies

$$\sum_{t=T_1}^{T_2-1} f(t, x_t, u_t) \leq \min\{U^f(T_1, T_2)+n, U^f(T_1, T_2, x_{T_1}, x_{T_2})+\delta(f^{(n)}, r^{(n)}, n)\}$$

we have

$$\rho_E(x_t, x_t^{f^{(n)}}) \leq 2^{-n}\Delta(f^{(n)}, r^{(n)}, n),$$

$$t = T_1 + L(f^{(n)}, r^{(n)}, n), \ldots, T_2 - L(f^{(n)}, r^{(n)}, n).$$

Property (d) implies that the following property holds:

(e) for each pair of integers $n, m \ge 1$, each pair of integers $T_1 \ge 0$,

$$T_2 \ge T_1 + 2\max\{L(f^{(n)}, r^{(n)}, n),\ L(f^{(m)}, r^{(m)}, m)\}$$

and each $(\{x_t\}_{t=T_1}^{T_2}, \{u_t\}_{t=T_1}^{T_2-1}) \in X(T_1, T_2)$ which satisfies

$$\sum_{t=T_1}^{T_2-1} f(t, x_t, u_t)$$

$$\le \min\{U^f(T_1, T_2)+1, U^h(T_1, T_2, x_{T_1}, x_{T_2})+\min\{\delta(f^{(i)}, r^{(i)}, i) :\ i = n, m\}\}$$

we have for $i = n, m$,

$$\rho_E(x_t, x_t^{f^{(i)}}) \le 2^{-i}\Delta(f^{(i)}, r^{(i)}, i),\ t = T_1 + L(f^{(i)}, r^{(i)}, i), \dots, T_2 - L(f^{(i)}, r^{(i)}, i)$$

and for all integers

$$t \in [T_1 + \max\{L(f^{(n)}, r^{(n)}, n),\ L(f^{(m)}, r^{(m)}, m)\},$$
$$T_2 - \max\{L(f^{(n)}, r^{(n)}, n),\ L(f^{(m)}, r^{(m)}, m)\}]$$

we have

$$\rho_E(x_t^{f^{(m)}}, x_t^{f^{(n)}}) \le 2^{-n}\Delta(f^{(n)}, r^{(n)}, n) + 2^{-m}\Delta(f^{(m)}, r^{(m)}, m).$$

Property (e) implies that the following property holds:

(f) for each pair of integers $n, m \ge 1$ and each integer

$$t \ge \max\{L(f^{(n)}, r^{(n)}, n),\ L(f^{(m)}, r^{(m)}, m)\}$$

we have

$$\rho_E(x_t^{f^{(m)}}, x_t^{f^{(n)}}) \le 2^{-n}\Delta(f^{(n)}, r^{(n)}, n) + 2^{-m}\Delta(f^{(m)}, r^{(m)}, m).$$

Construct a strictly increasing sequence of natural numbers $\{n_k\}_{k=1}^{\infty}$ such that

$$n_1 = 1 \tag{6.123}$$

and that for each integer $k \ge 1$,

$$n_{k+1} > L(f^{(n_k)}, r^{(n_k)}, n_k) + n_k(b_{f^{(n_k)}} + 1), \tag{6.124}$$

$$2^{-n_{k+1}}\Delta(f^{(n_{k+1})}, r^{(n_{k+1})}, n_{k+1}) < 2^{-n_k}\Delta(f^{(n_k)}, r^{(n_k)}, n_k). \tag{6.125}$$

Set

$$x_t^* = x_t^{f^{(n_1)}},\ t = 0, \dots, L(f^{(n_2)}, r^{(n_2)}, n_2),\ u_t^* = u_t^{f^{(n_1)}},$$

$$t = 0, \dots, L(f^{(n_2)}, r^{(n_2)}, n_2) - 1. \tag{6.126}$$

Assume that $k \geq 1$ is an integer and that we defined

$$(\{x_t^*\}_{t=0}^{L(f^{(n_{k+1})}, r^{(n_{k+1})}, n_{k+1})}, \{u_t^*\}_{t=0}^{L(f^{(n_{k+1})}, r^{(n_{k+1})}, n_{k+1})-1})$$

$$\in X(0, L(f^{(n_{k+1})}, r^{(n_{k+1})}, n_{k+1}))$$

such that

$$x^*_{L(f^{(n_{k+1})}, r^{(n_{k+1})}, n_{k+1})} = x^{f^{(n_k)}}_{L(f^{(n_{k+1})}, r^{(n_{k+1})}, n_{k+1})}. \tag{6.127}$$

(Clearly, our assumption holds for $k = 1$.) Property (f), Eqs. (6.124), (6.125) and the inequality

$$L(f^{(n_{k+1})}, r^{(n_{k+1})}, n_{k+1}) > n_{k+1}$$

imply that for each integer $t \geq L(f^{(n_{k+1})}, r^{(n_{k+1})}, n_{k+1})$,

$$\rho_E(x_t^{f^{(n_k)}}, x_t^{f^{(n_{k+1})}})$$

$$\leq 2^{-n_k}\Delta(f^{(n_k)}, r^{(n_k)}, n_k) + 2^{-n_{k+1}}\Delta(f^{(n_{k+1})}, r^{(n_{k+1})}, n_{k+1})$$

$$\leq 2^{-n_k+1}\Delta(f^{(n_k)}, r^{(n_k)}, n_k) \leq \Delta(f^{(n_k)}, r^{(n_k)}, n_k). \tag{6.128}$$

Property (a) and (6.128) imply that there exists

$$(\{x_t^*\}_{t=L(f^{(n_{k+1})}, r^{(n_{k+1})}, n_{k+1})}^{L(f^{(n_{k+1})}, r^{(n_{k+1})}, n_{k+1})+b_{f^{(n_k)}}}, \{u_t^*\}_{t=L(f^{(n_{k+1})}, r^{(n_{k+1})}, n_{k+1})}^{L(f^{(n_{k+1})}, r^{(n_{k+1})}, n_{k+1})+b_{f^{(n_k)}}-1})$$

$$\in X(L(f^{(n_{k+1})}, r^{(n_{k+1})}, n_{k+1}), L(f^{(n_{k+1})}, r^{(n_{k+1})}, n_{k+1}) + b_{f^{(n_k)}})$$

such that

$$x^*_{L(f^{(n_{k+1})}, r^{(n_{k+1})}, n_{k+1})} = x^{f^{(n_k)}}_{L(f^{(n_{k+1})}, r^{(n_{k+1})}, n_{k+1})}, \tag{6.129}$$

$$x^*_{L(f^{(n_{k+1})},r^{(n_{k+1})},n_{k+1})+b_{f^{(n_k)}}} = x^{f^{(n_{k+1})}}_{L(f^{(n_{k+1})},r^{(n_{k+1})},n_{k+1})+b_{f^{(n_k)}}}, \tag{6.130}$$

$$\sum\{f^{(n_k)}_{r^{(n_k)}}(t, x^*_t, u^*_t) :$$

$$t = L(f^{(n_{k+1})}, r^{(n_{k+1})}, n_{k+1}), \dots, L(f^{(n_{k+1})}, r^{(n_{k+1})}, n_{k+1}) + b_{f^{(n_k)}} - 1\}$$

$$\le \sum\{f^{(n_k)}_{r^{(n_k)}}(t, x^{f^{(n_k)}}_t, u^{f^{(n_k)}}_t) :$$

$$t = L(f^{(n_{k+1})}, r^{(n_{k+1})}, n_{k+1}), \dots, L(f^{(n_{k+1})}, r^{(n_{k+1})}, n_{k+1}) + b_{f^{(n_k)}} - 1\} + 2^{-n_k}, \tag{6.131}$$

$$\rho_E(x^*_t, x^{f^{(n_k)}}_t) \le 2^{-n_k},$$

$$t \in L(f^{(n_{k+1})}, r^{(n_{k+1})}, n_{k+1}), \dots, L(f^{(n_{k+1})}, r^{(n_{k+1})}, n_{k+1}) + b_{f^{(n_k)}}. \tag{6.132}$$

Set

$$x^*_t = x^{f^{(n_{k+1})}}_t,$$

$$t \in \{L(f^{(n_{k+1})}, r^{(n_{k+1})}, n_{k+1}) + b_{f^{(n_k)}}, \dots, L(f^{(n_{k+2})}, r^{(n_{k+2})}, n_{k+2})\}, \tag{6.133}$$

$$u^*_t = u^{f^{(n_{k+1})}}_t,$$

$$t \in \{L(f^{(n_{k+1})}, r^{(n_{k+1})}, n_{k+1}) + b_{f^{(n_k)}}, \dots, L(f^{(n_{k+2})}, r^{(n_{k+2})}, n_{k+2}) - 1\}. \tag{6.134}$$

It is not difficult to see that

$$(\{x^*_t\}_{t=0}^{L(f^{(n_{k+2})},r^{(n_{k+2})},n_{k+2})}, \{u^*_t\}_{t=0}^{L(f^{(n_{k+2})},r^{(n_{k+2})},n_{k+2}-1)})$$

$$\in X(0, L(f^{(n_{k+2})}, r^{(n_{k+2})}, n_{k+2})).$$

Therefore by induction we defined

$$(\{x^*_t\}_{t=0}^\infty, \{u^*_t\}_\infty) \in X(0, \infty)$$

such that (6.126) holds, for each integer $k \ge 1$ Eqs. (6.127) and (6.129)–(6.134) are true.

Assume that $k \ge 2$ and $t \ge L(f^{(n_k)}, r^{(n_k)}, n_k)$ are integers. By (6.124), there exists an integer

$$q \ge k \tag{6.135}$$

such that

$$t \in \{L(f^{(n_q)}, r^{(n_q)}, n_q), \dots, L(f^{(n_{q+1})}, r^{(n_{q+1})}, n_{q+1})\}. \tag{6.136}$$

It follows from (6.124), (6.125), (6.135), and (6.136) that

$$\rho_E(x_t^{f^{(n_k)}}, x_t^{f^{(n_q)}}) \leq 2^{-n_k+1}\Delta(f^{(n_k)}, r^{(n_k)}, n_k). \tag{6.137}$$

If

$$t \in \{L(f^{(n_q)}, r^{(n_q)}, n_q), \dots, L(f^{(n_q)}, r^{(n_q)}, n_q) + b_{f^{(n_{q-1})}}\}, \tag{6.138}$$

then it follows from Eq. (6.132) applied with $k = q - 1$, property (f) and Eqs. (6.124) and (6.138) that

$$\rho_E(x_t^*, x_t^{f^{(n_q)}}) \leq \rho_E(x_t^*, x_t^{f^{(n_{q-1})}}) + \rho_E(x_t^{f^{(n_{q-1})}}, x_t^{f^{(n_q)}})$$

$$\leq 2^{-n_q-1} + 2^{-n_{q-1}+1}\Delta(f^{(n_{q-1})}, r^{(n_{q-1})}, n_{q-1}) \leq 2^{-n_{q-1}+1}.$$

By (6.133) applied with $k = q - 1$, for each

$$t \in \{L(f^{(n_q)}, r^{(n_q)}, n_q) + b_{f^{(n_q)}}, \dots, L(f^{(n_{q+1})}, r^{(n_{q+1})}, n_{q+1})\}$$

we have

$$x_t^* = x_t^{f^{(n_q)}}.$$

Thus in all the cases

$$\rho_E(x_t^*, x_t^{f^{(n_q)}}) \leq 2^{-n_{q-1}+1}. \tag{6.139}$$

In view of (6.137) and (6.139),

$$\rho_E(x_t^*, x_t^{f^{(n_k)}}) \leq \rho_E(x_t^{f^{(n_k)}}, x_t^{f^{(n_q)}}) + \rho_E(x_t^{f^{(n_q)}}, x_t^*)$$

$$\leq 2^{-n_k+1}\Delta(f^{(n_k)}, r^{(n_k)}, n_k) + 2^{-n_{q-1}+1}$$

$$\leq 2^{-n_k+1}\Delta(f^{(n_k)}, r^{(n_k)}, n_k) + 2^{-n_{k-1}+1} \leq 3 \cdot 2^{-n_{k-1}}.$$

Thus we have shown that the following property holds:

(g) for each pair of integers $k \geq 2$ and $t \geq L(f^{(n_k)}, r^{(n_k)}, n_k)$ we have

$$\rho_E(x_t^*, x_t^{f^{(n_k)}}) \leq 3 \cdot 2^{-n_{k-1}}.$$

Assume that $M > 1$ and $\epsilon \in (0, 1)$. Choose an integer $k > 4$ such that

$$n_k > M,\ 4 \cdot 2^{-n_{k-1}} < \epsilon. \tag{6.140}$$

Assume that

$$g \in \mathcal{U}(f^{(n_k)}, r^{(n_k)}, n_k), \tag{6.141}$$

$$\delta = \delta(f^{(n_k)}, r^{(n_k)}, n_k),$$

$$T_1 \geq 0,\ T_2 \geq T_1 + 2L(f^{(n_k)}, r^{(n_k)}, n_k)$$

are integers and that $(\{x_t\}_{t=T_1}^{T_2}, \{u_t\}_{t=T_1}^{T_2-1}) \in X(T_1, T_2)$ satisfies

$$\sum_{t=T_1}^{T_2-1} g(t, x_t, u_t)$$

$$\leq \min\{U^g(T_1, T_2)+M, U^g(T_1, T_2, x_{T_1}, x_{T_2})+\delta(f^{(n_k)}, r^{(n_k)}, n_k)\}. \tag{6.142}$$

Property (b) and Eqs. (6.140) and (6.141) imply that for each integer

$$t \in [T_1 + L(f^{(n_k)}, r^{(n_k)}, n_k), T_2 - L(f^{(n_k)}, r^{(n_k)}, n_k)]$$

we have

$$\rho_E(x_t, x_t^{f^{(n_k)}}) \leq 2^{-n_k} \Delta(f^{(n_k)}, r^{(n_k)}, n_k). \tag{6.143}$$

Property (g) and Eqs. (6.140) and (6.143) imply that for each integer

$$t \in [T_1 + L(f^{(n_k)}, r^{(n_k)}, n_k), T_2 - L(f^{(n_k)}, r^{(n_k)}, n_k)]$$

we have

$$\rho_E(x_t, x_t^*) \leq \rho_E(x_t, x_t^{f^{(n_k)}}) + \rho_E(x_t^{f^{(n_k)}}, x_t^*)$$

$$\leq 2^{-n_k} \Delta(f^{(n_k)}, r^{(n_k)}, n_k) + 3 \cdot 2^{-n_{k-1}} \leq 4 \cdot 2^{-n_{k-1}} < \epsilon.$$

Theorem 6.29 is proved.

6.13 Examples of the Space $\mathfrak{M}$

Define $\phi : E \times E \to [0, 1]$ by

$$\phi(x, y) = \min\{1, \rho_E(x, y)\}, \ x, y \in E.$$

Clearly, the function ϕ is continuous and properties (i), (ii) hold. Set

$$\phi_0(x) = \min\{1, x\}, \ x \in R^1.$$

Then ϕ_0 is Lipschitz with a Lipschitz constant 1, $\phi = \phi_0 \circ \rho_E$ and ϕ is also a Lipschitz function on $E \times E$ with a Lipschitz constant 1. We equip the space $\mathfrak{M}_\psi$ with the uniformity determined by the base

$$E_s(\epsilon) = \{(h, g) \in \mathfrak{M}_\psi \times \mathfrak{M}_\psi : |h(t, x, u) - g(t, x.u)| \le \epsilon$$

$$\text{for each } (t, x, u) \in \mathcal{M}\}$$

$$\cap \{(h, g) \in \mathfrak{M}_\psi \times \mathfrak{M}_\psi : |(h - g)(t, x_1, u) - (h - g)(t, x_2, u)| \le \epsilon\rho_E(x_1, x_2)$$

$$\text{for each } (t, x_1, u), \ (t, x_2, u) \in \mathcal{M}\}$$

$$\cap \{(h, g) \in \mathfrak{M}_\psi \times \mathfrak{M}_\psi : |(h - g)(t, x, u_1) = (h - g)(t, x, u_2)$$

$$\text{for each } (t, x, u_i) \in \mathcal{M}, \ i = 1, 2\},$$

where $\epsilon > 0$. Clearly, the uniform space $\mathfrak{M}_\psi$ is Hausdorff and has a countable base. Therefore $\mathfrak{M}_\psi$ is metrizable (by a metric d_s). It is not difficult to see that the metric space $(\mathfrak{M}_\psi, d_s)$ is complete and the uniformity introduced above is stronger than the uniformity defined by (6.19).

Denote by $\mathfrak{M}_l$ the set of all $f \in \mathfrak{M}_\psi$ such that for each integer $t \ge 0$, the function $f(t, \cdot, \cdot)$ is lower semicontinuous on the set $\{(x.u) \in E \times F : \ (t, x, u) \in \mathcal{M}\}$, by $\mathfrak{M}_c$ the set of all $f \in \mathfrak{M}_\psi$ such that for each integer $t \ge 0$, the function $f(t, \cdot, \cdot)$ is continuous on the set $\{(x.u) \in E \times F : \ (t, x, u) \in \mathcal{M}\}$, by $\mathfrak{M}_L$ the set of all $f \in \mathfrak{M}_\psi$ such that for each integer $t \ge 0$, the function $f(t, \cdot, \cdot)$ is Lipschitz on the set $\{(x.u) \in E \times F : \ (t, x, u) \in \mathcal{M}\}$, by $\mathfrak{M}_{lL}$ the set of all $f \in \mathfrak{M}_\psi$ such that for each integer $t \ge 0$, the function $f(t, \cdot, \cdot)$ is Lipschitz on all bounded subsets of $\{(x.u) \in E \times F : \ (t, x, u) \in \mathcal{M}\}$. Clearly, all these sets are closed in $(\mathfrak{M}_\psi, d_s)$ and $\mathfrak{M}$ can be any of them.

6.14 Extensions of the Generic Results

Denote by $\tilde{\mathfrak{M}}_\psi$ the set of all functions $g : \{0, 1, \dots, \} \times E \times F \to R^1$, which satisfy

$$g(t, x, u) \geq \psi(\rho_E(x, \theta_0)) - a_0 \text{ for each } (t, x, u) \in \{0, 1, \dots, \} \times E \times F.$$

For each $g \in \tilde{\mathfrak{M}}_\psi$ denote by $\mathcal{R}(g)$ the restriction of g to $\mathcal{M}$. Clearly, $\mathcal{R}(g) \in \mathfrak{M}_\psi$ for every $g \in \tilde{\mathfrak{M}}_\psi$.

For each $g \in \tilde{\mathfrak{M}}_\psi$ and each pair of integers $T_2 > T_1 \geq 0$ and each pair of points $y, z \in E$ satisfying $(T_1, y), (T_2, z) \in \mathcal{A}$ we define

$$U^g(T_1, T_2, y, z) = U^{\mathcal{R}(g)}(T_1, T_2, y, z),$$

$$U^g(T_1, T_2, y) = U^{\mathcal{R}(g)}(T_1, T_2, y),$$

$$\widehat{U}^g(T_1, T_2, z) = \widehat{U}^{\mathcal{R}(g)}(T_1, T_2, z),$$

$$U^g(T_1, T_2) = U^{\mathcal{R}(g)}(T_1, T_2).$$

We equip the space $\tilde{\mathfrak{M}}_\psi$ with the uniformity determine by the base

$$\tilde{E}(N, \epsilon) = \{(g_1, g_2) \in \tilde{\mathfrak{M}}_\psi \times \tilde{\mathfrak{M}}_\psi : |g_1(t, x, u) - g_2(t, x.u)| \leq \epsilon$$

$$\text{for each } (t, x, u) \in \{0, 1, \dots, \} \times E \times F \text{ satisfying } \rho_E(x, \theta_0),\ \rho(u, \theta_1) \leq N\}$$

$$\cap \{(g_1, g_2) \in \tilde{\mathfrak{M}}_\psi \times \tilde{\mathfrak{M}}_\psi : |g_1(t, x, u) - g_2(t, x, u)| \leq \epsilon$$

$$\text{for each } (t, x, u) \in \{0, 1, \dots\} \times E \times F$$

$$\text{satisfying } \min\{g_1(t, x, u),\ g_2(t, x, u)\} \leq N\},$$

where $N, \epsilon > 0$. Clearly, the uniform space $\tilde{\mathfrak{M}}_\psi$ is Hausdorff and has a countable base. Therefore $\tilde{\mathfrak{M}}_\psi$ is metrizable (by a metric $\tilde{d}_\psi$). It is not difficult to see that the uniform space $\tilde{\mathfrak{M}}_\psi$ is complete. It is equipped with a topology induced by the uniformity which is called a weak topology. It is clear that the mapping $\mathcal{R} : \tilde{\mathfrak{M}}_\psi \to \mathfrak{M}_\psi$ is uniformly continuous.

Assume that a set $\tilde{\mathfrak{M}} \subset \tilde{\mathfrak{M}}_\psi$ is equipped with a metric $\tilde{d}_s$ which induces a uniformity which is stronger than the uniformity introduced above. The space $\tilde{\mathfrak{M}} \subset \mathfrak{M}_\psi$ is equipped with the topology induced by the metric d_s which is called the strong topology.

Denote by $\tilde{\mathfrak{M}}_{reg}$ the set of all $f \in \tilde{\mathfrak{M}}$ such that there exists an integer $b_f > 0$, $\Delta_f > 0$ and $(\{x_t^f\}_{t=0}^\infty, \{u_t^f\}_{t=0}^\infty) \in X(0, \infty)$ such that (A1), (A3) and Eqs. (6.11) and (6.12) hold.

Assume that $\phi : E \times E \to [0, 1]$ is a continuous function satisfying $\phi(x, x) = 0$ for all $x \in E$ and such that properties (i), (ii) of Sect. 6.11 hold.

Let $f \in \tilde{\mathfrak{M}}_{reg}$ and let $r \in (0, 1)$. Define

$$f_r(t, x, u) = f(t, x, u) + r\phi(x, x_t^f), \ (t, x, u) \in \{0, 1, \ldots, \} \times E \times F.$$

It was shown in Sect. 6.3 that (A1) and (A3) hold for f_r with $(x_t^{f_r}, u_t^{f_r}) = (x_t^f, u_t^f)$, $t = 0, 1, \ldots, b_{f_r} = b_f$ and $\Delta_{f_r} = \Delta_f$. Theorem 6.17 implies that f_r has properties TP, WTP, P1, P2. Assume that for every $f \in \tilde{\mathfrak{M}}_{reg}$ and every $r \in (0, 1)$, $f_r \in \tilde{\mathfrak{M}}$ and $f_r \to f$ as $r \to 0^+$ in the strong topology.

Denote by $\mathrm{cl}(\tilde{\mathfrak{M}}_{reg})$ the closure of $\tilde{\mathfrak{M}}_{reg}$ in the metric space $(\tilde{\mathfrak{M}}, d_s)$. The following analog of Theorem 6.28 is true.

Theorem 6.30 *Let $M > 0$ and $\epsilon \in (0, 1)$. Then there exists an open (in the weak topology) and everywhere dense (in the strong topology) set $\tilde{\mathcal{F}} \subset cl(\tilde{\mathfrak{M}}_{reg})$ such that*

$$\{f_r : f \in \tilde{\mathfrak{M}}_{reg}, \ r \in (0, 1)\} \subset \tilde{\mathcal{F}}$$

such that for each $g \in \tilde{\mathcal{F}}$ there exists an open set in the weak topology $V \subset \tilde{\mathfrak{M}}_\psi$ containing g, $\delta \in (0, 1)$, $L > 0$ and $(\{x_t^*\}_{t=0}^\infty, \{u_t^*\}_{t=0}^\infty) \in X(0, \infty)$ and the following assertion holds.

Assume that $h \in V$, $T_1 \geq 0$, $T_2 \geq T_1 + 2L$ are integers and a trajectory–control pair $(\{x_t\}_{t=T_1}^{T_2}, \{u_t\}_{t=T_1}^{T_2-1}) \in X(T_1, T_2)$ satisfies

$$\sum_{t=T_1}^{T_2-1} h(t, x_t, u_t) \leq \min\{U^h(T_1, T_2) + M, U^h(T_1, T_2, x_{T_1}, x_{T_2}) + \delta\}.$$

Then there exist integers $\tau_1 \in [T_1, T_1 + L]$, $\tau_2 \in [T_2 - L, T_2]$ such that

$$\rho_E(x_t, x_t^*) \leq \epsilon, \ t = \tau_1 \ldots, \tau_2.$$

Moreover, if $\rho_E(x_{T_2}, x_{T_2}^*) \leq \delta$, then $\tau_2 = T_2$ and if $\rho_E(x_{T_1}, x_{T_1}^*) \leq \delta$, $T_1 \geq L$, then $\tau_1 = T_1$.

Proof Our result follows from Theorem 6.28 applied for the space $\mathfrak{M} = \mathcal{R}(\tilde{\mathfrak{M}})$ equipped with the metric $d_s = d_\psi$. Let $\mathcal{F}$ be as guaranteed by Theorem 6.28 with $\mathfrak{M} = \mathcal{R}(\tilde{\mathfrak{M}})$. Set

$$\tilde{\mathcal{F}} = \mathcal{R}^{-1}(\mathcal{F}) \cap \mathrm{cl}(\tilde{\mathfrak{M}}_{reg}).$$

Clearly,

$$\{f_r : f \in \tilde{\mathfrak{M}}_{reg}, \ r \in (0, 1)\} \subset \tilde{\mathcal{F}}$$

which is dense in the strong topology. Note that the mapping $\mathcal{R} : \tilde{\mathfrak{M}}_\psi \to \mathfrak{M}_\psi$ is continuous in the weak topologies. This implies that

$$\mathcal{R}(\mathrm{cl}(\tilde{\mathfrak{M}})) \subset \bar{\mathfrak{M}}, \quad \mathcal{R}(\mathrm{cl}(\tilde{\mathfrak{M}}_{reg})) \subset \bar{\mathfrak{M}}_{reg}.$$

Thus

$$\tilde{\mathcal{F}} = \mathcal{R}^{-1}(\mathcal{F}) \cap (\mathrm{cl}(\tilde{\mathfrak{M}}_{reg}))$$

is an open set in $\mathrm{cl}(\tilde{\mathfrak{M}}_{reg})$ in the weak topology. Now the assertion of Theorem 6.30 follows from the assertion of Theorem 6.28.

The following analog of Theorem 6.29 is true.

Theorem 6.31 *Then there exists an everywhere dense (in the strong topology) set $\tilde{\mathcal{F}} \subset cl(\tilde{\mathfrak{M}}_{reg})$ which is a countable intersection of open (in the weak topology) subsets of $cl(\tilde{\mathfrak{M}}_{reg})$ such that*

$$\{f_r : f \in \tilde{\mathfrak{M}}_{reg}, \ r \in (0, 1)\} \subset \tilde{\mathcal{F}}$$

and that for each $f \in \tilde{\mathcal{F}}$ there exist $(\{x_t^*\}_{t=0}^\infty, \{u_t^*\}_{t=0}^\infty) \in X(0, \infty)$ such that the following assertion holds.

For each $M, \epsilon > 0$ there exists a neighborhood V of f in $\tilde{\mathfrak{M}}_\psi$ in the weak topology and $\delta, L > 0$ such that for each $g \in V$, each pair of integers $T_1 \geq 0$, $T_2 \geq T_1 + 2L$ and each $(\{x_t\}_{t=T_1}^{T_2}, \{u_t\}_{t=T_1}^{T_2-1}) \in X(T_1, T_2)$ satisfying

$$\sum_{t=T_1}^{T_2-1} g(t, x_t, u_t) \leq \min\{U^g(T_1, T_2) + M, U^g(T_1, T_2, x_{T_1}, x_{T_2}) + \delta\}$$

the inequality

$$\rho_E(x_t, x_t^*) \leq \epsilon, \ t = T_1 + L, \ldots, T_2 - L$$

is true.

Proof Our result follows from Theorem 6.29 applied for the space $\mathfrak{M} = \mathcal{R}(\tilde{\mathfrak{M}})$ equipped with the metric $d_s = d_\psi$. Let $\mathcal{F}$ be as guaranteed by Theorem 6.29 with $\mathfrak{M} = \mathcal{R}(\tilde{\mathfrak{M}})$. Set

$$\tilde{\mathcal{F}} = \mathcal{R}^{-1}(\mathcal{F}).$$

Clearly,

$$\{f_r : f \in \tilde{\mathfrak{M}}_{reg}, \ r \in (0, 1)\} \subset \tilde{\mathcal{F}}$$

which is dense in the strong topology. Note that the mapping $\mathcal{R} : \tilde{\mathfrak{M}}_{\psi} \to \mathfrak{M}_{\psi}$ is continuous in the weak topologies and

$$\mathcal{R}(\mathrm{cl}(\tilde{\mathfrak{M}})) \subset \mathrm{cl}(\mathcal{R}(\tilde{\mathfrak{M}})) = \bar{\mathfrak{M}}, \ \mathcal{R}(\mathrm{cl}(\tilde{\mathfrak{M}}_{reg})) \subset \bar{\mathfrak{M}}_{reg}.$$

Let $\mathcal{F}$ be as guaranteed by Theorem 6.29. Define

$$\tilde{\mathcal{F}} = \mathcal{R}^{-1}(\mathcal{F})(\mathrm{cl}(\tilde{\mathfrak{M}}_{reg})).$$

It is not difficult to see that

$$\{f_r : \ f \in \tilde{\mathfrak{M}}_{reg}, \ r \in (0, 1)\} \subset \tilde{\mathcal{F}},$$

$\tilde{\mathcal{F}}$ is dense in the strong topology and is a countable intersection of open sets in $\mathrm{cl}(\tilde{\mathfrak{M}}_{reg}))$ in the weak topology. Now the assertion of Theorem 6.31 follows from the assertion of Theorem 6.29.

6.15 Smooth Integrands

We use the convention that $\infty/\infty = 1$ and $\infty - \infty = 0$.

Let $(X, \|\cdot\|)$ and $(Y, \|\cdot\|)$ be Banach spaces. Denote by $C(X, Y)$ the set of all continuous mappings $f : X \to Y$.

Let $k \geq 1$ be an integer. For each mapping $f : X \to Y$ we denote by $f^{(k)}(x)$ its Frechet derivative of the order k at a point $x \in X$ if it exists. Denote by $C^k(X, Y)$ the set of all mappings $f : X \to Y$ such that $f^{(k)}(x)$ exists at any point $x \in X$ and the mapping $x \to f^{(k)}(x)$, $x \in X$ is continuous. We set $C^0(X, Y) = C(X, Y)$, $f^{(0)} = f$, $f \in C(X, Y)$.

Let $k \geq 0$ be an integer. For each $f, g \in C^{(k)}(X, Y)$ define

$$\tilde{d}^{(k)}_{X,Y}(f, g) = \sup\{\|f^{(j)}(x) - g^{(j)}(x)\| : \ j = 0, \dots, k, \ x \in X\},$$

$$d^{(k)}_{X,Y}(f, g) = \tilde{d}^{(k)}_{X,Y}(f, g)(1 + \tilde{d}^{(k)}_{X,Y}(f, g))^{-1}.$$

It is known that $(C^{(k)}(X, Y), d^{(k)}_{X,Y})$ is a complete metric space [41, 89].

Let $k \geq 1$ be an integer. We assume that there exists a function $\lambda_* \in C^k(X, R^1)$ such that

$$0 \leq \lambda_*(x) \leq 1, \ x \in X,$$

$$\{x \in X : \ \lambda_*(x) > 0\} \subset \{x \in X : \ \|x\| \leq 1\},$$

$$\lambda_*(z) > 0 \text{ for some } z \in X,$$

$$\sup\{\|\lambda_*^{(j)}(z)\| : \ z \in X, \ j = 0, \dots, k\} < \infty. \tag{6.144}$$

(Note that the function λ_* exists if the space X is Hilbert. It also exists if $k = 1$ and X has an equivalent Frechet differentiable norm. For more details concerning the existence of the function λ_* see [40].)

In view of Deville-Godefroy-Zizler variational principle [40] (see also Theorem 5.9 of [122]), there exists $\phi \in C^k(X; R^1)$ such that

$$\phi(0) = 0, \ \phi(z) \in [0, 1], \ z \in X,$$

$$\sup\{\|\phi^{(j)}(z)\| : \ z \in X, \ j = 0, \dots, k\} < \infty$$

and that the following property holds:

for each $\epsilon > 0$ there exists $\delta > 0$ such that if $x \in X$ and $\phi(x) \le \delta$, then $\|x\| \le \epsilon$.

Assume that $(E, \|\cdot\|)$ and $(F, \|\cdot\|)$ are Banach spaces, $\rho_E(x, y) = \|x - y\|$, $x, y \in E$, $\rho_F(x, y) = \|x - y\|$, $x, y \in F$. We consider the Banach space $E \times F$ equipped with the norm

$$\|(x, y)\| = \|x\| + \|y\|, \ x \in E, \ y \in F$$

and the space $\tilde{\mathfrak{M}}_\psi$ introduced in Sect. 6.14.

Assume that $k \ge 1$ is an integer and that there exists a function $\lambda_* \in C^k(E, R^1)$ satisfying (6.144) with $X = E$. Then there exists a function $\phi_0 \in C^k(E; R^1)$ such that

$$\phi_0(0) = 0, \ \phi_0(z) \in [0, 1], \ z \in E,$$

$$\sup\{\|\phi_0^{(j)}(z)\| : \ z \in E, \ j = 0, \dots, k\} < \infty$$

and that the following property holds:

for each $\epsilon > 0$ there exists $\delta > 0$ such that if $x \in E$ and $\phi(x) \le \delta$, then $\|x\| \le \epsilon$.

Denote by $\tilde{\mathfrak{M}}$ the set of all $f \in \tilde{\mathfrak{M}}_\psi$ such that for each integer $t \ge 0$, $f(t, \cdot, \cdot) \in C^k(E \times F)$. This set is equipped with the uniformity determined by the base

$$E(\epsilon) = \{(g_1, g_2) \in \tilde{\mathfrak{M}} \times \tilde{\mathfrak{M}} : \text{ for each integer } t \ge 0,$$

$$d^{(k)}_{E \times F, R^1}(g_1(t, \cdot, \cdot), g_2(t, \cdot, \cdot)) \le \epsilon\},$$

where $\epsilon > 0$. This uniformity is metrizable and complete and induces a topology which is called strong. Set $\phi(x, y) = \phi_0(x - y)$, $x, y \in E$. It is not difficult to see that Theorems 6.30 and 6.31 hold for the space $\tilde{\mathfrak{M}}$.

Chapter 7
Infinite Dimensional Control

The study of infinite dimensional optimal control has been a rapidly growing area of research [1–4, 15–17, 19–21, 37, 53, 61, 65, 67, 68, 73, 86, 87, 92, 101–104, 113]. In this chapter we present preliminaries which we need in order to study turnpike properties of infinite dimensional optimal control problems. We discuss unbounded operators, C_0 semigroups, evolution equations and admissible control operators.

7.1 Unbounded Operators

Let $(X, \|\cdot\|)$ and $(Y, \|\cdot\|)$ be Banach spaces. Denote by $\mathcal{L}(X, Y)$ the set of all linear continuous operators from X to Y. For every $A \in \mathcal{L}(X, Y)$, define

$$\|A\| = \sup\{\|Ax\| : x \in X, \ \|x\| \leq 1\}.$$

Set $\mathcal{L}(X) = \mathcal{L}(X, X)$, $X^* = \mathcal{L}(X, R^1)$, and $Y^* = \mathcal{L}(Y, R^1)$. For all $x \in X$, set $Ix = x$. In the space X, we always consider the norm convergence.

Let $\mathcal{D}(A)$ be a linear subspace of X (not necessarily closed), and let $A : \mathcal{D}(A) \to Y$ be a linear operator. We say that A is densely defined if $\mathcal{D}(A)$ is dense in X. We say that A is closed if the graph

$$G(A) = \{(x, y) \in X \times Y : x \in \mathcal{D}(A), \ y = Ax\}$$

of A is closed in $X \times Y$. We say that A is closable if there exists a closed operator $\bar{A} : \mathcal{D}(\bar{A}) \to Y$ such that

$$\mathcal{D}(A) \subset \mathcal{D}(\bar{A}) \subset X, \ \bar{A}x = Ax, \ x \in \mathcal{D}(A).$$

Let the symbol $\langle\cdot, \cdot\rangle$ be referred to as the duality pairing between X^* and X.

A. J. Zaslavski, *Turnpike Phenomenon in Metric Spaces*, Springer Optimization and Its Applications 201, https://doi.org/10.1007/978-3-031-27208-0_7

Similar to a bounded operator, for any linear operator (not necessarily bounded) $A : \mathcal{D}(A) \subset X \to X$, we still define the resolvent

$$\rho(A) = \{\lambda \in C^1 : (\lambda I - A)^{-1} \in \mathcal{L}(X)\},$$

the spectrum $\sigma(A) = C^1 \setminus \rho(A)$, the point spectrum (or the set of eigenvalues) $\sigma_p(A)$ of A, and we set $\text{Ran}(A) = \{Ax : x \in \mathcal{D}(A)\}$.

Let $A : \mathcal{D}(A) \subset X \to X$ be densely defined. Clearly, the map

$$x \to \langle Ax, y\rangle, \ x \in \mathcal{D}(A), \ y \in X^*$$

is well-defined. Suppose that $y \in X^*$ and

$$|\langle Ax, y\rangle| \le c_y \|x\| \text{ for all } x \in \mathcal{D}(A).$$

Then the functional $f_y(x) = \langle Ax, y\rangle, \ x \in \mathcal{D}(A)$, can be extended linearly and continuously to the whole X which is the closure of $\mathcal{D}(A)$. Such an extension (still denoted by itself) f_y is in X^* and unique. Hence we obtain

$$\langle Ax, y\rangle = f_y(x) = \langle x, f_y\rangle \text{ for all } x \in \mathcal{D}(A).$$

Define

$$\mathcal{D}(A^*) = \{y \in X^* : \exists c_y \ge 0 \text{ such that } |\langle Ax, y\rangle| \le c_y \|x\| \ \forall x \in \mathcal{D}(A)\}, \tag{7.1}$$

$$A^* y = f_y, \ y \in \mathcal{D}(A^*).$$

Clearly, $A^* : \mathcal{D}(A^*) \subset X^* \to X^*$ is a linear operator satisfying

$$\langle Ax, y\rangle = \langle x, A^* y\rangle \text{ for all } x \in \mathcal{D}(A) \text{ and all } y \in \mathcal{D}(A^*).$$

The mapping A^* is called the adjoint operator of A.

7.2 C_0 Semigroup

Let $(X, \|\cdot\|)$ be a Banach space, and let $\{T(t) : t \in [0, \infty)\} \subset \mathcal{L}(X)$. We call $T(\cdot)$ a C_0 semigroup (or a strongly continuous semigroup of operators) on X if the following properties hold:

$$T(0) = I, \tag{7.2}$$

$$T(t+s) = T(s)T(t) \text{ for all } s, t \ge 0, \tag{7.3}$$

$$\lim_{s \to 0} \|T(s)x - x\| = 0, \ x \in X. \tag{7.4}$$

In the case when $T(t)$ is defined for all $t \in R^1$ and (7.3) holds for all $t, s \in R^1$, we call $T(\cdot)$ as C_0 group or a strongly continuous group of operators. Equations (7.3) and (7.4) are usually referred to as the semigroup property and the strong continuity, respectively.

The following result is true (see Proposition 4.7, Chapter 2 of [73]).

Proposition 7.1 *Let $T(\cdot)$ be a C_0 semigroup on X. Then there exist constants $M \geq 1$ and $\omega \in R^1$ such that*

$$\|T(t)\| \leq Me^{\omega t}, \ t \in [0, \infty).$$

Let $T(\cdot)$ be a C_0 semigroup on X, and let

$$\mathcal{D}(A) = \{x \in X : \lim_{t \to 0} t^{-1}(T(t) - I)x \text{ exists}\},$$

$$Ax = \lim_{t \to 0} t^{-1}(T(t) - I)x, \ x \in \mathcal{D}(A).$$

The operator $A : \mathcal{D}(A) \to X$ is called the generator of the semigroup $T(\cdot)$. We also say that A generates the C_0 semigroup $T(\cdot)$. In general, the operator A is not bounded.

The following two results hold.

Proposition 7.2 (Proposition 4.10, Chapter 2 of [73]) *Let $T(\cdot)$ be a C_0 semigroup on X. Then the generator A of $T(\cdot)$ is a densely defined closed operator and $\cap_{n=1}^{\infty} \mathcal{D}(A^n)$ is dense in X. Furthermore, if $S(\cdot)$ is another C_0 semigroup on X with the same generator A as $T(\cdot)$, then $S(\cdot) = T(\cdot)$.*

Theorem 7.3 ([57, 106] and Theorem 4.11, Chapter 2 of [73]) *Let $A : \mathcal{D}(A) \subset X \to X$ be a linear operator. Then the following properties are equivalent:*

(i) *A generates a C_0 semigroup $T(\cdot)$ on X such that $\|T(t)\| \leq Me^{\omega t}$ for all $t \geq 0$ with some $M \geq 1$ and $\omega \in R^1$.*
(ii) *A is a densely defined and closed operator such that for the above $M \geq 1$ and $\omega \in R^1$, $\{\lambda \in C^1 : \ Re\,\lambda > \omega\} \subset \rho(A)$, and*

$$\|(\lambda I - A)^{-n}\| \leq M(Re\,\lambda - \omega)^{-n}$$

for all integers $n \geq 0$ and all $\lambda \in C^1$ with $Re\,\lambda > \omega$.
(iii) *A is a densely defined and closed operator such that for the above $M \geq 1$ and $\omega \in R^1$, there exists a sequence of positive numbers $\lambda_k \to \infty$ as $k \to \infty$ such that for all integers $k \geq 1$, $\lambda_k \in \rho(A)$, and*

$$\|(\lambda_k I - A)^{-n}\| \leq M(\lambda_k - \omega)^{-n} \text{ for all integers } n \geq 0 \text{ and all integers } k \geq 1.$$

Because the generator A determines the C_0 semigroup $T(\cdot)$ uniquely and in the case $A \in \mathcal{L}(X)$ the C_0 semigroup has an explicit expression $e^{At} =$

$\sum_{n=0}^{\infty}(n!)^{-1}t^n A^n$ (Proposition 4.9, Chapter 2 of [73]), we denote by e^{At} the C_0 semigroup generated by A.

7.3 Evolution Equations

Let $(X, \|\cdot\|)$ be a Banach space, $A : \mathcal{D}(A) \subset X \to X$ generate a C_0 semigroup e^{At} on X, $T > 0$, $f : [0, T] \to X$ be a Bochner integrable function, and $y_0 \in X$. We consider the following evolution equation:

$$y'(t) = Ay(t) + f(t),\ t \in [0, T], \qquad (P_0)$$

$$y_0 = y(0).$$

A continuous function $y : [0, T] \to X$ is called a (mild) solution of (P_0) if

$$y(t) = e^{At}y_0 + \int_0^t e^{A(t-s)}f(s)ds,\ t \in [0, T].$$

A continuous function $y : [0, T] \to X$ is called a weak solution of (P_0) if for any $x^* \in \mathcal{D}(A^*)$, $\langle y(\cdot), x^*\rangle$ is an absolutely continuous function on $[0, T]$ and that for all $t \in [0, T]$,

$$\langle y(t), x^*\rangle = \langle y_0, x^*\rangle + \int_0^t [\langle y(s), A^*x^*\rangle + \langle f(s), x^*\rangle]ds.$$

We have the following result:

Proposition 7.4 ([17], Proposition 5.2, Chapter 2 of [73]) *A continuous function $y : [0, T] \to X$ is a solution of (P_0) if and only if it is a weak solution of (P_0).*

The following useful result is also valid (see Proposition 4.14, Chapter 2 of [73]).

Proposition 7.5 *For any $x \in \mathcal{D}(A)$, $e^{At}x \in \mathcal{D}(A)$ for all $t \geq 0$ and*

$$(d/dt)(e^{At}x) = Ae^{At}x = e^{At}Ax \text{ for all } t \geq 0,$$

$$e^{At}x - e^{As}x = \int_s^t e^{Ar}Axdr \text{ for all } t > s \geq 0.$$

7.4 C_0 Groups

Let $(H, \langle\cdot,\cdot\rangle)$ be a Hilbert space equipped with an inner product $\langle\cdot,\cdot\rangle$ which induces the norm $\|\cdot\|$. A one-parametric family $S(t)$, $t \in R^1$, of continuous linear operators from H onto H is a strongly continuous group of continuous linear operators on H or C_0 group on H if

$$S(0)x = x \text{ for all } x \in H,$$

$$S(t_1 + t_2) = S(t_1)S(t_2) \text{ for all } t_1, t_2 \in R^1,$$

$$\lim_{t \to 0} S(t)x = x, \ x \in H.$$

Let $S(t)$, $t \in R^1$, be a C_0 group on H. Then the generator of S is the linear operator $A : \mathcal{D}(A) \subset H \to H$ defined by

$$\mathcal{D}(A) = \{x \in H : \lim_{t \to 0} t^{-1}(S(t) - I)x \text{ exists}\},$$

$$Ax = \lim_{t \to 0} t^{-1}(S(t) - I)x, \ x \in \mathcal{D}(A).$$

The following result is true.

Theorem 7.6 ([37]) *Let $S(\cdot)$ be a C_0 group on H. Then the following assertions hold:*

(i) *$\mathcal{D}(A)$ is dense in H and A is a closed linear operator.*
(ii) *For every $x^0 \in \mathcal{D}(A)$, there exists a unique function $x \in C^1(R^1; H) \cap C^0(R^1; \mathcal{D}(A))$ such that $x(0) = x^0$, $x(t) \in \mathcal{D}(A)$ for all $t \in R^1$ and $(dx/dt)(t) = Ax(t)$ for all $t \in R^1$; moreover, this solution satisfies $x(t) = S(t)x^0$ for all $t \in R^1$.*
(iii) *$S(t)^*$, $t \in R^1$ is a C_0 group on H and its generator is the adjoint operator A^* of A.*

It is clear that if $S(\cdot)$ is a C_0 group with the generator $A : \mathcal{D}(A) \to H$, then $S(-t), t \in R^1$, is also a C_0 group with the generator $-A : \mathcal{D}(A) \to H$.

Let $A : \mathcal{D}(A) \subset H \to H$. It is known that A and $-A$ are both generators of C_0 semigroups if and only if A is a generator of a C_0 group [104].

Note that a C_0 group is determined by its generator A uniquely and is denoted by e^{At}, $t \in R^1$. Clearly, for all $t \in R^1$, $e^{-At} = e^{A(-t)}$.

7.5 Admissible Control Operators

Let J be an open interval and U be a Hilbert space. Denote by $H^1(J; U)$ the Sobolev space of all locally absolutely continuous functions $z : J \to U$ for which $dz/dt \in L^2(J; U)$. The space $H^2(J; U)$ is the set of all locally absolutely continuous functions $z : J \to U$ for which $dz/dt \in H^1(J; U)$. The space $H_0^1(J; U)$ is the set of all functions in $H^1(J; U)$ which have limits equal to zero at the endpoints of J. Define

$$H_0^2(J;U) = \{h \in H^2(J;U) \cap H_0^1(J;U) : dh/dx \in H_0^1(J;U)\}.$$

In this section we collect several useful results of [104].

Proposition 7.7 (Proposition 2.1.2 of [104]) *Let e^{At}, $t \geq 0$, be a C_0 semigroup on a Hilbert space E. Then the mapping $(t, z) \to e^{At}z$, $t \in [0, \infty)$, $z \in E$, is continuous on $[0, \infty) \times E$ in the product topology.*

Let X and Z be Hilbert spaces and $\mathcal{D}(A)$ be a linear subspace of X. A linear operator $A : \mathcal{D}(A) \to Z$ is called closed if its graph $G(A) = \{(f, Af) : f \in \mathcal{D}(A)\}$ is closed in $X \times Z$. If A is closed, then $\mathcal{D}(A)$ is a Hilbert space with the graph norm $\|\cdot\|_{gr}$:

$$\|z\|_{gr}^2 = \|z\|_X^2 + \|Az\|_Z^2, \ z \in \mathcal{D}(A).$$

Proposition 7.8 (Proposition 2.10.1 of [104]) *Let X be a Hilbert space and $A : \mathcal{D}(A) \to X$ be a densely defined linear operator such that $\rho(A) \neq \emptyset$. Then, for every $\beta \in \rho(A) \cap R^1$, the space $\mathcal{D}(A)$ with the norm $\|z\|_1 = \|(\beta I - A)(z)\|$, $z \in \mathcal{D}(A)$, is a Hilbert space denoted by X_1. The norms generated as above for different $\beta \in \rho(A) \cap R^1$ are equivalent to the graph norm and the embedding $X_1 \subset X$ is continuous. If $L \in \mathcal{L}(X)$ satisfies $L\mathcal{D}(A) \subset \mathcal{D}(A)$, then $L \in \mathcal{L}(X_1)$.*

Note that the relation $\rho(A) \neq \emptyset$ implies that the operator A is closed. Let A be as in Proposition 7.8. It is clear that A^* has the same property. We define a Hilbert space $X_1^d = \mathcal{D}(A^*)$ equipped with the norm $\|z\|_1^d = \|(\beta I - A^*)z\|$, $z \in \mathcal{D}(A^*)$, where $\beta \in \rho(A) \cap R^1$, or equivalently, $\beta \in \rho(A^*) \cap R^1$.

Proposition 7.9 (Proposition 2.10.2 of [104]) *Let A be as in Proposition 7.8, and let $\beta \in \rho(A) \cap R^1$. We denote by X_{-1} the completion of X with respect to the norm*

$$\|z\|_{-1} = \|(\beta I - A)^{-1}z\|, \ z \in X. \tag{7.5}$$

The norms generated as above for different $\beta \in \rho(A) \cap R^1$ are equivalent (in particular, X_{-1} is independent of the choice of β). Moreover, X_{-1} is the dual of X_1^d with respect to the pivot space X. If $L \in \mathcal{L}(X)$ satisfies $L^*\mathcal{D}(A^*) \subset \mathcal{D}^*(A)$, then L has a unique extension to an operator $\tilde{L} \in \mathcal{L}(X_{-1})$.

Proposition 7.10 (Proposition 2.10.3 of [104]) *Let $A : \mathcal{D}(A) \to X$ be a densely defined linear operator such that $\rho(A) \neq \emptyset$, $\beta \in \rho(A) \cap R^1$, X_1 be as in Proposition 7.8, and X_{-1} be as in Proposition 7.9. Then $A \in \mathcal{L}(X_1, X)$ and A has a unique extension $\tilde{A} \in \mathcal{L}(X, X_{-1})$. Moreover, the operators $(\beta I - A)^{-1} \in \mathcal{L}(X, X_1)$ and $(\beta I - \tilde{A})^{-1} \in \mathcal{L}(X_{-1}, X)$ are unitary.*

Proposition 7.11 (Proposition 2.10.4 of [104]) *We use the notation from Proposition 7.10 and assume that A generates a C_0 semigroup $S(\cdot)$ on X. Then, for every*

$t \geq 0$, $S(t)$ has a unique extension $\tilde{S}(t) \in \mathcal{L}(X_{-1})$, $\tilde{S}(t)$, $t \geq 0$ is a C_0 semigroup on X_{-1}, and $\tilde{A}$ is its generator.

Let U and X be Hilbert spaces, $S(\cdot)$ be a C_0 semigroup on X, and $A : \mathcal{D}(A) \to X$ be its generator. We use the notation above and denote by A and $S(\cdot)$ also the extension of the generator to X and the extension of the semigroup to X_{-1}, respectively.

Consider the differential equation

$$z'(t) = Az(t) + f(t), \tag{7.6}$$

where $f \in L^1_{loc}([0, \infty); X_{-1})$. A solution of (7.6) in X_{-1} is a function $z \in L^1_{loc}([0, \infty); X) \cap C([0, \infty); X_{-1})$ which satisfies the following equations in X_{-1}:

$$z(t) - z(0) = \int_0^t [Az(s) + f(s)]ds \text{ for all } t \in [0, \infty). \tag{7.7}$$

It is also called a "strong" solution of (7.6) in X_{-1}. Equation (7.7) implies that z is an a.c. function with values in X_{-1} and (7.6) holds for almost every $t \geq 0$, with the derivative computed with respect to the norm of X_{-1}.

We can also define the concept of a "weak" solution of (7.6) in X_{-1} by regarding instead of (7.7) that for every $\phi \in X_1^d$ and every $t \geq 0$,

$$\langle z(t) - z(0), \phi \rangle_{X_{-1}, X_1^d} = \int_0^t [\langle z(s), A^*\phi \rangle_X + \langle f(s), \phi \rangle_{X_{-1}, X_1^d}]ds.$$

These two concepts are equivalent.

Proposition 7.12 (Proposition 4.1.4 of [104]) *Suppose that z is a solution of (7.6) in X_{-1}, and let $z_0 = z(0)$. Then z is given by*

$$z(t) = S(t)z_0 + \int_0^t S(t - \sigma)f(\sigma)d\sigma, \ t \geq 0. \tag{7.8}$$

In particular, for every $z_0 \in X$, there exists at most one solution in X_{-1} of (7.6) which satisfies initial condition $z(0) = z_0$.

Note that z satisfying (7.8) is called the mild solution of (7.6) with the initial state $z_0 \in X$.

It is not difficult to see that the following result is valid.

Proposition 7.13 *Let $f \in L^1_{loc}([0, \infty); X_{-1})$, $T > 0$, and $z \in C^0([0, T]; X_{-1})$. Then z satisfies (7.8) for all $t \in [0, T]$ in X_{-1} if and only if for all $\xi \in X_1^d = \mathcal{D}(A^*)$ and all $t \in [0, T]$,*

$$\langle z(t), \xi \rangle_{X_{-1}, X_1^d} = \langle S(t)z_0, \xi \rangle_{X_{-1}, X_1^d} + \int_0^t [\langle S(t-\sigma) f(\sigma), \xi \rangle_{X_{-1}, X_1^d}] d\sigma$$

$$= \langle S(t)z_0, \xi \rangle_{X_{-1}, X_1^d} + \int_0^t [\langle f(\sigma), S^*(t-\sigma)\xi \rangle_{X_{-1}, X_1^d}] d\sigma.$$

Proposition 7.14 (Proposition 12.1.2 of [104]) *Assume that Z_1, Z_2, and Z_3 are Hilbert spaces, $F \in \mathcal{L}(Z_1, Z_3)$, $G \in \mathcal{L}(Z_2, Z_3)$, and*

$$Ran(F) = \{F(z) : z \in Z_1\} \subset Ran(G) = \{G(z) : z \in Z_2\}.$$

Then there exists an operator $L \in \mathcal{L}(Z_1, Z_2)$ such that $F = GL$.

Corollary 7.15 *Assume that Z_1 and Z_2 are Hilbert spaces and $G \in \mathcal{L}(Z_1, Z_2)$ satisfies $Ran(G) = Z_2$. Then there exists an operator $L \in \mathcal{L}(Z_2, Z_1)$ such that $GL = I$—the identity operator in Z_2.*

Let $B \in \mathcal{L}(U, X_{-1})$ and $\tau \geq 0$. Define $\Phi_\tau \in \mathcal{L}(L^2(0, \infty; U), X_{-1})$ by

$$\Phi_\tau u = \int_0^\tau S(\tau - \sigma) B u(\sigma) d\sigma. \tag{7.9}$$

The operator $B \in \mathcal{L}(U, X_{-1})$ is called an admissible control operator for $S(\cdot)$ if for some $\tau > 0$,

$$\text{Ran}(\Phi_\tau) = \{\Phi_\tau u : u \in L^2(0, \infty; U)\} \subset X.$$

Proposition 7.16 (Proposition 4.2.2 and (4.2.5) of [104]) *Assume that $B \in \mathcal{L}(U, X_{-1})$ is an admissible control operator for $S(\cdot)$. Then, for every $t \geq 0$,*

$$\Phi_t \in \mathcal{L}(L^2(0, \infty; U), X),$$

and for all $T > t > 0$, $\|\Phi_t\| \leq \|\Phi_T\|$.

Proposition 7.17 (Proposition 4.2.5 of [104]) *Assume that $B \in \mathcal{L}(U, X_{-1})$ is an admissible control operator for $S(\cdot)$. Then, for every $z_0 \in X$ and every $u \in L^2_{loc}(0, \infty; U)$, the initial value problem*

$$z'(t) = Az(t) + Bu(t), \; z(0) = z_0,$$

has a unique solution in X_{-1}. This solution is given by

$$z(t) = S(t)z_0 + \Phi_t u = S(t)z_0 + \int_0^t S(t-\sigma) B u(\sigma) d\sigma,$$

and it satisfies

$$z \in C([0,\infty); X) \cap H^1_{loc}((0,\infty); X_{-1}).$$

Proposition 7.18 (Proposition 4.2.6 of [104]) *Assume that $B \in \mathcal{L}(U, X_{-1})$ is an admissible control operator for $S(\cdot)$. Then, for every $z_0 \in X$ and every $u \in L^2_{loc}(0,\infty; U)$, there exists a unique function $z \in C([0,\infty); X)$ such that for every $t \geq 0$ and every $\psi \in \mathcal{D}(A^*)$,*

$$\langle z(t) - z_0, \psi\rangle_X = \int_0^t [\langle z(\sigma), A^*\psi\rangle_X + \langle u(\sigma), B^*\psi\rangle_U] d\sigma.$$

Propositions 7.16 and 7.17 and Theorem 7.3 imply the following result.

Proposition 7.19 *Assume that $B \in \mathcal{L}(U, X_{-1})$ is an admissible control operator for $S(\cdot)$ and $T > 0$. Then there exists a constant $c_T > 0$ such that for every $z_0 \in X$ and every $u \in L^2(0, T; U)$, the unique solution z of the initial value problem*

$$z'(t) = Az(t) + Bu(t), \ t \in (0, T) \text{ a.e.}, z(0) = z_0,$$

satisfies $z \in C([0, T]; X)$ and that for all $t \in [0, T]$,

$$\|z(t)\| \leq c_T(\|z_0\| + \|u\|).$$

Proposition 7.16 and Corollary 7.15 imply the following result.

Proposition 7.20 *Assume that $B \in \mathcal{L}(U, X_{-1})$ is an admissible control operator for $S(\cdot)$, $T > 0$, and $Ran(\Phi_T) = X$. Then there exists $L \in \mathcal{L}(X, L^2(0, T; U))$ such that $\Phi_T Lx = x$ for all $x \in X$.*

The following result was proved in [148], Theorem 1.23.

Theorem 7.21 *Assume that $B \in \mathcal{L}(U, X_{-1})$ is an admissible control operator for a C_0 semigroup $S(\cdot)$, $T > 0$, $Ran(\Phi_T) = X$, $x_f \in X$, $u_f \in U$, $x^*(t) = x_f$, $u^*(t) = u_f$, $t \in [0,\infty)$, and $x^*(\cdot)$ is a unique solution in X_{-1} of the initial value problem*

$$z'(t) = Az(t) + Bu^*(t), \ z(0) = x_f.$$

Then there exists a constant $c > 0$ such that for each $z^0, z^1 \in X$, there exist $u \in L^2(0, T; U)$ and $z \in C^0([0, T]; X)$ which is a solution of the problem

$$z'(t) = Az(t) + Bu(t), \ t \in [0, T] \text{ a.e.}, \ z(0) = z^0$$

in X_{-1} and satisfy $z(T) = z^1$,

$$\|u(t) - u_f\| \le c(\|z^1 - x_f\| + \|z^0 - x_f\|),\ t \in [0, T],$$

$$\|z(t) - x_f\| \le c(\|z^1 - x_f\| + \|z^0 - x_f\|),\ t \in [0, T].$$

7.6 Examples

Let U and X be Hilbert spaces, $S(\cdot)$ be a C_0 semigroup on X, $A : \mathcal{D}(A) \to X$ be its generator, and $B \in \mathcal{L}(U, X_{-1})$ be an admissible control operator for $S(\cdot)$. We say that the pair (A, B) is exactly controllable in a time $\tau > 0$ if $\mathrm{Ran}(\Phi_\tau) = X$ [37, 104]. It is known that the exact controllability in time $\tau > 0$ is equivalent to the following property:

For each pair $z_0, z_1 \in X$, there exists $u \in L^2(0, \tau; U)$ such that the solution z of the initial value problem

$$z' = Az + Bu,\ z(0) = z_0 \tag{7.10}$$

satisfies $z(\tau) = z_1$.

In this section we consider examples of the pairs (A, B) which are exactly controllable in some time $\tau > 0$. All of these examples were discussed in [104] and [37].

It should be mentioned that in Chap. 8 our turnpike results will be established for large and general classes of infinite dimensional optimal control problems. One of these classes contains problems which are defined by an integrand and a pair of operators (A, B) introduced above, which is exactly controllable in some time $\tau > 0$. Therefore, each example of pairs (A, B) considered below gives us a subclass of infinite dimensional optimal control problems, for which the results of Chap. 8 hold.

Example 7.22 (Example 11.2.2 of [104]*)* We consider the problem of controlling the vibrations of an elastic membrane by a force field acting on a part of this membrane. More precisely, let n be a natural number, and let $\Omega \subset R^n$ be a bounded open set with $\partial\Omega$ of class C^2 or Ω be a rectangular domain. The physical problem described above can be modeled by the equations

$$\frac{\partial^2 w}{\partial t^2} - \Delta w = u \text{ in } \Omega \times (0, \infty),$$

$$w = 0 \text{ on } \partial\Omega \times (0, \infty),$$

$$w(x, 0) = f(x),\ \frac{\partial w}{\partial t}(x, 0) = g(x) \text{ for } x \in \Omega,$$

where f is the initial displacement and g is the initial velocity. Let $\mathcal{O}$ be a nonempty open subset of Ω and $u \in L^2(0, \infty; L^2(\mathcal{O}))$ be the input function. For any such u, we assume that $u(x,t) = 0$ for all $x \in \Omega \setminus \mathcal{O}$. The equation above can be written in the form (7.10) using the following spaces and operators: $X = H_0^1(\Omega) \times L^2(\Omega)$, $\mathcal{D}(A) = (H^2(\Omega) \cap H_0^1(\Omega)) \times H_0^1(\Omega)$, $U = L^2(\mathcal{O}) \subset L^2(\Omega)$,

$$A \begin{pmatrix} f \\ g \end{pmatrix} = \begin{pmatrix} g \\ \Delta f \end{pmatrix} \text{ for all } \begin{pmatrix} f \\ g \end{pmatrix} \in \mathcal{D}(A), \quad Bu = \begin{pmatrix} 0 \\ u \end{pmatrix} \text{ for all } u \in U.$$

Here $B \in \mathcal{L}(U, X)$. It was shown in Example 11.2.2 of [104] that B is an admissible control operator for the C_0 semigroup e^{At} and that the pair (A, B) is exactly controllable in time $\tau > 0$ if Γ and $\mathcal{O}$ satisfy the assumptions of Theorem 7.4.1 of [104], where Γ is a relatively open subset of $\partial\Omega$. In particular, (A, B) is exactly controllable in time $\tau > 0$ if there exist $x_0 \in R^n$ and $\epsilon > 0$ such that

$$N_\epsilon(\{x \in \partial\Omega : (x - x_0) \cdot \nu(x) > 0\}) \subset \operatorname{clos}\mathcal{O}, \quad \tau > 2r(x_0),$$

where $r(x_0) = \sup\{\|x - x_0\| : x \in \Omega\}$, $|\cdot|$ is the Euclidean norm in R^n, ν is the unit outward normal vector field in $\partial\Omega$, and $N_\epsilon(D) = \{x \in \Omega : d(x, D) < \epsilon\}$ for any $D \subset \Omega$ with $d(x, D) = \inf\{|x - y| : y \in D\}$.

Example 7.23 (Example 11.2.4 of [104]*)* Let n be a natural number, let $\Omega \subset R^n$ be a bounded open set with $\partial\Omega$ of class C^2 or let Ω be a rectangular domain, and let $\mathcal{O}$ be a nonempty open subset of Ω. We consider the problem of controlling the vibrations of an elastic plate occupying the domain Ω by a force field acting on $\mathcal{O}$. More precisely, we consider the following initial and boundary initial value problem:

$$\frac{\partial^2 w}{\partial t^2} + \Delta^2 w = u \text{ in } \Omega \times (0, \infty),$$

$$w = \Delta w = 0 \text{ on } \partial\Omega \times (0, \infty),$$

$$w(x, 0) = 0, \quad \frac{\partial w}{\partial t}(x, 0) = 0 \text{ for } x \in \Omega,$$

where $u \in L^2(0, \infty; L^2(\mathcal{O}))$ is the input function. As usual we assume that $u(x,t) = 0$ for all $x \in \Omega \setminus \mathcal{O}$. The equations above determine a system with the state space $X = (H^2(\Omega) \cap H_0^1(\Omega)) \times L^2(\Omega)$ and the input space $U = L^2(\Omega)$ which is exactly controllable in any time $\tau > 0$, as it was shown in Example 11.2.4 of [104], if the pair $(\Omega, \mathcal{O})$ satisfies one of the assumptions (A1) or (A2) in Example 11.2.3 of [104]. More precisely, let $H = L^2(\Omega)$, $\mathcal{D}(A_0) = H_1$ be the Sobolev space $H^2(\Omega) \cap H_0^1(\Omega)$, $A_0 : \mathcal{D}(A_0) \to H$ be defined by $A_0\phi = -\Delta\phi$, $\phi \in \mathcal{D}(A_0)$, and let $H_2 = \mathcal{D}(A_0^2)$ be endowed with the graph norm. Let χ be the Hilbert space $H_1 \times H$, consider the dense subset of χ defined by $\mathcal{D}(A) = H_2 \times H_1$, and let the linear operator $\mathcal{A} : \mathcal{D}(\mathcal{A}) \to \chi$ be defined by

$$\mathcal{A} = \begin{pmatrix} 0 & I \\ -A_0^2 & 0 \end{pmatrix}.$$

Then the equations above can be written in the form

$$z' = \mathcal{A}z + \mathcal{B}u, \ z(0) = 0,$$

where $\mathcal{B} \in \mathcal{L}(U, \chi)$ is defined by $\mathcal{B}u = \begin{pmatrix} 0 \\ u \end{pmatrix}$ for all $u \in U$. It was shown in Example 11.2.4 of [104] that the pair (A, B) is exactly controllable in any time $\tau > 0$ if the pair $(\Omega, \mathcal{O})$ satisfies one of the assumptions (A1) or (A2) in Example 11.2.3 of [104].

Example 7.24 (Example 11.2.6 of [104]*)* We consider the problem of controlling the vibrations of a string occupying the interval $[0, \pi]$ by means of a force $u(t)$ acting at its left end. The equations describing this problem are formulated as a well-posed boundary control system in subsection 10.2.2 of [104]:

$$\frac{\partial^2 w}{\partial t^2}(x, t) = \frac{\partial^2 w}{\partial x^2}(x, t), \ 0 < x < \pi, \ t \geq 0,$$

$$w(\pi, t) = 0, \ \frac{\partial w}{\partial x}(0, t) = u(t), \ t \geq 0,$$

$$w(x, 0) = f(x), \ \frac{\partial w}{\partial t}(x, 0) = g(x), 0 < x < \pi.$$

For this problem, $X = H_R^1(0, \pi) \times L^2(0, \pi)$, $A : \mathcal{D}(A) \to X$ is defined by $\mathcal{D}(A) = \{f \in H^2(0, \pi) \cap H_R^1(0, \pi) : \frac{\partial f}{\partial x}(0) = 0\} \times H_R^1(0, \pi)$,

$$A \begin{pmatrix} f \\ g \end{pmatrix} = \begin{pmatrix} g \\ \frac{\partial^2 f}{\partial x^2} \end{pmatrix} \text{ for all } \begin{pmatrix} f \\ g \end{pmatrix} \in \mathcal{D}(A).$$

The control operator B satisfies $B^* \begin{pmatrix} f \\ g \end{pmatrix} = -g(0)$ for all $\begin{pmatrix} f \\ g \end{pmatrix} \in \mathcal{D}(A)$. It was shown in Example 10.2.6 of [104] that the pair (A, B) is exactly controllable in any time $\tau \geq 2\pi$.

Example 7.25 (Example 11.2.7 of [104]*)* We consider the boundary control of the non-homogeneous elastic string. The model is described by the equation

$$\frac{\partial^2 w}{\partial t^2}(x, t) = \frac{\partial}{\partial x}(a(x)\frac{\partial w}{\partial x}(x, t)) - b(x)w(x, t), \ 0 < x < \pi, \ t > 0,$$

$$w(0,t) = u(t), \; w(\pi,t) = 0, \; w(\cdot,0) = f, \; \frac{\partial w}{\partial t}(\cdot,0) = g.$$

Here $a \in C^2[0,\pi]$, $b \in L^\infty(0,\pi)$, $a(x) \geq m > 0$, $b(x) \geq 0$ for all $x \in [0,\pi]$. These equations correspond to a well-posed boundary control system with state space $X = L^2(0,\pi) \times H^{-1}(0,\pi)$. The generator A is defined by

$$A\begin{pmatrix} f \\ g \end{pmatrix} = \begin{pmatrix} g \\ -A_0 f \end{pmatrix} \text{ for all } \begin{pmatrix} f \\ g \end{pmatrix} \in \mathcal{D}(A) = H_0^1(0,\pi) \times L^2(0,\pi),$$

where $A_0 \in \mathcal{L}(H_0^1(0,\pi), H^{-1}(0,\pi))$ is defined by

$$A_0 f = \frac{\partial}{\partial x}(a \frac{\partial f}{\partial x}) + bf \text{ for all } f \in H_0^1(0,\pi).$$

The control operator B of this system is determined by

$$B^* \begin{pmatrix} \phi \\ \psi \end{pmatrix} = a(0) \frac{\partial}{\partial x}(A_0^{-1}\psi)|_{x=0} \text{ for all } \begin{pmatrix} \phi \\ \psi \end{pmatrix} \in \mathcal{D}(A^*) = \mathcal{D}(A).$$

It was shown in Example 11.2.7 of [104] that the pair (A, B) is exactly controllable in any time $\tau \geq 2\int_0^\pi (a(x))^{-1/2} dx$.

Example 7.26 (Example 11.2.8 of [104]*)* We consider the problem of controlling the vibrations of a beam occupying the interval $[0,\pi]$ by means of a torque $u(t)$ acting at its left end. The model is described by the initial and boundary value problem

$$\frac{\partial^2 w}{\partial t^2}(x,t) = -\frac{\partial^4 w}{\partial x^4}(x,t), \; 0 < x < \pi, \; t > 0,$$

$$w(0,t) = 0, \; w(\pi,t) = 0, \; \frac{\partial^2 w}{\partial x^2}(0,t) = u(t), \; \frac{\partial^2 w}{\partial x^2}(\pi,t) = 0,$$

$$w(\cdot,0) = f, \; \frac{\partial w}{\partial t}(\cdot,0) = g.$$

These equations correspond to a control system where $H = L^2(0,\pi)$, $H_1 = H^2(0,\pi) \cap H_0^1(0,\pi)$, $A_0 : H_1 \to H$ is defined by $A_0 f = -\frac{\partial^2 f}{\partial x^2}$ for all $f \in H_1$, $H_{1/2} = H_0^1(0,\pi)$, $H_{-1/2} = H_0^{-1}(0,\pi)$. The unique extensions of A_0 to unitary operators from $H_{1/2}$ onto $H_{-1/2}$ and from H onto H_{-1} are still denoted by A_0. The space $H_{3/2} = A_0^{-1} H_{1/2}$ is

$$H_{3/2} = \{g \in H^3(0,\pi) \cap H_0^1(0,\pi) : \frac{\partial^2 g}{\partial x^2}(0) = \frac{\partial^2 g}{\partial x^2}(\pi) = 0\}.$$

We set $X = H_{1/2} \times H_{-1/2}$, $\mathcal{D}(A) = H_{3/2} \times H_{1/2}$,

$$A \begin{pmatrix} f \\ g \end{pmatrix} = \begin{pmatrix} g \\ -A_0^2 f \end{pmatrix} \text{ for all } \begin{pmatrix} f \\ g \end{pmatrix} \in \mathcal{D}(A).$$

The control operator B of this system is determined by

$$B^* \begin{pmatrix} f \\ g \end{pmatrix} = -\frac{d}{dx}(A_0^{-1} g)|_{x=0} \text{ for all } \begin{pmatrix} f \\ g \end{pmatrix} \in \mathcal{D}(A^*) = \mathcal{D}(A).$$

It was shown in Example 11.2.8 of [104] that the pair (A, B) is exactly controllable in any time $\tau > 0$.

Example 7.27 (Example 11.2.9 of [104]*)* We consider the problem of controlling the vibrations of a beam occupying the interval [0, 1] by means of an angular velocity $u(t)$ applied at its left end. The equations describing this problem have been formulated as a well-posed boundary control system in Section 10.5 of [104] as follows:

$$\frac{\partial^2 w}{\partial t^2}(x,t) = -\frac{\partial^4 w}{\partial x^4}(x,t), \; 0 < x < 1, \; t > 0,$$

$$w(0,t) = 0, \; w(1,t) = 0, \; \frac{\partial w}{\partial x}(0,t) = u(t), \; \frac{\partial w}{\partial x}(1,t) = 0,$$

$$w(\cdot, 0) = f, \; \frac{\partial f}{\partial t}(\cdot, 0) = g.$$

We denote $X = V \times L^2(0,1)$, where $V = \{h \in H^2(0,1) : \; h(0) = h(1) = \frac{\partial h}{\partial x}(1) = 0\}$. The norm on X is defined by

$$\|z\|^2 = \|z_1\|_V^2 + \|z_2\|_{L^2}^2, \text{ where } \|z_1\|_V^2 = \int_0^1 |\frac{\partial^2 z_1}{\partial x^2}|^2 dx.$$

Let $Z \subset X$ be defined by $Z = (V \cap H^4(0,1)) \times V$, $L : Z \to X$, $G : Z \to R^1$ be defined by

$$L \begin{pmatrix} 0 & I \\ -\frac{\partial^4}{\partial x^4} & 0 \end{pmatrix}, \; G \begin{pmatrix} z_1 \\ z_2 \end{pmatrix} = \frac{\partial z_2}{\partial x}(0),$$

$$\text{Ker} G = \{z \in Z : \; G(z) = 0\} = (V \cap H^4(0,1)) \times H_0^2(0,1), \; A = L|\text{Ker} G$$

and a control operator B be defined by

$$B^*\begin{pmatrix}\psi_1\\ \psi_2\end{pmatrix} = -\frac{\partial^2\psi_1}{\partial x^2}(0), \quad \text{for all } \begin{pmatrix}\psi_1\\ \psi_2\end{pmatrix} \in \mathcal{D}(A^*) = \mathcal{D}(A).$$

It was shown in Example 11.2.9 of [104] that the pair (A, B) is exactly controllable in any time $\tau > 0$.

The next two examples are discussed in [37].

Example 7.28 The transport equation. Let $L > 0$. We consider the linear control system

$$y_t + y_x = 0, \; t \in (0, T), \; x \in (0, L),$$

$$y(t, 0) = u(t), \; t \in (0, T),$$

where $u(t) \in R^1$, $y(t, \cdot) : (0, L) \to R^1$, with $X = L^2(0, L)$, $\mathcal{D}(A) = \{f \in H^1(0, L) : \; f(0) = 0\}$, $Af = -f_x$, $f \in \mathcal{D}(A)$, $U = R^1$ and $B : R^1 \to \mathcal{D}(A^*)'$ defined by $(Bu)z = uz(0)$ for all $u \in R^1$, and $z \in \mathcal{D}(A^*)$. It was shown in [37] that the pair (A, B) is controllable.

Example 7.29 The Korteweg–de Vries equation.

Let $L > 0$. We consider the linear control system

$$y_t + y_x + y_{xxx} = 0, \; t \in (0, T), \; x \in (0, L),$$

$$y(t, 0) = y(t, L) = 0, \; y_x(t, L) = u(t), \; t \in (0, T),$$

where $u(t) \in R^1$, $y(t, \cdot) : (0, L) \to R^1$, with $X = L^2(0, L)$, $\mathcal{D}(A) = \{f \in H^3(0, L) : \; f(0) = f(L) = f_x(L) = 0\}$, $Af = -f_x - f_{xxx}$, $f \in \mathcal{D}(A)$, $U = R^1$, and $B : R^1 \to \mathcal{D}(A^*)'$ is defined by $(Bu)z = uz_x(L)$ for all $u \in R^1$ and $z \in \mathcal{D}(A^*)$. It was shown in [37] that the pair (A, B) is controllable.

Chapter 8
Continuous-Time Nonautonomous Problems on Half-Axis

In this chapter, we establish sufficient and necessary conditions for the turnpike phenomenon for continuous-time optimal control problems on subintervals of half-axis in metric spaces. For these optimal control problems, the turnpike is not a singleton. The results of this chapter will be obtained for three large classes of problems that will be treated simultaneously.

8.1 Preliminaries

Let (E, ρ_E) be a complete metric space and (F, ρ_F) be a metric space. We suppose that $\mathcal{A}$ is a nonempty subset of $[0, \infty) \times E$ and $\mathcal{U} : \mathcal{A} \to 2^F$ is a point to set mapping with a graph

$$\mathcal{M} = \{(t, x, u) : (t, x) \in \mathcal{A}, \ u \in \mathcal{U}(t, x)\}.$$

We suppose that $\mathcal{M}$ is a Borel measurable subset of $[0, \infty) \times E \times F$. Assume that for each pair of numbers T_1, T_2 satisfying $0 \le T_1 < T_2 < \infty$ we are given a set $X(T_1, T_2)$ of pairs (x, u) (called as trajectory–control pairs) such that $x : [T_1, T_2] \to E$ is a continuous function and $u : [T_1, T_2] \to F$ is a Lebesgue function satisfying

$$(t, x(t)) \in \mathcal{A}, \ t \in [T_1, T_2], \tag{8.1}$$

$$u(t) \in \mathcal{U}(t, x(t)), \ t \in [T_1, T_2] \text{ almost everywhere (a.e.).} \tag{8.2}$$

We suppose that the following property holds:

if $0 \le T_1 \le S_1 < S_2 \le T_2 < \infty$, $(x, u) \in X(T_1, T_2)$, $\tilde{x}$ is the restriction of x to $[S_1, S_2]$, and $\tilde{u}$ is the restriction of u to $[S_1, S_2]$, then $(\tilde{x}, \tilde{u}) \in X(S_1, S_2)$;

A. J. Zaslavski, *Turnpike Phenomenon in Metric Spaces*, Springer Optimization and Its Applications 201, https://doi.org/10.1007/978-3-031-27208-0_8

for each triplet of nonnegative numbers $T_1 < T_2 < T_3$, each $(x_1, u_1) \in X(T_1, T_2)$, and each $(x_2, u_2) \in X(T_2, T_3)$ satisfying

$$x_1(T_2) = x_2(T_2),$$

a pair $(x_3, u_3) \in X(T_1, T_3)$, where

$$x_3(t) = x_1(t),\ t \in [T_1, T_2],\ x_3(t) = x_2(t),\ t \in (T_2, T_3],$$

$$u_3(t) = u_1(t),\ t \in [T_1, T_2],\ u_3(t) = u_2(t),\ t \in (T_2, T_3].$$

Let $a_0 > 0$, $\theta_0 \in E$, $\theta_1 \in F$, let $\psi : [0, \infty) \to [0, \infty)$ be an increasing function such that

$$\psi(t) \to \infty \text{ as } t \to \infty, \tag{8.3}$$

and let $\psi_1 : \mathcal{M} \to [0, \infty)$ satisfy

$$\psi_1(t, x, u) \geq \psi(\rho_E(x, \theta_0)),\ (t, x, u) \in \mathcal{M}. \tag{8.4}$$

Denote by $\mathfrak{M}_\psi$ the set of all Borelian functions $g : \mathcal{M} \to R^1$ such that for each $(t, x, u) \in \mathcal{M}$,

$$g(t, x, u) \geq \psi_1(t, x, u) - a_0,\ (t, x, u) \in \mathcal{M}. \tag{8.5}$$

For each pair of numbers $T_2 > T_1 \geq 0$, each $(x, u) \in X(T_1, T_2)$, and each $g \in \mathfrak{M}_\psi$, define

$$I^g(T_1, T_2, x, u) = \int_{T_1}^{T_2} g(t, x(t), u(t))dt \in (-\infty, \infty].$$

We consider functionals of the form $I^g(T_1, T_2, x, u)$, where $0 \leq T_1 < T_2$, $(x, u) \in X(T_1, T_2)$, and $g \in \mathfrak{M}_\psi$. Let $g \in \mathfrak{M}_\psi$. For each pair of numbers $T_2 > T_1 \geq 0$ and each pair of points $(T_1, y), (T_2, z) \in \mathcal{A}$, we define

$$U^g(T_1, T_2, y, z) = \inf\{I^g(T_1, T_2, x, u) :$$

$$(x, u) \in X(T_1, T_2),\ x(T_1) = y,\ x(T_2) = z\}, \tag{8.6}$$

$$U^g(T_1, T_2, y) = \inf\{I^g(T_1, T_2, x, u) : (x, u) \in X(T_1, T_2),\ x(T_1) = y\}, \tag{8.7}$$

$$\widehat{U}^g(T_1, T_2, z) = \inf\{I^g(T_1, T_2, x, u) : (x, u) \in X(T_1, T_2),\ x(T_2) = z\}, \tag{8.8}$$

$$U^g(T_1, T_2) = \inf\{I^g(T_1, T_2, x, u) : (x, u) \in X(T_1, T_2)\}. \tag{8.9}$$

We equip the space $\mathfrak{M}_\psi$ with the uniformity determined by the base

$$E(N, \epsilon, \lambda) = \{(f, g) \in \mathfrak{M}_\psi \times \mathfrak{M}_\psi : |f(t, x, u) - g(t, x.u)| \le \epsilon$$

$$\text{for each } (t, x, u) \in \mathcal{M} \text{ satisfying } \rho_E(x, \theta_0),\ \rho(u, \theta_1) \le N\}$$

$$\cap \{(f, g) \in \mathfrak{M}_\psi \times \mathfrak{M}_\psi : (|f(t, x, u)| + 1)(|g(t, x, u)| + 1)^{-1} \in [\lambda^{-1}, \lambda]$$

$$\text{for each } (t, x, u) \in \mathcal{M} \text{ satisfying } \rho_E(x, \theta_0) \le N\}, \tag{8.10}$$

where $N, \epsilon > 0, \lambda > 1$. Clearly, the uniform space $\mathfrak{M}_\psi$ is Hausdorff and has a countable base. Therefore, $\mathfrak{M}_\psi$ is metrizable (by a metric d_ψ). It is not difficult to see that the uniform space $\mathfrak{M}_\psi$ is complete. It is equipped with a topology induced by the uniformity.

In this chapter, we assume that the following assumption holds:

(A0) For each $L_0 > 0$, there exists $L_1 > 0$ such that for each $T_1 \ge 0$, each $T_2 \in (T_1, T_1 + L_0]$, each $(x, u) \in X(T_1, T_2)$, and each $g \in \mathfrak{M}_\psi$ that satisfy

$$\rho_E(x(T_1), \theta_0) \le L_0,\ I^g(T_1, T_2, x, u) \le L_0$$

the inequality

$$\rho_E(\theta_0, x(t)) \le L_1$$

is true for all $t \in [T_1, T_2]$.

Let $f \in \mathfrak{M}_\psi$. In some cases, we will use the following assumption:

(A0)' For each $L_0 > 0$, there exists $L_1 > 0$ such that for each $T_1 \ge 0$, each $T_2 \in (T_1, T_1 + L_0]$, and each $(x, u) \in X(T_1, T_2)$ that satisfy

$$I^f(T_1, T_2, x, u) \le L_0$$

and

$$\min\{\rho_E(\theta_0, x(t)) :\ t \in [T_1, T_2]\} \le L_0$$

the inequality

$$\rho_E(\theta_0, x(t)) \le L_1$$

is true for all $t \in [T_1, T_2]$.

Assumption (A0) and Eqs. (8.3)–(8.5) easily imply the following result.

Proposition 8.1 *Assume that $M_1 > 0$, $0 < \tau < \tau_0 < \tau_1$. Then there exists $M_2 > 0$ such that for each $T_1 \ge 0$, each $T_2 \in [T_1 + \tau_0, T_1 + \tau_1]$, each $g \in \mathfrak{M}_\psi$, and each $(x, u) \in X(T_1, T_2)$ that satisfy*

$$I^g(T_1, T_2, x, u) \le M_1$$

the inequality

$$\rho_E(\theta_0, x(t)) \le M_2$$

is true for all $t \in [T_1 + \tau, T_2]$.

8.2 The First Class of Problems

We begin with the description of the first class of problems. Let $(E, \|\cdot\|)$ be a Banach space and E^* be its dual. Let $\{A(t) : t \in [0, \infty)\}$ be the family of closed densely defined linear operators with the domain and range in the Banach space E. Let (F, ρ_F) be a metric space,

$$\theta_0 = 0, \ \theta_1 \in F,$$

$\rho_E(x, y) = \|x - y\|, \ x, y \in E$. We suppose that $\mathcal{A}$ is a nonempty subset of $[0, \infty) \times E$ and $\mathcal{U} : \mathcal{A} \to 2^F$ is a point to set mapping with a graph

$$\mathcal{M} = \{(t, x, u) : (t, x) \in \mathcal{A}, \ u \in \mathcal{U}(t, x)\}.$$

We suppose that $\mathcal{M}$ is a Borel measurable subset of $[0, \infty) \times E \times F$ and $G : \mathcal{M} \to E$ is a Borelian function. We consider the homogeneous Cauchy problem

$$x'(t) = A(t)x(t), \ t \in [0, \infty). \tag{8.11}$$

We assume that there exists a function $U : \{(t, s) \in R^2 : 0 \le s \le t < \infty\} \to L(E)$ that has the following properties [21]:

- (i) For each $x_0 \in E$, the function $(t, s) \to U(t, s)x_0$ is continuous on the set $\{(t, s) \in R^2 : 0 \le s \le t < \infty\}$.
- (ii) $U(s, s) = Id$ for all $s \in [0, \infty)$, where Id is the identity operator.
- (iii) $U(t, s)U(s, \tau) = U(t, \tau)$ for all numbers $t \ge s \ge \tau \ge 0$.
- (iv) For each $s \ge 0$, there exists a densely linear subspace E_s of E such that for each $x_0 \in E_s$ the function $t \to U(t, s)x_0$ is continuously differentiable on $[s, \infty)$ and

$$(\partial/\partial t)U(t, s)x_0 = A(t)U(t, s)x_0, \ t \in [s, \infty). \tag{8.12}$$

- (v) There exists an increasing function $\tau \to \Delta_\tau > 0, \tau > 0$ such that for each $\tau > 0$, each $s \ge 0$, and each $t \in [s, s + \tau]$,

$$\|U(t, s)\| \le \Delta_\tau.$$

In this case problem, (8.11) is called well-posed [21].

Let $0 \leq T_1 < T_2$, and consider the following equation:

$$x'(t) = A(t)x(t) + f(t),\ t \in [T_1, T_2],\ x(0) = x_{T_1}, \tag{8.13}$$

where $x_{T_1} \in E$ and $f \in L^1(T_1, T_2; E)$.

A continuous function $x : [T_1, T_2] \to E$ is a solution of (8.13) if

$$x(t) = U(t, T_1)x(T_1) + \int_{T_1}^{t} U(t, s)f(s)ds,\ t \in [T_1, T_2]. \tag{8.14}$$

Assume that (8.14) holds and $\tau \in [T_1, T_2]$. It is not difficult to see [148, p. 386] that $x : [\tau, T_2] \to E$ is a solution of the equation

$$y'(t) = A(t)y(t) + f(t),\ t \in [\tau, T_2],\ y(\tau) = x(\tau).$$

Let $0 \leq T_1 < T_2 < T_3$, $z_0 \in E$, $f \in L^1(T_1, T_3; E)$, a continuous function $x_1 : [T_1, T_2] \to E$ is a solution of the equation

$$x'(t) = A(t)x(t) + f(t),\ t \in [T_1, T_2],\ x(T_1) = z_0,$$

and a continuous function $x_2 : [T_2, T_3] \to E$ is a solution of the equation

$$x'(t) = A(t)x(t) + f(t),\ t \in [T_2, T_3],\ x(T_2) = x_1(T_2).$$

Set

$$x(t) = x_1(t),\ t \in [T_1, T_2],\ x(t) = x_2(t),\ t \in [T_2, T_3].$$

Clearly, the function $x : [T_1, T_3] \to E$ is continuous. It is not difficult to see [148, p. 387] that $x(\cdot)$ is a solution of the equation

$$x'(t) = A(t)x(t) + f(t),\ t \in [T_1, T_3],\ x(T_1) = z_0.$$

Let $0 \leq T_1 < T_2$. We consider the following equation:

$$x'(t) = A(t)x(t) + G(t, x(t), u(t)),\ t \in [T_1, T_2]. \tag{8.15}$$

A pair of functions $x : [T_1, T_2] \to E$, $u : [T_1, T_2] \to F$ is called a (mild) solution of (8.15) if $x : [T_1, T_2] \to E$ is a continuous function, $u : [T_1, T_2] \to F$ is a Lebesgue measurable function,

$$(t, x(t)) \in \mathcal{A},\ t \in [T_1, T_2], \tag{8.16}$$

$$u(t) \in \mathcal{U}(t, x(t)),\ t \in [T_1, T_2] \text{ almost everywhere (a.e.)}, \tag{8.17}$$

$G(s, x(s), u(s)),\ s \in [T_1, T_2]$ is Bochner integrable, and for every $t \in [T_1, T_2]$,

$$x(t) = U(t, T_1)x(T_1) + \int_{T_1}^{t} U(t, s)G(s, x(s), u(s))ds. \tag{8.18}$$

The set of all pairs (x, u), which are solutions of (8.15), is denoted by $X(T_1, T_2)$.

Let $T_1 \geq 0$. A pair of functions $x : [T_1, \infty) \to E$, $u : [T_1, \infty) \to F$ is called a (mild) solution of the system

$$x'(t) = A(t)x(t) + G(t, x(t), u(t)),\ t \in [T_1, \infty) \tag{8.19}$$

if for every $T_2 > T_1$, $x : [T_1, T_2] \to E$, $u : [T_1, T_2] \to F$ is a solution of (8.15). The set of all such pairs (x, u), which are solutions of the equation above, is denoted by $X(T_1, \infty)$.

A function $x : I \to E$, where I is either $[T_1, T_2]$ or $[T_1, \infty)$ $(0 \leq T_1 < T_2)$, is called a trajectory if there exists a Lebesgue measurable function $u : I \to F$ (referred to as a control) such that $(x, u) \in X(T_1, T_2)$ or $(x, u) \in X(T_1, \infty)$, respectively.

For this class of problem for each $(t, x, u) \in \mathcal{M}$, set

$$\psi_1(t, x, u) = \max\{\psi(\|x\|),\ \psi(\|u\|),$$

$$\psi((\|G(t, x, u)\| - a_0\|x\|)_+)(\|G(t, x, u)\| - a_0\|x\|)_+\}. \tag{8.20}$$

In Sect. 8.4, we show that assumption (A0) holds for this class of problems.

8.3 The Second Class of Problems

Let $(E, \langle\cdot,\cdot\rangle)_E$ be a Hilbert space equipped with an inner product $\langle\cdot,\cdot\rangle_E$ that induces the norm $\|\cdot\|_E$, and let $(F, \langle\cdot,\cdot\rangle_F)$ be a Hilbert space equipped with an inner product $\langle\cdot,\cdot\rangle_F$ that induces the norm $\|\cdot\|_F$. For simplicity, we set $\langle\cdot,\cdot\rangle_E = \langle\cdot,\cdot\rangle$, $\|\cdot\|_E = \|\cdot\|$, $\langle\cdot,\cdot\rangle_F = \langle\cdot,\cdot\rangle$, $\|\cdot\|_F = \|\cdot\|$, if E, F are understood. Set

$$\theta_0 = 0,\ \theta_1 = 0,$$

$\rho_E(x, y) = \|x - y\|,\ x, y \in E$, and $\rho_F(x, y) = \|x - y\|,\ x, y \in F$.

We suppose that $\mathcal{A}_0$ is a nonempty subset of E and $\mathcal{U}_0 : \mathcal{A}_0 \to 2^F$ is a point to set mapping with a graph

$$\mathcal{M}_0 = \{(x, u) :\ x \in \mathcal{A}_0,\ u \in \mathcal{U}_0(x)\}.$$

We suppose that $\mathcal{M}_0$ is a Borel measurable subset of $E \times F$. Define

$$\mathcal{A} = [0, \infty) \times \mathcal{A}_0,$$

$\mathcal{U} : \mathcal{A} \to 2^F$ by

$$\mathcal{U}(t, x) = \mathcal{U}_0(x), \ (t, x) \in \mathcal{A},$$

$$\mathcal{M} = [0, \infty) \times \mathcal{M}_0.$$

Let a linear operator $A : \mathcal{D}(A) \to E$ generate a C_0 semigroup $S(t) = e^{At}$, $t \in [0, \infty)$ on E. As usual, we denote by $S(t)^*$ the adjoint of $S(t)$. Then $S(t)^*$, $t \in [0, \infty)$, is a C_0 semigroup, and its generator is the adjoint A^* of A. The domain $\mathcal{D}(A^*)$ is a Hilbert space equipped with the graph norm $\| \cdot \|_{\mathcal{D}(A^*)}$:

$$\|z\|^2_{\mathcal{D}(A^*)} = \|z\|^2_E + \|A^* z\|^2_E, \ z \in \mathcal{D}(A^*).$$

Let $\mathcal{D}(A^*)'$ be the dual of $\mathcal{D}(A^*)$ with the pivot space E. In particular,

$$E_1^d := \mathcal{D}(A^*) \subset E \subset \mathcal{D}(A^*)' = E_{-1}.$$

(Here we use the notation of Chap. 7.)

Let $G : \mathcal{M} \to \mathcal{D}(A^*)' = E_{-1}$, $B \in \mathcal{L}(F, E_{-1})$ be an admissible control operator for e^{At}, $t \geq 0$, and for all $(t, x, u) \in \mathcal{M}$,

$$G(t, x, u) = Bu.$$

Let $0 \leq T_1 < T_2$. We consider the following equation:

$$x'(t) = Ax(t) + Bu(t), \ t \in [T_1, T_2] \text{ a.e.} \tag{8.21}$$

A pair of functions $x : [T_1, T_2] \to E$, $u : [T_1, T_2] \to F$ is called a (mild) solution of (8.21) if $x : [T_1, T_2] \to E$ is a continuous function, $u : [T_1, T_2] \to F$ is a Lebesgue measurable function, $u \in L^2(T_1, T_2; F)$,

$$(t, x(t)) \in \mathcal{A}, \ t \in [T_1, T_2], \tag{8.22}$$

$$u(t) \in \mathcal{U}(t, x(t)), \ t \in [T_1, T_2] \text{ a.e.}, \tag{8.23}$$

and for each $t \in [T_1, T_2]$,

$$x(t) = e^{A(t-T_1)} x(T_1) + \int_{T_1}^t e^{A(t-s)} Bu(s) ds \tag{8.24}$$

in E_{-1}. The set of all pairs (x, u), which are solutions of (8.21), is denoted by $X(T_1, T_2, A, G)$. In the sequel for simplicity, we use the notation $X(T_1, T_2) = X(T_1, T_2, A, G)$ if the pair (A, G) is understood.

Let $T_1 \geq 0$. A pair of functions $x : [T_1, \infty) \to E$, $u : [T_1, \infty) \to F$ is called a (mild) solution of the system

$$x'(t) = Ax(t) + Bu(t), \ t \in [T_1, \infty) \tag{8.25}$$

if for every $T_2 > T_1$, $x : [T_1, T_2] \to E$, and $u : [T_1, T_2] \to F$ is a solution of (8.21). The set of all such pairs (x, u), which are solutions of (8.25), is denoted by $X(T_1, \infty)$.

A function $x : I \to E$, where I is either $[T_1, T_2]$ or $[T_1, \infty)$ $(0 \leq T_1 < T_2)$, is called a trajectory if there exists a Lebesgue measurable function $u : I \to F$ (referred to as a control) such that $(x, u) \in X(T_1, T_2)$ or $(x, u) \in X(T_1, \infty)$, respectively.

For this class of problems, for each $(t, x, u) \in \mathcal{M}$, set

$$\psi_1(t, x, u) = \max\{\psi(\|x\|), \ K_0\|u\|^2\} - a_0, \tag{8.26}$$

where K, a_0 are positive constants.

In Sect. 8.4, we show that assumption (A0) holds for this class of problems.

8.4 A0 and the First and Second Classes of Problems

In this section, we prove that assumption (A0) holds for the first and second classes of problems. Namely, we prove the following result where the two classes are treated simultaneously.

Proposition 8.2 *Let $M_0, M_1, \tau_0 > 0$. Then there exists $M_2 > M_1$ such that for each $g \in \mathfrak{M}_\psi$, each $T_1 \geq 0$, each $T_2 \in (T_1, T_1 + \tau_0]$, and each $(x, u) \in X(T_1, T_2)$ satisfying*

$$\|x(T_1)\| \leq M_1, \tag{8.27}$$

$$I^g(T_1, T_2, x, u) \leq M_0, \tag{8.28}$$

the following inequality holds:

$$\|x(t)\| \leq M_2 \text{ for all } t \in [T_1, T_2]. \tag{8.29}$$

Proof Let us consider the first class of problems. Fix a positive number

$$\delta < \min\{16^{-1}\tau_0, 8^{-1}(a_0 + 1)^{-1}, 8^{-1}(a_0 + 1)^{-1}\Delta_\delta^{-1}\}. \tag{8.30}$$

By (8.5) and (8.20), there exist $h_0 > M_1 + 1$ and $\gamma_0 > 0$ such that for each $g \in \mathfrak{M}_\psi$,

$$g(t, x, u) \geq 4(M_0 + a_0\tau_0)\delta^{-1} \text{ for each } (t, x, u) \in \mathcal{M} \text{ satisfying } \|x\| \geq h_0, \tag{8.31}$$

$$g(t, x, u) \geq 8(\|G(t, x, u)\| - a_0\|x\|)_+ \text{ for each } (t, x, u) \in \mathcal{M} \text{ satisfying}$$

$$\|G(t, x, u)\| - a_0\|x\| \geq \gamma_0. \tag{8.32}$$

Choose a number

$$M_2 > (8M_1 + 8 + 2h_0 + 2\gamma_0\delta + M_0 + a_0\tau_0)(\Delta_\delta + 1). \tag{8.33}$$

Let

$$g \in \mathfrak{M}_\psi,\ T_1 \geq 0,\ T_2 \in (T_1, T_1 + \tau_0],\ (x, u) \in X(T_1, T_2),$$

and Eqs. (8.27) and (8.28) hold. We show that (8.29) holds. Assume the contrary. Then there exists $t_0 \in [T_1, T_2]$ such that

$$\|x(t_0)\| > M_2. \tag{8.34}$$

By the choice of h_0 (see (8.31)) and (8.27), (8.28), there exists $t_1 \in [T_1, t_0]$ such that

$$\|x(t_1)\| \leq h_0,\ |t_1 - t_0| \leq \delta. \tag{8.35}$$

There exists $t_2 \in [t_1, t_0]$ such that

$$\|x(t_2)\| \geq \|x(t)\| \text{ for all } t \in [t_1, t_0]. \tag{8.36}$$

In view of (8.33)–(8.36),

$$t_2 > t_1. \tag{8.37}$$

By (8.18), (8.36), and (8.37),

$$\|x(t_2) - U(t_2, t_1)x(t_1)\| = \|\int_{t_1}^{t_2} U(t_2, s)G(x(s), u(s))ds\|. \tag{8.38}$$

It follows from property (v) from Sect. 8.2 and (8.35) and (8.38) that

$$\|x(t_2)\| - \Delta_\delta h_0 \leq \|x(t_2)\| - \Delta_\delta\|x(t_1)\| \leq \Delta_\delta \int_{t_1}^{t_2} \|G(s, x(s), u(s))\|ds. \tag{8.39}$$

Define

$$E_1 = \{t \in [t_1, t_2] : \ \|G(t, x(t), u(t))\| \ge a_0\|x(t)\| + \gamma_0\}, \tag{8.40}$$

$$E_2 = [t_1, t_2] \setminus E_1. \tag{8.41}$$

By (8.28), (8.32), (8.35), (8.36), and (8.39)–(8.41),

$$\Delta_\delta^{-1}\|x(t_2)\| \le h_0 + \int_{t_1}^{t_2} \|G(s, x(s), u(s))\|ds$$

$$\le h_0 + a_0 \int_{t_1}^{t_2} \|x(t)|dt + \int_{t_1}^{t_2} (\|G(s, x(s), u(s))\| - a_0\|x(s)\|)_+ds$$

$$\le h_0 + a_0\delta\|x(t_2)\| + \int_{E_1} (\|G(t, x(t), u(t))\| - a_0\|x(t)\|)_+dt$$

$$+ \int_{E_2} (\|G(t, x(t), u(t))\| - a_0\|x(t)\|)_+dt$$

$$\le h_0 + a_0\|x(t_2)\|\delta + \gamma_0\delta + \int_{E_1} (\|G(t, x(t), u(t))\| - a_0\|x(t)\|)_+dt$$

$$\le h_0 + a_0\|x(t_2)\|\delta + \gamma_0\delta + 8^{-1} \int_{E_1} g(t, x(t), u(t))dt$$

$$\le h_0 + a_0\|x(t_2)\|\delta + \gamma_0\delta + 8^{-1}(M_0 + a_0\tau_0). \tag{8.42}$$

It follows from (8.30), (8.34), (8.36), and (8.42) that

$$2^{-1}M_2\Delta_\delta^{-1} \le 2^{-1}\Delta_\delta^{-1}\|x(t_2)\| \le h_0 + \gamma_0\delta + 8^{-1}(M_0 + a_0\tau_0)$$

and

$$M_2 \le (2h_0 + 2\gamma_0\delta + 4^{-1}(M_0 + a_0\tau_0))\Delta_\delta.$$

This contradicts (8.33). The contradiction we have reached proves Proposition 8.2 for the first class of problems.

Consider the second class of problems. Recall (see Proposition 7.1) that there exist $M_* \ge 1$ and $\omega_* \in R^1$ such that

$$\|e^{At}\| \le M_*e^{\omega_* t}, \ t \in [0, \infty) \tag{8.43}$$

and that for every $\tau \ge 0$,

$$\Phi_\tau u = \int_0^\tau e^{A(\tau-s)} Bu(s)ds, \; u \in L^2(0,\tau;F), \tag{8.44}$$

where $\Phi_\tau \in \mathcal{L}(L^2(0,\tau;F),E)$ (see Proposition 7.16). Choose a number

$$\delta \in (0,1).$$

By (8.5) and (8.26), there exist

$$h_0 > M_1 + 1 \tag{8.45}$$

and $\gamma > 1$ such that

$$g(t,x,u) \geq 4(M_0 + a_0\tau_0)\delta^{-1} \text{ for each } (t,x,u) \in \mathcal{M} \text{ satisfying } \|x\| \geq h_0, \tag{8.46}$$

$$g(t,x,u) \geq K_0\|u\|^2/2 \text{ for each } u \in F \text{ satisfying } \|u\| \geq \gamma. \tag{8.47}$$

Choose

$$M_2 > M_1 + M_* e^{|\omega_*|\delta} h_0 + \|\Phi_1\|\gamma\delta^{1/2} + \|\Phi_1\|(2K_0^{-1}(M_0 + a_0\tau_0))^{1/2}. \tag{8.48}$$

Let $g \in \mathfrak{M}_\psi$, $T_1 \geq 0$, $T_2 \in (T_1, T_1 + \tau_0]$, $(x,u) \in X(T_1,T_2)$, and (8.27) and (8.28) hold. We show that (8.29) holds. Assume the contrary. Then there exists $t_0 \in [T_1, T_2]$ such that (8.34) holds. By the choice of h_0, (8.45), (8.46), there exists

$$t_1 \in [T_1, t_0]$$

satisfying (8.35). There exists $t_2 \in [t_1, t_0]$ such that

$$\|x(t_2)\| \geq \|x(t)\| \text{ for all } t \in [t_1, t_0]. \tag{8.49}$$

Clearly, $(x,u) \in X(t_1,t_2)$,

$$x(t_2) = e^{A(t_2-t_1)}x(t_1) + \int_{t_1}^{t_2} e^{A(t_2-s)} Bu(s)ds \tag{8.50}$$

in $E_{-1} = \mathcal{D}(A^*)'$. Proposition 7.16, (8.35), (8.43), (8.44), and (8.50) imply that

$$\|x(t_2)\| \leq \|x(t_1)\| M_* e^{|\omega_*|\delta} + \|\int_{t_1}^{t_2} e^{A(t_2-s)} Bu(s)ds\|$$

$$\leq M_* e^{|\omega_*|\delta} h_0 + \|\int_0^{t_2-t_1} e^{A(t_2-t_1-s)} Bu(t_1+s)ds\|$$

$$\le M_* e^{|\omega_*|\delta} h_0 + \|\Phi_{t_2-t_1}\| \|u(t_1+\cdot)\|_{L^2(0,t_2-t_1;F)}$$

$$\le M_* e^{|\omega_*|\delta} h_0 + \|\Phi_1\| (\int_{t_1}^{t_2} \|u(s)\|^2 ds)^{1/2}. \tag{8.51}$$

Define

$$\Omega_1 = \{t \in [t_1, t_2] : \|u(t)\| \ge \gamma\},\ \Omega_2 = [t_1, t_2] \setminus \Omega_1. \tag{8.52}$$

By (8.26), (8.28), (8.34), (8.47)–(8.49), (8.51), and (8.52),

$$M_2 < \|x(t_0)\| \le \|x(t_2)\| \le M_* e^{|\omega_*|\delta} h_0$$

$$+ \|\Phi_1\| [(\int_{\Omega_1} \|u(s)\|^2 ds)^{1/2} + (\int_{\Omega_2} \|u(s)\|^2 ds)^{1/2}]$$

$$\le M_* e^{|\omega_*|\delta} h_0 + \|\Phi_1\| (2K_0^{-1} \int_{\Omega_1} g(s, x(s), u(s)) ds)^{1/2} + \|\Phi_1\| \gamma \delta^{1/2}$$

$$\le M_* e^{|\omega_*|\delta} h_0 + \|\Phi_1\| \gamma \delta^{1/2} + \|\Phi_1\| (2K_0^{-1}(M_0 + a_0 \tau_0))^{1/2} < M_2.$$

The contradiction we have reached proves Proposition 8.2 for the second class of problems.

8.5 The Third Class of Problems

Let (E, ρ_E) be a complete metric space, (F, ρ_F) be a metric space, and

$$\theta_0 \in E,\ \theta_1 \in F.$$

We suppose that $\mathcal{A}$ is a nonempty subset of $[0, \infty) \times E$ and $\mathcal{U} : \mathcal{A} \to 2^F$ is a point to set mapping with a graph

$$\mathcal{M} = \{(t, x, u) : (t, x) \in \mathcal{A},\ u \in \mathcal{U}(t, x)\}.$$

We suppose that $\mathcal{M}$ is a Borel measurable subset of $[0, \infty) \times E \times F$ and $G : \mathcal{M} \to [0, \infty)$ is a Borelian function.

Let $0 \le T_1 < T_2 < \infty$. Denote by $X(T_1, T_2)$ the set of all pairs of functions $x : [T_1, T_2] \to E$ and $u : [T_1, T_2] \to F$ such that x is continuous on $[T_1, T_2]$, u is Lebesgue measurable on function $[T_1, T_2]$,

$$(t, x(t)) \in \mathcal{A},\ t \in [T_1, T_2], \tag{8.53}$$

$$u(t) \in \mathcal{U}(t, x(t)),\ t \in [T_1, T_2] \text{ almost everywhere (a.e.)}, \tag{8.54}$$

$G(s, x(s), u(s)),\ s \in [T_1, T_2]$ is integrable, and for every pair $S_1, S_2 \in [T_1, T_2]$ satisfying $S_1 < S_2$,

$$\rho_E(x(S_1), x(S_2)) \le \int_{S_1}^{S_2} G(t, x(t), u(t))dt. \tag{8.55}$$

(See Section 1.1 of [5].) Pairs $(x, u) \in X(T_1, T_2)$ are called trajectory–control pairs.

Let $T_1 \ge 0$. Denoted by $X(T_1, \infty)$ the set of all pairs of functions $x : [T_1, \infty) \to E$, $u : [T_1, \infty) \to F$ such that for every $T_2 > T_1$, the pair $x : [T_1, T_2] \to E$, $u : [T_1, T_2] \to F$ belongs to $X(T_1, T_2)$. Pairs $(x, u) \in X(T_1, \infty)$ are called trajectory–control pairs.

A function $x : I \to E$, where I is either $[T_1, T_2]$ or $[T_1, \infty)$ $(0 \le T_1 < T_2)$, is called a trajectory if there exists a Lebesgue measurable function $u : I \to F$ (referred to as a control) such that $(x, u) \in X(T_1, T_2)$ or $(x, u) \in X(T_1, \infty)$, respectively.

For this class of problems for each $(t, x, u) \in \mathcal{M}$, set

$$\psi_1(t, x, u) = \max\{\psi(\rho_E(x, \theta_0)),\ \psi(\rho_F(u, \theta_1)),$$

$$\psi((G(t, x, u) - a_0\rho_E(x, \theta_0))_+)(G(t, x, u) - a_0\rho_E(x, \theta_0))_+\}. \tag{8.56}$$

For each pair of numbers $T_2 > T_1 \ge 0$, each $(x, u) \in X(T_1, T_2)$, and each $g \in \mathfrak{M}_\psi$, define

$$I^g(T_1, T_2, x, u) = \int_{T_1}^{T_2} g(t, x(t), u(t))dt \in (-\infty, \infty].$$

We show now that assumption (A0) holds for this class of problems.

Proposition 8.3 *Assumption (A0) holds for the third class of problems. Moreover, for each $L_0 > 0$, there exists $L_1 > 0$ such that for each $g \in \mathfrak{M}_\psi$, each $T_1 \ge 0$, each $T_2 \in (T_1, T_1 + L_0]$, and each $(x, u) \in X(T_1, T_2)$ satisfying*

$$I^g(T_1, T_2, x, u) \le L_0$$

and

$$\min\{\rho_E(x(t), \theta_0) :\ t \in [T_1, T_2]\} \le L_0$$

the following inequality holds:

$$\rho_E(x(t), \theta_0) \le L_1 \text{ for all } t \in [T_1, T_2].$$

Proof In view of (8.5) and (8.56), for each $g \in \mathfrak{M}_\psi$ and each $(t, x, u) \in \mathcal{M}$,

$$g(t, x, u) \geq \max\{\psi(\rho_E(x, \theta_0)), \ \psi(\rho_F(u, \theta_1)),$$

$$\psi((G(t, x, u) - a_0\rho_E(x, \theta_0))_+)(G(t, x, u) - a_0\rho_E(x, \theta_0))_+\} - a_0. \tag{8.57}$$

Let $L_0 > 0$. Fix a positive number

$$\delta < \min\{16^{-1}L_0, 8^{-1}(a_0 + 1)^{-1}\}. \tag{8.58}$$

By (8.3) and (8.57), there exist $h_0 > L_0 + 1$ and $\gamma_0 > 0$ such that for each $g \in \mathfrak{M}_\psi$,

$$g(t, x, u) \geq 4(L_0 + a_0\tau_0)\delta^{-1} \text{ for each } (t, x, u) \in \mathcal{M} \text{ satisfying } \rho_E(x, \theta_0) \geq h_0, \tag{8.59}$$

$$g(t, x, u) \geq 8(G(t, x, u) - a_0\rho_E(x, \theta_0))_+ \text{ for each } (t, x, u) \in \mathcal{M} \text{ satisfying}$$

$$G(t, x, u) - a_0\rho_E(x, \theta_0) \geq \gamma_0. \tag{8.60}$$

Choose a number

$$L_1 > 8L_0 + 8 + 2h_0 + 8\gamma_0\delta + 4L_0(1 + a_0) + 4. \tag{8.61}$$

Let

$$g \in \mathfrak{M}_\psi, \ T_1 \geq 0, \ T_2 \in (T_1, T_1 + L_0], \ (x, u) \in X(T_1, T_2),$$

$$I^g(T_1, T_2, x, u) \leq L_0, \tag{8.62}$$

and

$$\min\{\rho_E(x(t), \theta_0) : \ t \in [T_1, T_2]\} \leq L_0. \tag{8.63}$$

We show that

$$\rho_E(x(t), \theta_0) \leq L_1 \text{ for all } t \in [T_1, T_2]. \tag{8.64}$$

Assume the contrary. Then there exists $t_0 \in [T_1, T_2]$ such that

$$\rho_E(x(t_0), \theta_0) > L_1. \tag{8.65}$$

By the choice of h_0 (see (8.59)) and (8.62), (8.63), there exists $t_1 \in [T_1, T_2]$ such that

$$\rho_E(x(t_1), \theta_0) \leq h_0, \ |t_1 - t_0| \leq \delta. \tag{8.66}$$

There exists

$$t_2 \in [\min\{t_1, t_0\},\ \max\{t_1, t_0\}] \tag{8.67}$$

such that

$$\rho_E(x(t_2), \theta_0) \geq \rho_E(x(t), \theta_0),\ t \in [\min\{t_1, t_0\},\ \max\{t_1, t_0\}]. \tag{8.68}$$

In view of (8.61), (8.65), (8.66), and (8.68),

$$\begin{aligned}\rho_E(x(t_2), x(t_1)) &\geq \rho_E(x(t_2), \theta_0) - \rho_E(x(t_1), \theta_0)\\ &\geq \rho_E(x(t_2), \theta_0) - h_0 \geq 2^{-1}\rho_E(x(t_2), \theta_0).\end{aligned} \tag{8.69}$$

By (8.55) and (8.69),

$$2^{-1}\rho_E(x(t_2), \theta_0) \leq \rho_E(x(t_2), x(t_1)) \leq \int_{\min\{t_1,t_2\}}^{\max\{t_1,t_2\}} G(t, x(t), u(t))dt. \tag{8.70}$$

Define

$$\begin{aligned}E_1 &= \{t \in [\min\{t_1, t_2\},\ \max\{t_1, t_2\}] :\ G(t, x(t), u(t)) \geq a_0\rho_E(x(t), \theta_0) + \gamma_0\},\\ E_2 &= [\min\{t_1, t_2\},\ \max\{t_1, t_2\}] \setminus E_1.\end{aligned} \tag{8.71}$$

By (8.60), (8.62), (8.66), (8.67), (8.70), and (8.71),

$$\begin{aligned}2^{-1}\rho_E(x(t_2), \theta_0) &\leq \int_{E_1} G(t, x(t), u(t))dt + \int_{E_2} G(t, x(t), u(t))dt\\ &\leq \int_{E_1} (G(t, x(t), u(t)) - a_0\rho_E(x(t), \theta_0))_+ dt\\ &\quad + \int_{E_2} (G(t, x(t), u(t)) - a_0\rho_E(x(t), \theta_0))_+ dt\\ &\quad + a_0\rho_E(x(t_2), \theta_0)\delta\\ &\leq a_0\rho_E(x(t_2), \theta_0)\delta + \gamma_0\delta + 8^{-1}\int_{E_1} g(t, x(t), u(t))dt\\ &\leq a_0\delta\rho_E(x(t_2), \theta_0) + \gamma_0\delta + 8^{-1}(L_0 + a_0 L_0).\end{aligned} \tag{8.72}$$

It follows from (8.58), (8.65), (8.68), and (8.72) that

$$4^{-1}L_1 \le 4^{-1}\rho_E(x(t_2), \theta_0) \le \gamma_0\delta + 8^{-1}(L_0 + a_0 L_0),$$

$$L_1 \le 4\gamma_0\delta + L_0(1 + a_0).$$

This contradicts (8.61). The contradiction we have reached proves (8.64) and Proposition 8.3.

8.6 Boundedness Results

We consider the general class of problems introduced in Sect. 8.1 and use the notation, definitions, and assumptions introduced there. We suppose that $(x_f, u_f) \in X(0, \infty)$ satisfies

$$\sup\{\rho_E(x_f(t), \theta_0) : t \in [0, \infty)\} < \infty, \tag{8.73}$$

$$\Delta_f := \sup\{|I^f(j, j+1, x_f, u_f)| : j = 0, 1, \dots\} < \infty. \tag{8.74}$$

We suppose that there exists a number $b_f > 0$ and the following assumptions hold:

(A1) For each $S_1 > 0$, there exist $S_2 > 0$ and $c > 0$ such that

$$I^f(T_1, T_2, x_f, u_f) \le I^f(T_1, T_2, x, u) + S_2$$

for each $T_1 \ge 0$, each $T_2 \ge T_1 + c$, and each $(x, u) \in X(T_1, T_2)$ satisfying $\rho_E(x(T_j), \theta_0) \le S_1$, $j = 1, 2$.

(A2) For each $\epsilon > 0$, there exists $\delta > 0$ such that for each $(T_i, z_i) \in \mathcal{A}$, $i = 1, 2$, satisfying $\rho_E(z_i, x_f(T_i)) \le \delta$, $i = 1, 2$, and $T_2 \ge b_f$, there exist $\tau_1, \tau_2 \in (0, b_f]$ and $(x_1, u_1) \in X(T_1, T_1 + \tau_1)$, $(x_2, u_2) \in X(T_2 - \tau_2, T_2)$, which satisfies

$$x_1(T_1) = z_1, \ x_1(T_1 + \tau_1) = x_f(T_1 + \tau_1),$$

$$I^f(T_1, T_1 + \tau_1, x_1, u_1) \le I^f(T_1, T_1 + \tau_1, x_f, u_f) + \epsilon,$$

$$x_2(T_2) = z_2, \ x_2(T_2 - \tau_2) = x_f(T_2 - \tau_2),$$

$$I^f(T_2 - \tau_2, T_2, x_2, u_2) \le I^f(T_2 - \tau_2, T_2, x_f, u_f) + \epsilon.$$

Relations (8.5) and (8.74) imply the following result.

Lemma 8.4 *Let $c > 0$. Then*

$$\Delta_f(c) = \sup\{|I^f(T_1, T_2, x_f, u_f)| : T_1 \ge 0, \ T_2 \in (T_1, T_1 + c]\} < \infty.$$

The following result is proved in Sect. 8.10.

Theorem 8.5

1. *There exists* $S > 0$ *such that for each pair of numbers* $T_2 > T_1 \geq 0$ *and each* $(x, u) \in X(T_1, T_2)$,

$$I^f(T_1, T_2, x, u) + S \geq I^f(T_1, T_2, x_f, u_f).$$

2. *For each* $(x, u) \in X(0, \infty)$, *either*

$$I^f(0, T, x, u) - I^f(0, T, x_f, u_f) \to \infty \text{ as } T \to \infty$$

or

$$\sup\{|I^f(0, T, x, u) - I^f(0, T, x_f, u_f)| : T \in (0, \infty)\} < \infty. \tag{8.75}$$

Moreover, if (8.75) *holds, then* $\sup\{\rho_E(x(t), \theta_0)) : t \in (0, \infty)\} < \infty$.

We say that $(x, u) \in X(0, \infty)$ is (f)-good if (8.75) holds. The next boundedness result is proved in Sect. 8.11.

Theorem 8.6 *Let* $M_0 > 0$, $c > 0$, $c_0 \in (0, c)$. *Then there exists* $M_1 > 0$ *such that for each* $T_1 \geq 0$, *each* $T_2 \geq T_1 + c$, *and each* $(x, u) \in X(T_1, T_2)$ *satisfying*

$$I^f(T_1, T_2, x, u) \leq I^f(T_1, T_2, x_f, u_f) + M_0$$

the inequality

$$\rho_E(x(t), \theta_0) \leq M_1$$

holds for all $t \in [T_1 + c_0, T_2]$ *and* $\rho_E(x(t), \theta_0) \leq M_1$ *for all* $t \in [T_1, T_2]$ *if assumption (A0)' holds.*

Let $L > 0$. Denote by $\mathcal{A}_L$ the set of all $(s, z) \in \mathcal{A}$ for which there exist $\tau \in (0, L]$ and $(x, u) \in X(s, s + \tau)$ such that

$$x(s) = z, \ x(s + \tau) = x_f(s + \tau), \ I^f(s, s + \tau, x, u) \leq L.$$

Denote by $\widehat{\mathcal{A}}_L$ the set of all $(s, z) \in \mathcal{A}$ for which $s \geq L$ and there exist $\tau \in (0, L]$ and $(x, u) \in X(s - \tau, s)$ such that

$$x(s - \tau) = x_f(s - \tau), \ x(s) = z, \ I^f(s - \tau, s, x, u) \leq L.$$

Theorems 8.7–8.9 stated below are also boundedness results. They follow easily from Theorem 8.6. For details, see Chapter 6 of [148].

Theorem 8.7 *Let* $L > 0$, $M_0 > 0$, *and* $c \in (0, L)$. *Then there exists* $M_1 > 0$ *such that for each* $T_1 \geq 0$, *each* $T_2 \geq T_1 + 2L$, *and each* $(x, u) \in X(T_1, T_2)$ *satisfying*

$$(T_1, x(T_1)) \in \mathcal{A}_L, \ (T_2, x(T_2)) \in \widehat{\mathcal{A}}_L,$$

$$I^f(T_1, T_2, x, u) \le U^f(T_1, T_2, x(T_1), x(T_2)) + M_0$$

the inequality

$$\rho_E(x(t), \theta_0) \le M_1$$

holds for all $t \in [T_1 + c, T_2]$ *and* $\rho_E(x(t), \theta_0) \le M_1$ *for all* $t \in [T_1, T_2]$ *if assumption (A0)' holds.*

Theorem 8.8 *Let* $L > 0$, $M_0 > 0$, *and* $c \in (0, L)$. *Then there exists* $M_1 > 0$ *such that for each* $T_1 \ge 0$, *each* $T_2 \ge T_1 + L$, *and each* $(x, u) \in X(T_1, T_2)$ *satisfying*

$$(T_1, x(T_1)) \in \mathcal{A}_L, \ I^f(T_1, T_2, x, u) \le U^f(T_1, T_2, x(T_1)) + M_0$$

the inequality $\rho_E(x(t), \theta_0) \le M_1$ *holds for all* $t \in [T_1 + c, T_2]$ *and* $\rho_E(x(t), \theta_0) \le M_1$ *for all* $t \in [T_1, T_2]$ *if assumption (A0)' holds.*

Theorem 8.9 *Let* $L > 0$, $M_0 > 0$, *and* $c \in (0, L)$. *Then there exists* $M_1 > 0$ *such that for each* $T_1 \ge 0$, *each* $T_2 \ge T_1 + L$, *and each* $(x, u) \in X(T_1, T_2)$ *satisfying*

$$(T_2, x(T_2)) \in \widehat{\mathcal{A}}_L, \ I^f(T_1, T_2, x, u) \le \widehat{U}^f(T_1, T_2, x(T_2)) + M_0$$

the inequality

$$\rho_E(x(t), \theta_0) \le M_1$$

holds for all $t \in [T_1 + c, T_2]$ *and* $\rho_E(x(T_1), \theta_0) \le M_1$ *for all* $t \in [T_1, T_2]$ *if assumption (A0)' holds.*

The next result is proved in Sect. 8.12.

Theorem 8.10 *Let* $M_0 > 0$ *and* $c > 0$. *Then there exists* $M_1 > 0$ *such that for each* $T_1 \ge 0$, *each* $T_2 \ge T_1 + c$, *and each* $(x, u) \in X(T_1, T_2)$ *satisfying*

$$\rho_E(x(T_1), \theta_0) \le M_0, \ I^f(T_1, T_2, x, u) \le I^f(T_1, T_2, x_f, u_f) + M_0$$

the inequality $\rho_E(x(t), \theta_0) \le M_1$ *holds for all* $t \in [T_1, T_2]$.

The next three theorems follow from Theorem 8.10. For details, see Chapter 6 of [148].

Theorem 8.11 *Let* $L > 0$, $M_0 > 0$. *Then there exists* $M_1 > 0$ *such that for each* $T_1 \ge 0$, *each* $T_2 \ge T_1 + 2L$, *and each* $(x, u) \in X(T_1, T_2)$ *satisfying*

$$\rho_E(x(T_1), \theta_0) \le M_0, \ (T_1, x(T_1)) \in \mathcal{A}_L, \ \ (T_2, x(T_2)) \in \widehat{\mathcal{A}}_L,$$

$$I^f(T_1, T_2, x, u) \le U^f(T_1, T_2, x(T_1), x(T_2)) + M_0$$

the inequality $\rho_E(x(t), \theta_0) \le M_1$ *holds for all* $t \in [T_1, T_2]$.

Theorem 8.12 *Let* $L > 0$, $M_0 > 0$. *Then there exists* $M_1 > 0$ *such that for each* $T_1 \ge 0$, *each* $T_2 \ge T_1 + L$, *and each* $(x, u) \in X(T_1, T_2)$ *satisfying*

$$\rho_E(x(T_1), \theta_0) \le M_0, \ (T_1, x(T_1)) \in \mathcal{A}_L,$$

$$I^f(T_1, T_2, x, u) \le U^f(T_1, T_2, x(T_1)) + M_0$$

the inequality $\rho_E(x(T_1), \theta_0) \le M_1$ *holds for all* $t \in [T_1, T_2]$.

Theorem 8.13 *Let* $L > 0$, $M_0 > 0$. *Then there exists* $M_1 > 0$ *such that for each* $T_1 \ge 0$, *each* $T_2 \ge T_1 + L$, *and each* $(x, u) \in X(T_1, T_2)$ *satisfying*

$$\rho_E(x(T_1), \theta_0) \le M_0, \ (T_2, x(T_2)) \in \widehat{\mathcal{A}}_L,$$

$$I^f(T_1, T_2, x, u) \le \widehat{U}^f(T_1, T_2, x(T_2)) + M_0$$

the inequality $\rho_E(x(T_1), \theta_0) \le M_1$ *holds for all* $t \in [T_1, T_2]$.

8.7 Turnpike Results

We say that the integrand f possesses the turnpike property (or TP for short) if for each $\epsilon > 0$ and each $M > 0$ there exist $\delta > 0$ and $L > 0$ such that for each $T_1 \ge 0$, each $T_2 \ge T_1 + 2L$, and each $(x, u) \in X(T_1, T_2)$ that satisfy

$$I^f(T_1, T_2, x, u) \le \min\{U^f(T_1, T_2) + M, \ U^f(T_1, T_2, x(T_1), x(T_2)) + \delta\}$$

there exist $\tau_1, \tau_2 \in [0, L]$ such that

$$\rho_E(x(t), x_f(t)) \le \epsilon \text{ for all } t \in [T_1 + \tau_1, T_2 - \tau_2].$$

Moreover, if $\rho_E(x(T_2), x_f(T_2)) \le \delta$, then $\tau_2 = 0$, and if $T_1 \ge L$ and $\rho_E(x(T_1), x_f(T_1)) \le \delta$, then $\tau_1 = 0$.

We say that the integrand f possesses the strong turnpike property (or STP for short) if for each $\epsilon > 0$ and each $M > 0$ there exist $\delta > 0$ and $L > 0$ such that for each $T_1 \ge 0$, each $T_2 \ge T_1 + 2L$, and each $(x, u) \in X(T_1, T_2)$ that satisfy

$$I^f(T_1, T_2, x, u) \le \min\{U^f(T_1, T_2) + M, \ U^f(T_1, T_2, x(T_1), x(T_2)) + \delta\}$$

there exist $\tau_1, \tau_2 \in [0, L]$ such that

$$\rho_E(x(t), x_f(t)) \le \epsilon \text{ for all } t \in [T_1 + \tau_1, T_2 - \tau_2].$$

Moreover, if $\rho_E(x(T_1), x_f(T_1)) \le \delta$, then $\tau_1 = 0$, and if $\rho_E(x(T_2), x_f(T_2)) \le \delta$ then $\tau_2 = 0$.

Theorem 8.5 implies the following two results.

Theorem 8.14 *Assume that f has TP and that $\epsilon, M > 0$. Then there exist $\delta > 0$ and $L > 0$ such that for each $T_1 \ge 0$, each $T_2 \ge T_1 + 2L$, and each $(x, u) \in X(T_1, T_2)$ that satisfy*

$$I^f(T_1, T_2, x, u) \le \min\{I^f(T_1, T_2, x_f, u_f) + M,\ U^f(T_1, T_2, x(T_1), x(T_2)) + \delta\}$$

there exist $\tau_1 \in [T_1, T_1 + L]$, $\tau_2 \in [T_2 - L, T_2]$ such that

$$\rho_E(x(t), x_f(t)) \le \epsilon \text{ for all } t \in [\tau_1, \tau_2].$$

Moreover, if $\rho_E(x(T_2), x_f(T_2)) \le \delta$, then $\tau_2 = T_2$, and if $T_1 \ge L$ and $\rho_E(x(T_1), x_f(T_1)) \le \delta$, then $\tau_1 = T_1$.

Theorem 8.15 *Assume that f has STP and that $\epsilon, M > 0$. Then there exist $\delta > 0$ and $L > 0$ such that for each $T_1 \ge 0$, each $T_2 \ge T_1 + 2L$, and each $(x, u) \in X(T_1, T_2)$ that satisfy*

$$I^f(T_1, T_2, x, u) \le \min\{I^f(T_1, T_2, x_f, u_f) + M,\ U^f(T_1, T_2, x(T_1), x(T_2)) + \delta\}$$

there exist $\tau_1 \in [T_1, T_1 + L]$, $\tau_2 \in [T_2 - L, T_2]$ such that

$$\rho_E(x(T_1), x_f(T_1)) \le \epsilon \text{ for all } t \in [\tau_1, \tau_2].$$

Moreover, for $i = 1, 2$, if $\rho_E(x(T_i), x_f(T_i)) \le \delta$, then $\tau_i = T_i$.

It is easy to see that the following result holds (for details, see [148]).

Proposition 8.16 *Let $L > 0$. Then for each $T_1 \ge 0$, each $T_2 \ge T_1 + 2L$, and each pair of points $(T_1, z_1) \in \mathcal{A}_L$, $(T_2, z_2) \in \widehat{\mathcal{A}}_L$,*

$$U^f(T_1, T_2, z_1, z_2) \le I^f(T_1, T_2, x_f, u_f) + 2L(1 + a_0).$$

Proposition 8.16 and Theorems 8.14 and 8.15 imply the following two results.

Theorem 8.17 *Assume that f has TP and that $\epsilon, L_0 > 0$. Then there exist $\delta > 0$ and $L > L_0$ such that for each $T_1 \ge 0$, each $T_2 \ge T_1 + 2L$, and each $(x, u) \in X(T_1, T_2)$ that satisfy*

$$(T_1, x(T_1)) \in \mathcal{A}_{L_0},\ \ (T_2, x(T_2)) \in \widehat{\mathcal{A}}_{L_0},$$

$$I^f(T_1, T_2, x, u) \le U^f(T_1, T_2, x(T_1), x(T_2)) + \delta$$

there exist $\tau_1 \in [T_1, T_1 + L]$, $\tau_2 \in [T_2 - L, T_2]$ *such that*

$$\rho_E(x(t), x_f(t)) \le \epsilon \textit{ for all } t \in [\tau_1, \tau_2].$$

Moreover, if $\rho_E(x(T_2), x_f(T_2)) \le \delta$, *then* $\tau_2 = T_2$, *and if* $T_1 \ge L$ *and* $\rho_E(x(T_1), x_f(T_1)) \le \delta$, *then* $\tau_1 = T_1$.

Theorem 8.18 *Assume that* f *has STP and that* $\epsilon, L_0 > 0$. *Then there exist* $\delta > 0$ *and* $L > L_0$ *such that for each* $T_1 \ge 0$, *each* $T_2 \ge T_1 + 2L$, *and each* $(x, u) \in X(T_1, T_2)$ *that satisfy*

$$(T_1, x(T_1)) \in \mathcal{A}_{L_0}, \ (T_2, x(T_2)) \in \widehat{\mathcal{A}}_{L_0},$$

$$I^f(T_1, T_2, x, u) \le U^f(T_1, T_2, x(T_1), x(T_2)) + \delta$$

there exist $\tau_1 \in [T_1, T_1 + L]$, $\tau_2 \in [T_2 - L, T_2]$ *such that*

$$\rho_E(x(t), x_f(t)) \le \epsilon \textit{ for all } t \in [\tau_1, \tau_2].$$

Moreover, if $i \in \{1, 2\}$ *and* $\rho_E(x(T_i), x_f(T_i)) \le \delta$, *then* $\tau_i = T_i$.

It is easy to see that the following result holds.

Proposition 8.19 *Let* $L > 0$. *Then for each* $(T_1, z) \in \mathcal{A}_L$, *each* $T_2 \ge T_1 + L$,

$$U^f(T_1, T_2, z) \le I^f(T_1, T_2, x_f, u_f) + L(1 + a_0).$$

Proposition 8.19 implies the following two results.

Theorem 8.20 *Assume that* f *has TP and that* $\epsilon, L_0 > 0$. *Then there exist* $\delta > 0$ *and* $L > L_0$ *such that for each* $T_1 \ge 0$, *each* $T_2 \ge T_1 + 2L$, *and each* $(x, u) \in X(T_1, T_2)$ *that satisfy*

$$(T_1, x(T_1)) \in \mathcal{A}_{L_0},$$

$$I^f(T_1, T_2, x, u) \le U^f(T_1, T_2, x(T_1)) + \delta$$

there exist $\tau_1 \in [T_1, T_1 + L]$, $\tau_2 \in [T_2 - L, T_2]$ *such that*

$$\rho_E(x(t), x_f(t)) \le \epsilon \textit{ for all } t \in [\tau_1, \tau_2].$$

Moreover, if $\rho_E(x(T_2), x_f(T_2)) \le \delta$, *then* $\tau_2 = T_2$, *and if* $T_1 \ge L$ *and* $\rho_E(x(T_1), x_f(T_1)) \le \delta$, *then* $\tau_1 = T_1$.

Theorem 8.21 *Assume that* f *has STP and that* $\epsilon, L_0 > 0$. *Then there exist* $\delta > 0$ *and* $L > L_0$ *such that for each* $T_1 \ge 0$, *each* $T_2 \ge T_1 + 2L$, *and each* $(x, u) \in X(T_1, T_2)$ *that satisfy*

$$(T_1, x(T_1)) \in \mathcal{A}_{L_0},$$

$$I^f(T_1, T_2, x, u) \le U^f(T_1, T_2, x(T_1)) + \delta$$

there exist $\tau_1 \in [T_1, T_1 + L]$, $\tau_2 \in [T_2 - L, T_2]$ *such that*

$$\rho_E(x(t), x_f(t)) \le \epsilon \text{ for all } t \in [\tau_1, \tau_2].$$

Moreover, if $i \in \{1, 2\}$ *and* $\rho_E(x(T_2), x_f(T_2)) \le \delta$, *then* $\tau_i = T_i$.

It is easy to see that the following result holds.

Proposition 8.22 *Let* $L > 0$. *Then for each* $T_1 \ge 0$ *and each* $(T_2, z) \in \widehat{\mathcal{A}_L}$ *satisfying* $T_2 \ge T_1 + L$,

$$\widehat{U}^f(T_1, T_2, z) \le I^f(T_1, T_2, x_f, u_f) + L(1 + a_0).$$

Proposition 8.22 implies the following two results.

Theorem 8.23 *Assume that* f *has TP and that* $\epsilon, L_0 > 0$. *Then there exist* $\delta > 0$ *and* $L > L_0$ *such that for each* $T_1 \ge 0$, *each* $T_2 \ge T_1 + 2L$, *and each* $(x, u) \in X(T_1, T_2)$ *that satisfy*

$$(T_2, x(T_2)) \in \widehat{\mathcal{A}}_{L_0},$$

$$I^f(T_1, T_2, x, u) \le \widehat{U}^f(T_1, T_2, x(T_2)) + \delta$$

there exist $\tau_1 \in [T_1, T_1 + L]$, $\tau_2 \in [T_2 - L, T_2]$ *such that*

$$\rho_E(x(t), x_f(t)) \le \epsilon \text{ for all } t \in [\tau_1, \tau_2].$$

Moreover, if $\rho_E(x(T_2), x_f(T_2)) \le \delta$, *then* $\tau_2 = T_2$, *and if* $T_1 \ge L$ *and* $\rho_E(x(T_1), x_f(T_1)) \le \delta$, *then* $\tau_1 = T_1$.

Theorem 8.24 *Assume that* f *has STP and that* $\epsilon, L_0 > 0$. *Then there exist* $\delta > 0$ *and* $L > L_0$ *such that for each* $T_1 \ge 0$, *each* $T_2 \ge T_1 + 2L$, *and each* $(x, u) \in X(T_1, T_2)$ *that satisfy*

$$(T_2, x(T_2)) \in \widehat{\mathcal{A}}_{L_0},$$

$$I^f(T_1, T_2, x, u) \le \widehat{U}^f(T_1, T_2, x(T_2)) + \delta$$

there exist $\tau_1 \in [T_1, T_1 + L]$, $\tau_2 \in [T_2 - L, T_2]$ *such that*

$$\rho_E(x(t), x_f(t)) \le \epsilon \text{ for all } t \in [\tau_1, \tau_2].$$

Moreover, if $i \in \{1, 2\}$ *and* $\rho_E(x(T_2), x_f(T_2)) \le \delta$, *then* $\tau_i = T_i$.

The next theorem is the main result of this chapter. It is proved in Sect. 8.13.

Theorem 8.25 *f has TP if and only if the following properties hold:*

(P1) For each (f)-good pair $(x, u) \in X(0, \infty)$,

$$\lim_{t \to \infty} \rho_E(x(t), x_f(t)) = 0.$$

(P2) For each $\epsilon > 0$ and each $M > 0$, there exist $\delta > 0$ and $L > 0$ such that for each $T \geq 0$ and each $(x, u) \in X(T, T + L)$ that satisfy

$$I^f(T, T + L, x, u) \leq \min\{U^f(T, T + L, x(T), x(T + L)) + \delta,$$

$$I^f(T, T + L, x_f, u_f) + M\}$$

there exists $s \in [T, T + L]$ such that $\rho_E(x(s), x_f(s)) \leq \epsilon$.

We say that f possesses the weak turnpike property (or WTP for short) if for each $\epsilon > 0$ and each $M > 0$ there exist a natural number Q and $l > 0$ such that for each $T_1 \geq 0$, each $T_2 \geq T_1 + lQ$, and each $(x, u) \in X(T_1, T_2)$ that satisfy

$$I^f(T_1, T_2, x, u) \leq I^f(T_1, T_2, x_f, u_f) + M$$

there exist finite sequences $\{a_i\}_{i=1}^q$, $\{b_i\}_{i=1}^q \subset [T_1, T_2]$ such that an integer $q \leq Q$,

$$0 \leq b_i - a_i \leq l, \; i = 1, \ldots, q,$$

$$b_i \leq a_{i+1} \text{ for all integers } i \text{ satisfying } 1 \leq i < q,$$

$$\rho_E(x(t), x_f(t)) \leq \epsilon \text{ for all } t \in [T_1, T_2] \setminus \cup_{i=1}^q [a_i, b_i].$$

The next result is proved in Sect. 8.14.

Theorem 8.26 *f has WTP if and only if f has (P1) and (P2).*

Theorem 8.27 *Consider the three classes of problems. For the second class of problems, we assume that the following assumption holds:*

(A3) For each $\epsilon > 0$, there exist $\delta, L > 0$ such that for each $S_1 \geq L$ and each $S_2 \in [S_1, S_1 + \delta]$ the inequality $I^f(S_1, S_2, x_f, u_f) \leq \epsilon$.

The integrand f has (P1) and (P2) if and only if the following property holds:

(P3) For each $\epsilon > 0$ and each $M > 0$, there exists $L > 0$ such that for each $T_1 \geq 0$, each $T_2 \geq T_1 + L$, and each $(x, u) \in X(T_1, T_2)$ that satisfy

$$I^f(T_1, T_2, x, u) \leq I^f(T_1, T_2, x_f, u_f) + M$$

the inequality

$$mes(\{t \in [T_1, T_2] : \rho_E(x(t), x_f(t)) > \epsilon\}) \leq L$$

is valid.

In view of Theorem 8.26, (P1) and (P2) imply (P3). Clearly, (P3) implies (P2). Therefore, Theorem 8.27 follows from the next result that is proved in Sect. 8.16.

Proposition 8.28 *We consider three classes of problems. For the second class of problems, we assume that (A3) holds. Let f have (P3). Then f has (P1).*

8.8 Lower Semicontinuity Property

We continue to consider the general class of problems studied in Sect. 8.7.

We say that f possesses lower semicontinuity property (or LSC property for short) if for each $T_2 > T_1 \geq 0$ and each sequence $(x_j, u_j) \in X(T_1, T_2)$, $j = 1, 2, \dots$, which satisfy

$$\sup\{I^f(T_1, T_2, x_j, u_j) : j = 1, 2, \dots\} < \infty,$$

there exist a subsequence $\{(x_{j_k}, u_{j_k})\}_{k=1}^{\infty}$ and $(x, u) \in X(T_1, T_2)$ such that

$$x_{j_k}(t) \to x(t) \text{ as } k \to \infty \text{ for every } t \in [T_1, T_2],$$

$$I^f(T_1, T_2, x, u) \leq \liminf_{j \to \infty} I^f(T_1, T_2, x_j, u_j).$$

LSC property plays an important role in the calculus of variations and optimal control theory [35].

Let $S \geq 0$. A pair $(x, u) \in X(S, \infty)$ is called (f)-overtaking optimal if for every $(y, v) \in X(S, \infty)$ satisfying $x(S) = y(S)$,

$$\limsup_{T \to \infty}[I^f(S, T, x, u) - I^f(S, T, y, v)] \leq 0.$$

A pair $(x, u) \in X(S, \infty)$ is called f-weakly optimal if for every $(y, v) \in X(S, \infty)$ satisfying $x(S) = y(S)$,

$$\liminf_{T \to \infty}[I^f(S, T, x, u) - I^f(S, T, y, v)] \leq 0.$$

A pair $(x, u) \in X(S, \infty)$ is called (f)-minimal if for every $T > S$

$$I^f(S, T, x, u) = U^f(S, T, x(S), x(T)).$$

The next result is proved in Sect. 8.18.

Theorem 8.29 *Assume that f has (P1), (P2), and (LSC) property and $S \geq 0$. Then the following assertions hold:*

1. *Assume that $(x, u) \in X(S, \infty)$ is (f)-good. Then there exists an (f)-overtaking optimal pair $(x_*, u_*) \in X(S, \infty)$ such that $x_*(S) = x(S)$.*
2. *Assume that $(\tilde{x}, \tilde{u}) \in X(S, \infty)$ is (f)-good and that $(x_*, u_*) \in X(S, \infty)$ satisfies $x_*(S) = \tilde{x}(S)$. Then the following conditions are equivalent:*

   (i) *(x_*, u_*) is (f)-overtaking optimal.*
   (ii) *(x_*, u_*) is (f)-weakly optimal.*
   (iii) *(x_*, u_*) is (f)-minimal and (f)-good.*
   (iv) *(x_*, u_*) is (f)-minimal and satisfies $\lim_{t\to\infty} \rho_E(x_*(t), x_f(t)) = 0$.*
   (v) *(x_*, u_*) is (f)-minimal and satisfies $\liminf_{t\to\infty} \rho_E(x_*(t), x_f(t)) = 0$.*

The next result easily follows from Theorems 8.25 and 8.26.

Theorem 8.30 *Assume that f has (P1), (P2) and LSC property and $\epsilon, L > 0$. Then there exists $\tau_0 > 0$ such that for every $T_0 \geq 0$ and every (f)-overtaking optimal pair $(x, u) \in X(T_0, \infty)$ that satisfy $(T_0, x(T_0)) \in \mathcal{A}_L$ the inequality $\rho_E(x(t), x_f(t)) \leq \epsilon$ holds for all $t \geq T_0 + \tau_0$.*

In the sequel, we use the following result that can be easily proved.

Proposition 8.31 *Let f have (P1) and $(T_0, z_0) \in \mathcal{A}$. There exists an (f)-good pair $(x, u) \in X(T_0, \infty)$ satisfying $x(T_0) = z_0$ if and only if $(T_0, z_0) \in \cup\{\mathcal{A}_L : L > 0\}$.*

LSC property implies the following result.

Proposition 8.32 *Let f have LSC property, $T_2 > T_2 \geq 0$, $L > 0$. Then the following assertions hold:*

1. *There exists $(x, u) \in X(T_1, T_2)$ such that*

$$I^f(T_1, T_2, x, u) \leq I^f(T_1, T_2, y, v) \text{ for all } (y, v) \in X(T_1, T_2).$$

2. *If $T_2 - T_1 \geq 2L$, $(T_1, z_1) \in \mathcal{A}_L$, $(T_2, z_2) \in \widehat{\mathcal{A}}_L$, then there exists $(x, u) \in X(T_1, T_2)$ such that $x(T_i) = z_i$, $i = 1, 2$ and*

$$I^f(T_1, T_2, x, u) = U^f(T_1, T_2, z_1, z_2).$$

3. *If $T_2 - T_1 \geq L$, $(T_1, z) \in \mathcal{A}_L$, then there exists $(x, u) \in X(T_1, T_2)$ such that $x(T_1) = z$ and*

$$I^f(T_1, T_2, x, u) = U^f(T_1, T_2, z).$$

4. *If $T_2 - T_1 \geq L$, $(T_2, z) \in \widehat{\mathcal{A}}_L$, then there exists $(x, u) \in X(T_1, T_2)$ such that $x(T_2) = z$ and*

$$I^f(T_1, T_2, x, u) = \widehat{U}^f(T_1, T_2, z).$$

8.9 Perturbed Problems

In this section, we use the following assumption:

(A4) For each $\epsilon > 0$, there exists $\delta > 0$ such that for each $(T_i, z_i) \in \mathcal{A}$, $i = 1, 2$, satisfying $\rho_E(z_i, x_f(t)) \le \delta$, $i = 1, 2$, and $T_2 \ge b_f$, there exist $\tau_1, \tau_2 \in (0, b_f]$ and $(x_1, u_1) \in X(T_1, T_1 + \tau_1)$, $(x_2, u_2) \in X(T_2 - \tau_2, T_2)$ that satisfy

$$x_1(T_1) = z_1, \ x_1(T_1 + \tau_1) = x_f(T_1 + \tau_1),$$

$$I^f(T_1, T_1 + \tau_1, x_1, u_1) \le I^f(T_1, T_1 + \tau_1, x_f, u_f) + \epsilon,$$

$$\rho_E(x_1(t), x_f(t)) \le \epsilon, \ t \in [T_1, T_1 + \tau_1],$$

$$x_2(T_2) = z_2, \ x_2(T_2 - \tau_2) = x_f(T_2 - \tau_2),$$

$$I^f(T_2 - \tau_2, T_2, x_2, u_2) \le I^f(T_2 - \tau_2, T_2, x_f, u_f) + \epsilon,$$

$$\rho_E(x_2(t), x_f(t)) \le \epsilon, \ t \in [T_2 - \tau_2, T_2].$$

Clearly, (A4) implies (A2). Assume that $\phi : E \times E \to [0, 1]$ is a continuous function satisfying $\phi(0, 0) = 0$ and such that the following property holds:

(i) For each $\epsilon > 0$, there exists $\delta > 0$ such that for each $x, y \in E$ satisfying $\phi(x, y) \le \delta$, we have $\rho_E(x, y) \le \epsilon$.

For each $r \in (0, 1)$, set

$$f_r(t, x, u) = f(t, x, u) + r\phi(x, x_f(t)), \ (t, x, u) \in \mathcal{M}.$$

Clearly, for any $r \in (0, 1)$, f_r is a Borelian function; if (A4) holds, then (A1) and (A4) hold for f_r with $(x_{f_r}, u_{f_r}) = (x_f, u_f)$.

In the next result, we consider each of the three classes of problems. For the second class, we assume that (A3) holds.

Theorem 8.33 *Let (A4) hold and $r \in (0, 1)$. Then f_r has (P1) and (P2). Moreover, if (x_f, u_f) is (f)-minimal, then (x_f, u_f) is a unique (f_r)-overtaking optimal with an initial point $x_f(0)$.*

8.10 Auxiliary Results for Theorems 8.5 and Its Proof

The next result follows easily from (A2).

Proposition 8.34 *Let $\gamma > 0$. Then there exists $\delta > 0$ such that for each $(T, z_1), (T + 2b_f, z_2) \in \mathcal{A}$ satisfying*

$$\rho_E(z_1, x_f(T)) \le \delta, \ \rho_E(z_2, x_f(T + 2b_f)) \le \delta,$$

there exists $(x, u) \in X(T, T + 2b_f)$*, which satisfies*

$$x(T) = z_1, \ x(T + 2b_f) = z_2,$$

$$I^f(T, T + 2b_f, x, u) \le I^f(T, T + 2b_f, x_f, u_f) + \gamma.$$

Lemma 8.35 *There exist numbers* $S > 0$*,* $c_0 > 1$ *such that for each* $T_1 \ge 0$*, each* $T_2 \ge T_1 + c_0$*, and each* $(x, u) \in X(T_1, T_2)$*,*

$$I^f(T_1, T_2, x_f, u_f) \le I^f(T_1, T_2, x, u) + S. \tag{8.76}$$

Proof In view of (8.3), (8.5), and (8.74), there exists $S_1 > 0$ such that

$$\psi(S_1) > a_0 + 1 + \sup\{|I^f(j, j + 1, x_f, u_f)| : \ j = 0, 1, \dots\}. \tag{8.77}$$

By (A1), there exist $S_2 > 0$, $c_0 > 1$ such that

$$I^f(T_1, T_2, x_f, u_f) \le I^f(T_1, T_2, x, u) + S_2 \tag{8.78}$$

for each $T_1 \ge 0$, each $T_2 \ge T_1 + c_0$, and each $(x, u) \in X(T_1, T_2)$ satisfying

$$\rho_E(x(T_1), \theta_0), \ \rho_E(x(T_2), \theta_0) \le S_1.$$

Fix a number

$$S \ge S_2 + 2 + 2a_0(4 + 2c_0) + 4(c_0 + 4)$$
$$\times \sup\{|I^f(j, j + 1, x_f, u_f)| : \ j = 0, 1, \dots\}. \tag{8.79}$$

Assume that

$$T_1 \ge 0, \ T_2 \ge T_1 + c_0, \ (x, u) \in X(T_1, T_2).$$

We show that (8.76) is true. If

$$\rho_E(x(t), \theta_0) \ge S_1, \ t \in [T_1, T_2],$$

then by (8.4), (8.5), (8.77), and (8.79) for all $t \in [T_1, T_2]$,

$$f(t, x(t), u(t)) \ge -a_0 + \psi(S_1)$$

and

$$I^f(T_1, T_2, x, u) \geq (T_2 - T_1)(\psi(S_1) - a_0) \geq I^f(T_1, T_2, x_f, u_f) - S$$

and (8.76) holds.

Assume that

$$\inf\{\rho_E(x(t), \theta_0) : t \in [T_1, T_2]\} < S_1. \tag{8.80}$$

Set

$$\tau_1 = \inf\{t \in [T_1, T_2] : \rho_E(x(t), \theta_0) \leq S_1\}, \tag{8.81}$$

$$\tau_2 = \sup\{t \in [T_1, T_2] : \rho_E(x(t), \theta_0) \leq S_1\}. \tag{8.82}$$

Clearly, τ_1, τ_2 are well-defined and

$$\rho_E(x(\tau_i), \theta_0) \leq S_1, \; i = 1, 2. \tag{8.83}$$

By (8.4), (8.5), (8.81), and (8.82) for each $t \in [T_1, \tau_1] \cup [T_2, \tau_2]$,

$$\rho_E(x(t), \theta_0) \geq S_1, \; f(t, x(t), u(t)) \geq -a_0 + \psi(S_1). \tag{8.84}$$

It follows from (8.74), (8.77), and (8.84) that

$$I^f(T_1, \tau_1, x, u) \geq I^f(T_1, \tau_1, x_f, u_f) - 2\Delta_f - 2a_0,$$

$$I^f(\tau_2, T_2, x, u) \geq I^f(\tau_2, T_2, x_f, u_f) - 2\Delta_f - 2a_0. \tag{8.85}$$

If $\tau_2 - \tau_1 < c_0$, then by (8.5), (8.74), and (8.85),

$$I^f(T_1, T_2, x, u) \geq I^f(T_1, \tau_1, x_f, u_f) + I^f(\tau_2, T_2, x_f, u_f) - 4\Delta_f - 4a_0 - a_0c_0$$

$$\geq I^f(T_1, T_2, x_f, u_f) - 4a_0 - a_0c_0 - \Delta_f c_0 - 5\Delta_f - 2a_0 \geq I^f(T_1, T_2, x_f, u_f) - S.$$

Assume that

$$\tau_2 - \tau_1 \geq c_0.$$

It follows from (8.78) and (8.83) that

$$I^f(\tau_1, \tau_2, x_f, u_f) \leq I^f(\tau_1, \tau_2, x, u) + S_2.$$

Together with (8.79) and (8.85), this implies (8.76). Lemma 8.35 is proved.

Proof of Theorem 8.5

Assertion 1 follows from Lemma 8.35. Let $(x, u) \in X(0, \infty)$. Assumption 1 implies that either

$$I^f(0, T, x, u) - I^f(0, T, x_f, u_f) \to \infty \text{ as } T \to \infty$$

or

$$\sup\{|I^f(0, T, x, u) - I^f(0, T, x_f, u_f)| : T \in (0, \infty)\} < \infty. \tag{8.86}$$

Assume that (8.86) holds. Then the boundedness of x follows from Proposition 8.1.

8.11 Proof of Theorem 8.6

We may assume that $c < 1/2$ and $c_0 < c/4$. By Theorem 8.5, there exists $S_0 > 0$ such that the following property holds:

(i) For each $T_2 > T_1 \geq 0$ and each $(x, u) \in X(T_1, T_2)$,

$$I^f(T_1, T_2, x, u) + S_0 \geq I^f(T_1, T_2, x_f, u_f).$$

Proposition 8.1 implies that there exists $M_1 > 0$ such that the following property holds:

(ii) For each $T \geq 0$ and each $(x, u) \in X(T, T + c/2)$ satisfying

$$I^f(T, T + c/2, x, u) \leq M_0 + 2S_0 + 2 + 2a_0 + 2\Delta_f,$$

we have

$$\rho_E(x(t), \theta_0) \leq M_1 \text{ for all } t \in [c_0 + T, T + c/2],$$

and if (A0)' holds, then

$$\rho_E(x(t), \theta_0) \leq M_1 \text{ for all } t \in [T, T + c/2].$$

Assume that

$$T_1 \geq 0,\ T_2 \geq T_1 + c,\ (x, u) \in X(T_1, T_2) \tag{8.87}$$

and

$$I^f(T_1, T_2, x, u) \leq I^f(T_1, T_2, x_f, u_f) + M_0. \tag{8.88}$$

Property (i) and (8.88) imply that for each $a \in R^1$ satisfying

$$[a, a + 2^{-1}c] \subset [T_1, T_2],$$

we have

$$I^f(a, a + c/2, x, u) = I^f(T_1, T_2, x, u) - I^f(T_1, a, x, u) - I^f(a + c/2, T_2, x, u)$$

$$\leq I^f(T_1, T_2, x, u) + M_0 - I^f(T_1, a, x_f, u_f) + S_0 - I^f(a + c/2, T_2, x_f, u_f) + S_0$$

$$= I^f(a, a + c/2, x_f, u_f) + 2S_0 + M_0 \leq 2\Delta_f + 2a_0 + 2S_0 + M_0,$$

and combined with property (ii), this implies that

$$\rho_E(x(t), \theta_0) \leq M_1 \text{ for all } t \in [c_0 + a, a + c/2],$$

and if (A0)' holds, then

$$\rho_E(x(t), \theta_0) \leq M_1 \text{ for all } t \in [a, a + c/2].$$

Since a is any number satisfying (8.87), this completes the proof of Theorem 8.6.

8.12 Proof of Theorem 8.10

We may assume that $c < 1$. Theorem 8.5 implies that there exists $S_0 > 0$ such that the following property holds:

(i) For each $T_2 > T_1 \geq 0$ and each $(x, u) \in X(T_1, T_2)$, we have

$$I^f(T_1, T_2, x, u) + S_0 \geq I^f(T_1, T_2, x_f, u_f).$$

By Theorem 8.6, there exists $M_2 > 0$ such that the following property holds:

(ii) For each $T_1 \geq 0$, each $T_2 \geq T_1 + c$, and each $(x, u) \in X(T_1, T_2)$, satisfying

$$I^f(T_1, T_2, x, u) \leq I^f(T_1, T_2, x_f, u_f) + M_0,$$

the inequality $\rho_E(x(t), \theta_0) \leq M_2$ holds for all $t \in [T_1 + c/4, T_2]$.

Assumption (A0) implies that there exists $M_1 > M_2$ such that the following property holds:

(iii) For each $T \geq 0$ and each $(x, u) \in X(T, T + c/4)$ satisfying

$$\rho_E(x(T), \theta_0) \le M_0, \; I^f(T, T + c/4, x, u) \le M_0 + S_0 + a_0 c + 2|\Delta_f| + 2a_0,$$

we have

$$\rho_E(x(t), \theta_0) \le M_1 \text{ for all } t \in [T, T + c/4].$$

Assume that $T_1 \ge 0$, $T_2 \ge T_1 + c$, $(x, u) \in X(T_1, T_2)$,

$$\rho_E(x(T_1), \theta_0) \le M_0, \tag{8.89}$$

$$I^f(T_1, T_2, x, u) \le I^f(T_1, T_2, x_f, u_f) + M_0. \tag{8.90}$$

Property (ii) and (8.90) imply that

$$\rho_E(x(t), \theta_0) \le M_2 < M_1, \; t \in [T_1 + c/4, T_2]. \tag{8.91}$$

It follows from (8.90) and property (i) that

$$I^f(T_1, T_1 + c/4, x, u) = I^f(T_1, T_2, x, u) - I^f(T_1 + c/4, T_2, x, u)$$

$$\le I^f(T_1, T_2, x_f, u_f) + M_0 - I^f(T_1 + c/4, T_2, x, u) + S_0$$

$$\le S_0 + M_0 + I^f(T_1, T_1 + c/4, x_f, u_f) \le M_0 + S_0 + 2\Delta_f + 2a_0. \tag{8.92}$$

Property (iii) and Eqs. (8.89) and (8.92) imply

$$\rho_E(x(t), \theta_0) \le M_1, \; t \in [T_1, T_1 + c/4].$$

This completes the proof of Theorem 8.10.

8.13 Proof of Theorem 8.25

First we show that TP implies (P1) and (P2). In view of Theorem 8.5, TP implies (P2). We show that TP implies (P1). Assume that TP holds and $(x, u) \in X(0, \infty)$ is (f)-good. Let $\epsilon > 0$. There exists $S > 0$ such that for each $T > 0$,

$$|I^f(0, T, x, u) - I^f(0, T, x_f, u_f)| < S.$$

This implies that for each pair of numbers $T_2 > T_1 \ge 0$,

$$|I^f(T_1, T_2, x, u) - I^f(T_1, T_2, x_f, u_f)| < 2S. \tag{8.93}$$

Let $\delta > 0$. We show that there exists $T_\delta > 0$ such that for each $T > T_\delta$,

$$I^f(T_\delta, T, x, u) \le U^f(T_\delta, T, x(T_\delta), x(T)) + \delta.$$

Assume the contrary. Then for each $T \ge 0$, there exists $S > T$ such that

$$I^f(T, S, x, u) > U^f(T, S, x(T), x(S)) + \delta.$$

This implies that there exists a strictly increasing sequence of numbers $\{T_i\}_{i=0}^{\infty}$ such that

$$T_0 = 0,$$

and for every integer $i \ge 0$,

$$I^f(T_i, T_{i+1}, x, u) > U^f(T_i, T_{i+1}, x(T_i), x(T_{i+1})) + \delta.$$

By the relation above, there exists $(y, v) \in X(0, \infty)$ such that for every integer $i \ge 0$,

$$y(T_i) = x(T_i),$$

$$I^f(T_i, T_{i+1}, x, u) > I^f(T_i, T_{i+1}, y, v) + \delta.$$

Combined with (8.93), this implies that for each integer $k \ge 1$,

$$I^f(0, T_k, y, v) - I^f(0, T_k, x_f, u_f)$$

$$= I^f(0, T_k, y, v) - I^f(0, T_k, x, u) + I^f(0, T_k, x, u) - I^f(0, T_k, x_f, u_f)$$

$$\le -k\delta + 2S \to -\infty \text{ as } k \to \infty.$$

This contradicts Theorem 8.5. The contradiction we have reached proves that the following property holds:

(i) For each $\delta > 0$, there exists $T_\delta > 0$ such that for each $T > T_\delta$,

$$I^f(T_\delta, T, x, u) \le U^f(T_\delta, T, x(T_\delta), x(T)) + \delta.$$

Theorem 8.14 implies that there exist $\delta > 0$, $L > 0$ such that the following property holds:

(ii) For each $S_1 \ge 0$, each $S_2 \ge S_1 + 2L$, and each $(z, \xi) \in X(S_1, S_2)$ that satisfy

$$I^f(S_1, S_2, z, \xi) \le \min\{I^f(S_1, S_2, x_f, u_f) + 2S,\ U^f(S_1, S_2, z(S_1), z(S_2)) + \delta\},$$

we have

$$\rho_E(z(t), x_f(t)) \le \epsilon,\ t \in [S_1 + L, S_2 - L].$$

Let $T_\delta > 0$ be as guaranteed by (i). Properties (i) and (ii), (8.93), and the choice of T_δ imply that for each $T \ge T_\delta + 2L$,

$$\rho_E(x(t), x_f(t)) \le \epsilon,\ t \in [T_\delta, T - L]$$

and (P1) holds. Thus TP implies (P1) and (P2).

Now we show that (P1) and (P2) imply TP. We begin with the following auxiliary result.

Lemma 8.36 *Assume that (P1) holds and that $\epsilon > 0$. Then there exists $\delta, L > 0$ such that for each $T_1 \ge L$, each $T_2 \ge T_1 + 2b_f$, and each $(x, u) \in X(T_1, T_2)$ that satisfy*

$$\rho_E(x(T_i), x_f(T_i)) \le \delta,\ i = 1, 2,$$

$$I^f(T_1, T_2, x, u) \le U^f(T_1, T_2, x(T_1), x(T_2)) + \delta$$

the inequality $\rho_E(x(t), x_f(t)) \le \epsilon$ holds for all $t \in [T_1, T_2]$.

Proof By (A2), for each integer $k \ge 1$, there exists $\delta_k \in (0, 2^{-k})$ such that the following property holds:

(iii) For each $(T_i, z_i) \in \mathcal{A}$, $i = 1, 2$, satisfying $\rho_E(z_i, x_f(T_i)) \le \delta_k$, $i = 1, 2$, and $T_2 \ge b_f$, there exist $\tau_1, \tau_2 \in (0, b_f]$ and $(x_1, u_1) \in X(T_1, T_1 + \tau_1)$, $(x_2, u_2) \in X(T_2 - \tau_2, T_2)$ that satisfy

$$x_1(T_1) = z_1,\ x_1(T_1 + \tau_1) = x_f(T_1 + \tau_1),\ x_2(T_2) = z_2,\ x_2(T_2 - \tau_2)$$
$$= x_f(T_2 - \tau_2),$$

$$I^f(T_1, T_1 + \tau_1, x_1, u_1) \le I^f(T_1, T_1 + \tau_1, x_f, u_f) + 2^{-k},$$

$$I^f(T_2 - \tau_2, T_2, x_2, u_2) \le I^f(T_2 - \tau_2, T_2, x_f, u_f) + 2^{-k}.$$

We may assume without loss of generality that $\{\delta_k\}_{k=1}^{\infty}$ is a decreasing sequence. Assume that the lemma does not hold. Then for each integer $k \ge 1$, there exist

$$T_{k,1} \ge k + b_f,\ T_{k,2} \ge T_{k,1} + 2b_f, \tag{8.94}$$

and $(x_k, u_k) \in X(T_{k,1}, T_{k,2})$ such that

$$\rho_E(x_k(T_{k,i}), x_f(T_{k,i})) \le \delta_k,\ i = 1, 2, \tag{8.95}$$

$$I^f(T_{k,1}, T_{k,2}, x_k, u_k) \le U^f(T_{k,1}, T_{k,2}, x_k(T_{k,1}), x_k(T_{k,2})) + \delta_k, \tag{8.96}$$

$$\sup\{\rho_E(x_k(t), x_f(t)) : t \in [T_{k,1}, T_{k,2}]\} > \epsilon. \tag{8.97}$$

Extracting a subsequence and re-indexing, we may assume without loss of generality that for each integer $k \ge 1$,

$$T_{k+1,1} \ge T_{k,2} + 4b_f. \tag{8.98}$$

Let $k \ge 1$ be an integer. Property (iii), (8.94) and (8.95) imply that there exist $\tau_{k,1}, \tau_{k,2} \in (0, b_f]$ and $(\tilde{x}_k, \tilde{u}_k) \in X(T_{k,1} - \tau_{k,1}, T_{k,2} + \tau_{k,2})$ such that

$$(\tilde{x}_k(t), \tilde{u}_k(t)) = (x_k(t), u_k(t)),\ t \in [T_{k,1}, T_{k,2}], \tag{8.99}$$

$$\tilde{x}_k(T_{k,1} - \tau_{k,1}) = x_f(T_{k,1} - \tau_{k,1}),\ \tilde{x}_k(T_{k,2} + \tau_{k,2}) = x_f(T_{k,2} + \tau_{k,2}), \tag{8.100}$$

$$I^f(T_{k,1} - \tau_{k,1}, T_{k,1}, \tilde{x}_k, \tilde{u}_k) \le I^f(T_{k,1} - \tau_{k,1}, T_{k,1}, x_f, u_f) + 2^{-k}, \tag{8.101}$$

$$I^f(T_{k,2}, T_{k,2} + \tau_{k,2}, \tilde{x}_k, \tilde{u}_k) \le I^f(T_{k,2}, T_{k,2} + \tau_{k,2}, x_f, u_f) + 2^{-k}. \tag{8.102}$$

Property (iii) and (8.94) and (8.95) imply that there exist $\tau_{k,3}, \tau_{k,4} \in (0, b_f]$ and $(\widehat{x}_k, \widehat{u}_k) \in X(T_{k,1}, T_{k,2})$ such that

$$\widehat{x}_k(T_{k,1}) = x_k(T_{k,1}),\ \widehat{x}_k(T_{k,2}) = x_k(T_{k,2}), \tag{8.103}$$

$$(\widehat{x}_k(t), \widehat{u}_k(t)) = (x_f(t), u_f(t)),\ t \in [T_{k,1} + \tau_{k,3}, T_{k,2} - \tau_{k,4}], \tag{8.104}$$

$$I^f(T_{k,1}, T_{k,1} + \tau_{k,3}, \widehat{x}_k, \widehat{u}_k) \le I^f(T_{k,1}, T_{k,1} + \tau_{k,3}, x_f, u_f) + 2^{-k}, \tag{8.105}$$

$$I^f(T_{k,2} - \tau_{k,4}, T_{k,2}, \widehat{x}_k, \widehat{u}_k) \le I^f(T_{k,2} - \tau_{k,4}, T_{k,2}, x_f, u_f) + 2^{-k}. \tag{8.106}$$

In view of (8.103)–(8.106),

$$U^f(T_{k,1}, T_{k,2}, x_k(T_{k,1}), x_k(T_{k,2})) \le I^f(T_{k,1}, T_{k,2}, \widehat{x}_k, \widehat{u}_k)$$
$$\le I^f(T_{k,1}, T_{k,2}, x_f, u_f) + 2^{-k+1}. \tag{8.107}$$

By (8.96), (8.107) and the inequality $\delta_k < 2^{-k}$,

$$I^f(T_{k,1}, T_{k,2}, x_k, u_k) \le I^f(T_{k,1}, T_{k,2}, x_f, u_f) + 2^{-k+2}. \tag{8.108}$$

It follows from (8.99)–(8.102) and (8.108) that

$$I^f(T_{k,1} - \tau_{k,1}, T_{k,2} + \tau_{k,2}, \tilde{x}_k, \tilde{u}_k)$$

$$\leq I^f(T_{k,1} - \tau_{k,1}, T_{k,1}, x_f, u_f)$$

$$+ I^f(T_{k,1}, T_{k,2}, x_k, u_k) + I^f(T_{k,2}, T_{k,2} + \tau_{k,2}, x_f, u_f) + 2^{-k+1}$$

$$\leq I^f(T_{k,1} - \tau_{k,1}, T_{k,2} + \tau_{k,2}, x_f, u_f) + 2^{-k+3}. \quad (8.109)$$

By (8.98) and (8.100), there exists $(x, u) \in X(0, \infty)$ such that for each integer $k \geq 1$,

$$(x(t), u(t)) = (\tilde{x}_k(t), \tilde{u}_k(t)), \ t \in [T_{k,1} - \tau_{k,1}, T_{k,2} + \tau_{k,2}], \quad (8.110)$$

$$(x(t), u(t)) = (x_f(t), u_f(t)), \ t \in [0, \infty) \setminus \cup_{k=1}^{\infty} [T_{k,1} - \tau_{k,1}, T_{k,2} + \tau_{k,2}]. \quad (8.111)$$

It follows from (8.111) that for each integer $k \geq 1$,

$$I^f(0, T_{k,2} + \tau_{k,2}, x, u) - I^f(0, T_{k,2} + \tau_{k,2}, x_f, u_f)$$

$$= \sum_{i=1}^{k} (I^f(T_{i,1} - \tau_{i,1}, T_{i,2} + \tau_{i,2}, \tilde{x}_i, \tilde{u}_i) - I^f(T_{i,1} - \tau_{i,1}, T_{i,2} + \tau_{i,2}, x_f, u_f))$$

$$\leq \sum_{i=1}^{k} 2^{-i+3} \leq 8.$$

Combined with Theorem 8.5 implies that (x, u) is (f)-good. Together with (P1), this implies that

$$\lim_{t \to \infty} \rho_E(x(t), x_f(t)) = 0.$$

On the other hand, in view of (8.97), (8.99), and (8.110),

$$\limsup_{t \to \infty} \rho_E(x(t), x_f(t)) \geq \epsilon.$$

The contradiction we have reached completes the proof of Lemma 8.36.

Completion of the Proof of Theorem 8.25
Assume that properties (P1) and (P2) hold. Let $\epsilon, M > 0$. By Lemma 8.36, there exist $\delta_0, L_0 > 0$ such that the following property holds:

(iv) For each $T_1 \geq L_0$, each $T_2 \geq T_1 + 2b_f$, and each $(x, u) \in X(T_1, T_2)$ that satisfy

$$\rho_E(x(T_i), x_f(T_i)) \le \delta_0, \ i = 1, 2,$$

$$I^f(T_1, T_2, x, u) \le U^f(T_1, T_2, x(T_1), x(T_2)) + \delta_0$$

the inequality $\rho_E(x(t), x_f(t)) \le \epsilon$ holds for all $t \in [T_1, T_2]$.

By Theorem 8.5, there exists $S_0 > 0$ such that the following property holds:

(v) For each $T_2 > T_1 \ge 0$ and each $(x, u) \in X(T_1, T_2)$, we have

$$I^f(T_1, T_2, x, u) + S_0 > I^f(T_1, T_2, x_f, u_f).$$

In view of (P2), there exist $\delta \in (0, \delta_0)$ and $L_1 > 0$ such that the following property holds:

(vi) For each $T \ge 0$ and each $(x, u) \in X(T, T + L_1)$ that satisfy

$$I^f(T, T + L_1, x, u) \le \min\{U^f(T, T + L_1, x(T), x(T + L_1)) + \delta,$$

$$I^f(T, T + L_1, x_f, u_f) + 2S_0 + M\},$$

there exists $S \in [T, T + L_1]$ such that $\rho_E(x(S), x_f(S)) \le \delta_0$.

Set

$$L = L_0 + L_1 + b_f. \tag{8.112}$$

Assume that $T_1 \ge 0$, $T_2 \ge T_1 + 2L$ and that $(x, u) \in X(T_1, T_2)$ satisfies

$$I^f(T_1, T_2, x, u) \le U^f(T_1, T_2) + M, \tag{8.113}$$

$$I^f(T_1, T_2, x, u) \le U^f(T_1, T_2, x(T_1), x(T_2)) + \delta. \tag{8.114}$$

By (8.113),

$$I^f(T_1, T_2, x, u) \le I^f(T_1, T_2, x_f, u_f) + M. \tag{8.115}$$

Property (v) and (8.115) imply that for each pair of numbers $Q_1, Q_2 \in [T_1, T_2]$ satisfying $Q_1 < Q_2$,

$$I^f(Q_1, Q_2, x, u) = I^f(T_1, T_2, x, u) - I^f(T_1, Q_1, x, u) - I^f(Q_2, T_2, x, u)$$

$$\le I^f(T_1, T_2, x_f, u_f) + M - I^f(T_1, Q_1, x_f, u_f) + S_0 - I^f(Q_2, T_2, x_f, u_f) + S_0$$

$$= I^f(Q_1, Q_2, x_f, u_f) + M + 2S_0. \tag{8.116}$$

It follows from (8.112) and (8.116) that

$$I^f(\max\{T_1, L_0\}, \max\{T_1, L_0\} + L_1, x, u)$$

$$\leq I^f(\max\{T_1, L_0\}, \max\{T_1, L_0\} + L_1, x_f, u_f) + 2S_0 + M, \quad (8.117)$$

$$I^f(T_2 - L_1, T_2, x, u) \leq I^f(T_2 - L_1, T_2, x_f, u_f) + 2S_0 + M. \quad (8.118)$$

By (8.114), (8.117), (8.118), and property (vi), there exist

$$\tau_1 \in [\max\{T_1, L_0\}, \max\{T_1, L_0\} + L_1], \ \tau_2 \in [T_2 - L_1, T_2] \quad (8.119)$$

such that

$$\rho_E(x(\tau_i), x_f(\tau_i)) \leq \delta_0, \ i = 1, 2. \quad (8.120)$$

If $\rho_E(x(T_2), x_f(T_2)) \leq \delta$, then we may assume that $\tau_2 = T_2$, and if $T_1 \geq L_0$ and $\rho_E(x(T_1), x_f(T_1)) \leq \delta$, then we may assume that $\tau_1 = T_1$. In view of (8.112) and (8.119),

$$\tau_2 - \tau_1 \geq T_2 - T_1 - L_1 - L_1 - L_0 \geq L_0 + 2b_f. \quad (8.121)$$

Property (iv), (8.114), (8.120), and (8.121) imply that

$$\rho_E(x(t), x_f(t)) \leq \epsilon, \ t \in [\tau_1, \tau_2].$$

Thus TP holds and Theorem 8.25 is proved.

8.14 Proof of Theorem 8.26

Assume that f has properties (P1) and (P2). Theorem 8.25 implies that f has TP. Let $\epsilon, M > 0$. By Theorem 8.5, there exists $S_0 > 0$ such that the following property holds:

(i) For each $\tau_2 > \tau_1 \geq 0$ and each $(y, v) \in X(\tau_1, \tau_2)$,

$$I^f(\tau_1, \tau_2, y, v) + S_0 \geq I^f(\tau_1, \tau_2, x_f, u_f).$$

TP implies that there exist $\delta \in (0, \epsilon)$ and $L_0 > 0$ such that the following property holds:

(ii) For each $\tau_1 \geq 0$, each $\tau_2 \geq \tau_1 + 2L_0$, and each $(y, v) \in X(\tau_1, \tau_2)$ that satisfy

$$I^f(\tau_1, \tau_2, y, v) \le \min\{U^f(\tau_1, \tau_2) + M + 3S_0, U^f(\tau_1, \tau_2, y(\tau_1), y(\tau_2)) + \delta\},$$

we have

$$\rho_E(y(t), x_f(t)) \le \epsilon,\ t \in [\tau_1 + L_0, \tau_2 - L_0].$$

Set

$$l = 2L_0 + 1. \tag{8.122}$$

Choose a natural number

$$Q \ge 2 + 2(M + S_0)\delta^{-1}. \tag{8.123}$$

Assume that $T_1 \ge 0$, $T_2 \ge T_1 + lQ$ and $(x, u) \in X(T_1, T_2)$ satisfies

$$I^f(T_1, T_2, x, u) \le I^f(T_1, T_2, x_f, u_f) + M. \tag{8.124}$$

Property (i) and (8.124) imply that for each pair of numbers $\tau_1, \tau_2 \in [T_1, T_2]$ satisfying $\tau_1 < \tau_2$,

$$\begin{aligned} I^f(\tau_1, \tau_2, x, u) &= I^f(T_1, T_2, x, u) - I^f(T_1, \tau_1, x, u) - I^f(\tau_2, T_2, x, u) \\ &\le M + I^f(T_1, T_2, x_f, u_f) - I^f(T_1, \tau_1, x_f, u_f) + S_0 - I^f(\tau_2, T_2, x_f, u_f) + S_0 \\ &\le I^f(\tau_1, \tau_2, x_f, u_f) + M + 2S_0. \end{aligned} \tag{8.125}$$

Set

$$t_0 = T_1. \tag{8.126}$$

If

$$I^f(T_1, T_2, x, u) \le U^f(T_1, T_2, x(T_1), x(T_2)) + \delta,$$

then we set

$$t_1 = T_2,\ s_1 = T_2. \tag{8.127}$$

Assume that

$$I^f(T_1, T_2, x, u) > U^f(T_1, T_2, x(T_1), x(T_2)) + \delta. \tag{8.128}$$

It is easy to see that for each $t \in (T_1, T_2)$ such that $t - T_1$ is sufficiently small, we have

$$I^f(T_1, t, x, u) - U^f(T_1, t, x(T_1), x(t)) < \delta.$$

Set

$$\tilde{t}_1 = \inf\{t \in (T_1, T_2] : \ I^f(T_1, t, x, u) - U^f(T_1, t, x(T_1), x(t)) > \delta\}.$$

Clearly, $\tilde{t}_1 \in (T_1, T_2]$ is well-defined. There exist $s_1, t_1 \in [T_1, T_2]$ such that

$$t_0 < s_1 < \tilde{t}_1, \ s_1 \geq \tilde{t}_1 - 1/4, \tag{8.129}$$

$$I^f(t_0, s_1, x, u) - U^f(t_0, s_1, x(t_0), x(s_1)) \leq \delta, \tag{8.130}$$

$$\tilde{t}_1 \leq t_1 < \tilde{t}_1 + 1/4, \ t_1 \leq T_2, \tag{8.131}$$

$$I^f(t_0, t_1, x, u) - U^f(t_0, t_1, x(t_0), x(t_1)) > \delta. \tag{8.132}$$

Assume that $k \geq 1$ is an integer, and we defined finite sequences of numbers $\{t_i\}_{i=0}^k \subset [T_1, T_2]$, $\{s_i\}_{i=1}^k \subset [T_1, T_2]$ such that

$$T_1 = t_0 < t_1 \cdots < t_k, \tag{8.133}$$

for each integer $i = 1, \ldots, k$,

$$t_{i-1} < s_i \leq t_i, \ t_i - s_i \leq 2^{-1}, \tag{8.134}$$

$$I^f(t_{i-1}, s_i, x, u) - U^f(t_{i-1}, s_i, x(t_{i-1}), x(s_i)) \leq \delta, \tag{8.135}$$

and if $t_i < T_2$, then

$$I^f(t_{i-1}, t_i, x, u) - U^f(t_{i-1}, t_i, x(t_{i-1}), x(t_i)) > \delta. \tag{8.136}$$

(Note that in view of (8.126), (8.127), (8.129), (8.132), our assumption holds for $k = 1$.)

By (8.136), there exists $(y, v) \in X(T_1, T_2)$ such that

$$y(t_i) = x(t_i), \ i = 1, \ldots, k,$$

$$I^f(t_{i-1}, t_i, x, u) - I^f(t_{i-1}, t_i, y, v) > \delta \text{ for all intergers } i \in [1, k]\setminus\{k\}, \tag{8.137}$$

$$(y(t), v(t)) = (x(t), u(t)), \ t \in [t_{k-1}, T_2]. \tag{8.138}$$

Property (i), (8.124), (8.137), and (8.138) imply that

$$I^f(T_1, T_2, x_f, u_f) + M \geq I^f(T_1, T_2, x, u) \geq I^f(T_1, T_2, y, v) + \delta(k-1)$$

$$\geq I^f(T_1, T_2, x_f, u_f) - S_0 + \delta(k-1),$$

$$k \leq 1 + \delta^{-1}(M + S_0). \tag{8.139}$$

If $t_k = T_2$, then the construction of the sequence is completed. Assume that $t_k < T_2$. If

$$I^f(t_k, T_2, x, u) \leq U^f(t_k, T_2, x(t_k), x(T_2)) + \delta,$$

then we set $t_{k+1} = T_2$, $s_{k+1} = T_2$, and the construction of the sequence is completed. Assume that

$$I^f(t_k, T_2, x, u) > U^f(t_k, T_2, x(t_k), x(T_2)) + \delta.$$

It is easy to see that for each $t \in (t_k, T_2)$ such that $t - t_k$ is sufficiently small, we have

$$I^f(t_k, t, x, u) - U^f(t_k, t, x(t_k), x(t)) < \delta.$$

Set

$$\tilde{t} = \inf\{t \in (t_k, T_2] : \ I^f(t_k, t, x, u) - U^f(t_k, t, x(t_k), x(t)) > \delta\}.$$

Clearly, $\tilde{t}$ is well-defined and $\tilde{t} \in (t_k, T_2]$. There exist $s_{k+1}, t_{k+1} \in [T_1, T_2]$ such that

$$t_k < s_{k+1} < \tilde{t}, \ s_{k+1} \geq \tilde{t} - 1/4, \ \tilde{t} + 4^{-1} > t_{k+1} \geq \tilde{t},$$

$$I^f(t_k, s_{k+1}, x, u) - U^f(t_k, s_{k+1}, x(t_k), x(s_{k+1})) \leq \delta,$$

$$I^f(t_k, t_{k+1}, x, u) - U^f(t_k, t_{k+1}, x(t_k), x(t_{k+1})) > \delta.$$

It is not difficult to see that the assumption made for k also holds for $k+1$. In view of (8.139), by induction, we constructed finite sequences $\{t_i\}_{i=0}^q \subset [T_1, T_2]$, $\{s_i\}_{i=1}^q \subset [T_1, T_2]$ such that

$$q \leq 1 + \delta^{-1}(M + S_0), \tag{8.140}$$

$$T_1 = t_0 < t_1 \cdots < t_q = T_2,$$

for each integer $i = 1, \dots, q$,

$$t_{i-1} < s_i \leq t_i, \ t_i - s_i \leq 2^{-1}, \tag{8.141}$$

$$I^f(t_{i-1}, s_i, x, u) - U^f(t_{i-1}, s_i, x(t_{i-1}), x(s_i)) \le \delta, \tag{8.142}$$

and if $t_i < T_2$, then

$$I^f(t_{i-1}, t_i, x, u) - U^f(t_{i-1}, t_i, x(t_{i-1}), x(t_i)) > \delta. \tag{8.143}$$

Assume that

$$i \in \{0, \dots, q-1\},\ t_{i+1} - t_i \ge 2L_0 + 1. \tag{8.144}$$

By (8.141) and (8.144),

$$s_{i+1} - t_i \ge 2L_0. \tag{8.145}$$

In view of (8.142),

$$I^f(t_i, s_{i+1}, x, u) - U^f(t_i, s_{i+1}, x(t_i), x(s_{i+1})) \le \delta. \tag{8.146}$$

Property (i) and (8.124) imply that

$$I^f(t_i, s_{i+1}, x, u) = I^f(T_1, T_2, x, u) - I^f(T_1, t_i, x, u) - I^f(s_{i+1}, T_2, x, u)$$

$$\le M + I^f(T_1, T_2, x_f, u_f) - I^f(T_1, t_i, x_f, u_f) + S_0 - I^f(s_{i+1}, T_2, x_f, u_f) + S_0$$

$$\le I^f(t_i, s_{i+1}, x_f, u_f) + M + 2S_0 \le M + 3S_0 + U^f(t_i, s_{i+1}). \tag{8.147}$$

It follows from (8.145)–(8.147) and property (ii) that

$$\rho_E(x(t), x_f(t)) \le \epsilon,\ t \in [t_i + L_0, s_{i+1} - L_0],$$

and together with (8.141), this implies that

$$\rho_E(x(t), x_f(t)) \le \epsilon,\ t \in [t_i + L_0, t_{i+1} - L_0 - 1]. \tag{8.148}$$

By (8.148),

$$\{t \in [T_1, T_2] :\ \rho_E(x(t), x_f(t)) > \epsilon\}$$

$$\subset \cup\{[t_i, t_i + L_0] \cup [t_{i+1} - 1 - L_0, t_{i+1}] :\ i \in \{0, \dots, q-1\} \text{ and } t_{i+1} - t_i \ge 2L_0 + 1\}$$

$$\cup \{[t_i, t_{i+1}] :\ i \in \{0, \dots, q-1\} \text{ and } t_{i+1} - t_i < 2L_0 + 1\}.$$

In view of (8.123) and (8.140), the maximal length of intervals in the right-hand side of the relation above does not exceed $2L_0 + 1 = l$, and their number does not exceed $2q \le Q$. Thus WTP holds. It is not difficult to see that WTP implies (P1) Theorem 8.26 is proved.

8.15 An Auxiliary Result

Lemma 8.37 *Let $\epsilon > 0$. Then there exist $\delta, L > 0$ such that for each $T_1 \geq L$, each $T_2 > T_1$, and each $(x, u) \in X(T_1, T_2)$ that satisfy*

$$\rho_E(x(T_i), x_f(T_i)) \leq \delta, \ i = 1, 2,$$

the following inequality holds:

$$I^f(T_1, T_2, T, x, u) \geq I^f(T_1, T_2, T, x_f, u_f) - \epsilon.$$

Proof By (A2), there exists $\delta > 0$ such that the following property holds:

(i) For each $(T, z) \in \mathcal{A}$ satisfying $\rho_E(z, x_f(T)) \leq \delta$, there exist $\tau_1 \in (0, b_f]$, $(\tilde{x}_1, \tilde{u}_1) \in X(T, T + \tau_1)$ such that

$$\tilde{x}_1(T) = z \ , \tilde{x}_1(T + \tau_1) = x_f(T + \tau_1),$$

$$I^f(T, T + \tau_1, \tilde{x}_1, \tilde{u}_1) \leq I^f(T, T + \tau_1, x_f, u_f) + \epsilon/4,$$

and if $T \geq b_f$, then there exist $\tau_2 \in (0, b_f]$ and $(\tilde{x}_2, \tilde{u}_2) \in X(T - \tau_2, T)$ such that

$$\tilde{x}_2(T - \tau_2) = x_f(T - \tau_2) \ , \tilde{x}_2(T) = z,$$

$$I^f(T - \tau_2, T, \tilde{x}_2, \tilde{u}_2) \leq I^f(T - \tau_2, T, x_f, u_f) + \epsilon/4.$$

Theorem 8.5 implies that there exists $L > 2b_f$ such that for each $T_1 \geq L - b_f$ and each $T_2 > T_1$

$$I^f(T_1, T_2, x_f, u_f) < U^f(T_1, T_2, x_f(T_1), x_f(T_2)) + \epsilon/4. \tag{8.149}$$

Assume that $T_1 \geq L$, $T_2 > T_1$, $(x, u) \in X(T_1, T_2)$, and

$$\rho_E(x(T_1), x_f(T_1)) \leq \delta, \ \ \rho_E(x(T_2), x_f(T_2)) \leq \delta. \tag{8.150}$$

By (8.150) and property (i), there exist $\tau_1, \tau_2 \in (0, b_f]$ and $(y, v) \in X(T_1 - \tau_1, T_2 + \tau_2)$ such that

$$y(T_1 - \tau_1) = x_f(T_1 - \tau_1), \tag{8.151}$$

$$(y(t), v(t)) = (x(t), u(t)) \text{ for all } t \in [T_1, T_2], \tag{8.152}$$

$$I^f(T_1 - \tau_1, T_1, y, v) \leq I^f(T_1 - \tau_1, T_1, x_f, u_f) + \epsilon/4, \tag{8.153}$$

$$y(T_2 + \tau_2) = x_f(T_2 + \tau_2), \tag{8.154}$$

$$I^f(T_2, T_2 + \tau_2, y, v) \le I^f(T_2, T_2 + \tau_2, x_f, u_f) + \epsilon/4. \tag{8.155}$$

It follows from (8.149), (8.151), and (8.154) that

$$I^f(T_1 - \tau_1, T_2 + \tau_2, x_f, u_f) < I^f(T_1 - \tau_1, T_2 + \tau_2, y, v) + \epsilon/4. \tag{8.156}$$

Relations (8.152), (8.153), (8.155), and (8.156) imply that

$$I^f(T_1, T_2, x, u) = I^f(T_1 - \tau_1, T_2 + \tau_2, y, v)$$

$$- I^f(T_1 - \tau_1, T_1, y, v) - I^f(T_2, T_2 + \tau_2, y, v)$$

$$\ge I^f(T_1 - \tau_1, T_2 + \tau_2, x_f, u_f) - \epsilon/4 - I^f(T_1 - \tau_1, T_1, x_f, u_f)$$

$$- \epsilon/4 - I^f(T_2, T_2 + \tau_2, x_f, u_f) - \epsilon/4$$

$$= I^f(T_1, T_2, x_f, u_f) - 3\epsilon/4.$$

Lemma 8.37 is proved.

8.16 Proof of Proposition 8.28

Proposition 7.1 implies that there exist $M_* \ge 1$, $\omega_* \in R^1$ such that

$$\|e^{At}\| \le M_* e^{\omega_* t}, \ t \in [0, \infty),$$

in the case of the second class of problems.

We show that (P1) holds.

Assume that $(x, u) \in X(0, \infty)$ is (f)-good. We show that

$$\lim_{t \to \infty} \rho_E(x(t), x_f(t)) = 0.$$

Assume the contrary. Then there exist $\epsilon > 0$ and a sequence of numbers $\{t_k\}_{k=1}^{\infty}$ such that

$$t_1 \ge 8, \ t_{k+1} \ge t_k + 8 \text{ for all integers } k \ge 1, \tag{8.157}$$

$$\rho_E(x(t_k), x_f(t_k)) > \epsilon \text{ for all integers } k \ge 1. \tag{8.158}$$

Theorem 8.5 implies that there exists $M_1 > 0$ such that

$$\rho_E(x_f(t), \theta_0) \le M_1, \ \ \rho_E(x(t), \theta_0) \le M_1, \ \ t \in [0, \infty), \tag{8.159}$$

$$|I^f(0, t, x, u) - I^f(0, t, x_f, u_f)| < M_1, \ \ t \in (0, \infty).$$

By the relation above, for all $S > T \ge 0$,

$$|I^f(T, S, x, u) - I^f(T, S, x_f, u_f)| \le 2M_1. \tag{8.160}$$

By (8.16), there exists $M_2 > 0$ such that

$$I^f(S_1, S_2, x_f, u_f) \le M_2 \text{ for all } S_1 \ge 0 \text{ and all } S_2 \in (S_1, S_1 + 1]. \tag{8.161}$$

From now, we consider the three classes of problems separately and first consider the first class of problems. Recall that in this case $\theta_0 = 0$, $\theta_1 = 0$, and $\rho_E(z_1, z_2) = \|z_1 - z_2\|$, $z_1, z_2 \in E$. Fix

$$\epsilon_1 = 4^{-1}\epsilon(2\Delta_1 + 1)^{-1}. \tag{8.162}$$

By (8.5) and (8.20), there exists $h_0 > 0$ such that

$$f(t, x, u) \ge 4\epsilon_1^{-1}(M_1 + M_2 + a_0 + 8)(\|G(t, x, u)\| - a_0\|x\|_+) \tag{8.163}$$

for each $(t, x, u) \in \mathcal{M}$ satisfying $\|G(t, x, u)\| - a_0\|x\| \ge h_0$.

There exists $\delta \in (0, 1)$ such that

$$\delta < \epsilon/8, \ \ \Delta_1\delta < \epsilon_1/32, \ \ \delta(a_0 M_1 + h_0) < \epsilon_1/4. \tag{8.164}$$

Let $k \ge 1$ be an integer. We show that for all $t \in [t_k - \delta, t_k]$,

$$\|x(t) - x_f(t)\| \ge \delta. \tag{8.165}$$

Assume the contrary. Then there exists

$$\tau \in [t_k - \delta, t_k] \tag{8.166}$$

such that

$$\|x(\tau) - x_f(\tau)\| < \delta. \tag{8.167}$$

By (8.18) and (8.166),

$$x(t_k) = U(t_k, \tau)x(\tau) + \int_{\tau}^{t_k} U(t_k, s)G(s, x(s), u(s))ds, \tag{8.168}$$

$$x_f(t_k) = U(t_k, \tau)x_f(\tau) + \int_{\tau}^{t_k} U(t_k, s)G(s, x_f(s), u_f(s))ds. \tag{8.169}$$

Property (v) (from Sect. 8.2), (8.166), (8.167), (8.169) imply that

$$\begin{aligned}
\|x(t_k) - x_f(t_k)\| &= \|U(t_k, \tau)\| \|x_f(\tau) - x(\tau)\| \\
&+ \int_{\tau}^{t_k} \|U(t_k, s)\| \|G(s, x(s), u(s))\| ds \\
&+ \int_{\tau}^{t_k} \|U(t_k, s)\| \|G(s, x_f(s), u_f(s))\| ds \\
&\le \Delta_\delta \|x_f(\tau) - x(\tau)\| + \Delta_\delta \int_{\tau}^{t_k} \|G(s, x(s), u(s))\| ds \\
&+ \Delta_\delta \int_{\tau}^{t_k} \|G(s, x_f(s), u_f(s))\| ds \\
&\le \Delta_1 \delta + \Delta_\delta \int_{\tau}^{t_k} \|G(s, x(s), u(s))\| ds + \Delta_\delta \int_{\tau}^{t_k} \|G(s, x_f(s), u_f(s))\| ds.
\end{aligned} \tag{8.170}$$

Let

$$(y, v) \in \{(x, u),\ (x_f, u_f)\}. \tag{8.171}$$

Set

$$\Omega_1 = \{t \in [\tau, t_k] :\ \|G(t, y(t), v(t))\| - a_0\|y(t)\| \ge h_0\},\ \Omega_2 = [\tau, t_k] \setminus \Omega_1. \tag{8.172}$$

By (8.159)–(8.161), (8.163), (8.164), (8.166), (8.171), and (8.172),

$$\begin{aligned}
\int_{\tau}^{t_k} \|G(s, y(s), v(s))\| ds &\le a_0 \int_{\tau}^{t_k} \|y(s)\| ds \\
&+ \int_{\tau}^{t_k} (\|G(s, y(s), v(s))\| - a_0\|y(s)\|)_+ ds \\
&\le a_0 \delta M_1 + h_0 \delta + \int_{\Omega_1} (\|G(s, y(s), v(s))\| - a_0\|y(s)\|)_+ ds
\end{aligned}$$

$$\leq a_0\delta M_1 + h_0\delta + 4^{-1}\epsilon_1(M_1 + M_2 + a_0 + 8)^{-1}\int_{\Omega_1} f(s, y(s), v(s))ds$$

$$\leq a_0\delta M_1 + h_0\delta + (4^{-1}\epsilon_1(M_1 + M_2 + a_0 + 8)^{-1})(I^f(\tau, t_k, y, v) + a_0\delta)$$

$$\leq \delta(a_0 M_1 + h_0) + (4^{-1}\epsilon_1(M_1 + M_2 + a_0 + 8)^{-1})(I^f(\tau, t_k, x_f, u_f) + 2M_1 + a_0\delta)$$

$$\leq \delta(a_0 M_1 + h_0) + (4^{-1}\epsilon_1(M_1 + M_2 + a_0 + 8)^{-1})(M_2 + 2M_1 + a_0\delta) < \epsilon_1/4 + \epsilon_1/2.$$

In view of (8.158), (8.164), (8.170) and the equation above,

$$\epsilon \leq \|x(t_k) - x_f(t_k)\| \leq \Delta_1\delta + 2\Delta_\delta(\epsilon_1/4 + \epsilon_1/2)$$

$$\leq \epsilon_1/32 + 2\Delta_1\epsilon_1 \leq \epsilon_1(2\Delta_1 + 1) = 4^{-1}\epsilon,$$

a contradiction. The contradiction we have reached proves that (8.165) holds for all $t \in [t_k - \delta, t_k]$. Together with (8.157) and (8.158), this implies that

$$\text{mes}(\{t \in [0, T] : \|x(t) - x_f(t)\| \geq \delta\}) \to \infty \text{ as } T \to \infty.$$

On the other hand, property (P3) and (8.160) imply that there exists $L > 0$ such that for all $T > 0$,

$$\text{mes}(\{t \in [0, T] : \|x(t) - x_f(t)\| \geq \delta\}) \leq L.$$

The contradiction we have reached proves that $\lim_{t\to\infty}(x(t) - x_f(t)) = 0$, and (P1) holds for the first class of problems.

Consider the second class of problems. Recall (see Proposition 7.16 and (7.9)) that for each $\tau \geq 0$, $\Phi_\tau \in \mathcal{L}(L^2(0, \tau; , F), E)$ is defined by

$$\Phi_\tau u = \int_0^\tau e^{A(\tau - s)}Bu(s)ds, \; u \in L^2(0, \tau; F). \tag{8.173}$$

Lemma 8.37 and (A2) and (A3) imply that there exist $\delta \in (0, 1)$, $\bar{L} > 0$ such that

$$M_* e^{|\omega_*|}\delta < \epsilon/8, \; \delta < b_f/16, \; 64a_0\delta < K_0\epsilon^2(\|\Phi_1\| + 1)^{-2} \tag{8.174}$$

and the following properties hold:

(i) For each $(T_i, z_i) \in \mathcal{A}$, $i = 1, 2$, satisfying $\|z_i - x_f(T_i)\| \leq \delta$, $i = 1, 2$ and $T_2 \geq b_f$, there exist $\tau_1, \tau_2 \in (0, b_f]$ and $(x_1, u_1) \in X(T_1, T_1 + \tau_1)$, $(x_2, u_2) \in X(T_2 - \tau_2, T_2)$ satisfying

$$x_1(T_1) = z_1 \;, x_1(T_1+\tau_1) = x_f(T_1+\tau_1), \;\; x_2(T_2) = z_2 \;, x_2(T_2-\tau_2) = x_f(T_2-\tau_2),$$

$$I^f(T_1, T_1+\tau_1, x_1, u_1) \le I^f(T_1, T_1+\tau_1, x_f, u_f) + 32^{-1}\epsilon^2 K_0(\|\Phi_1\|+1)^{-2},$$

$$I^f(T_2-\tau_2, T_2, x_2, u_2) \le I^f(T_2-\tau_2, T_2, x_f, u_f) + 32^{-1}\epsilon^2 K_0(\|\Phi_1\|+1)^{-2}.$$

(ii) For each $t_1 \ge \bar{L}$, each $t_2 \in (t_1, t_1+4\delta]$,

$$I^f(t_1, t_2, x_f, u_f) \le 32^{-1}\epsilon^2 K_0(\|\Phi_1\|+1)^{-2}.$$

(iii) For each $S_1 \ge \bar{L}$, each $S_2 > S_1$, and each $(\xi, \eta) \in X(S_1, S_2)$ that satisfy $\|\xi(S_i) - x_f(S_i)\| \le \delta$, $i = 1, 2$, we have

$$I^f(S_1, S_2, \xi, \eta) \ge I^f(S_1, S_2, x_f, u_f) - 32^{-1}\epsilon^2 K_0(\|\Phi_1\|+1)^{-2}.$$

Property (P3) and (8.160) imply that there exists $L > 0$ such that for all $T > 0$,

$$\mathrm{mes}(\{t \in [0, T] : \; \|x(t) - x_f(t)\| \ge \delta\}) \le L.$$

Therefore, there exists $T_0 > 0$ such that for each $S > T_0$,

$$\mathrm{mes}(\{t \in [T_0, S] : \; \|x(t) - x_f(t)\| > \delta\}) \le \delta/2. \tag{8.175}$$

Since the pair (x, u) is (f)-good, it follows from Theorem 8.6 that there exists $T_1 > 0$ such that for each $S > T_1$,

$$I^f(T_1, S, x, u) \le U^f(T_1, S, x(T_1), x(S)) + 32^{-1}\epsilon^2 K_0(\|\Phi_1\|+1)^{-2}, \tag{8.176}$$

$$I^f(T_1, S, x_f, u_f) \le U^f(T_1, S, x_f(T_1), x_f(S)) + 32^{-1}\epsilon^2 K_0(\|\Phi_1\|+1)^{-2}. \tag{8.177}$$

Let $k \ge 1$ be an integer such that

$$t_k > T_0 + T_1 + 1 + \bar{L}. \tag{8.178}$$

Assume that

$$\inf\{\|x(t) - x_f(t)\| : \; t \in [t_k - \delta, t_k]\} < \delta. \tag{8.179}$$

Therefore, there exists

$$\tau \in [t_k - \delta, t_k] \tag{8.180}$$

such that

$$\|x(\tau) - x_f(\tau)\| < \delta. \tag{8.181}$$

It is clear that $(x, u), (x_f, u_f) \in X(\tau, t_k)$. In view of (8.24), (8.173), and (8.180),

$$x(t_k) = e^{A(t_k-\tau)}x(\tau) + \int_\tau^{t_k} e^{A(t_k-s)}Bu(s)ds = e^{A(t_k-\tau)}x(\tau) + \Phi_{t_k-\tau}u(\tau + \cdot), \tag{8.182}$$

$$x_f(t_k) = e^{A(t_k-\tau)}x_f(\tau) + \int_\tau^{t_k} e^{A(t_k-s)}Bu_f(s)ds$$

$$= e^{A(t_k-\tau)}x_f(\tau) + \Phi_{t_k-\tau}u_f(\tau + \cdot). \tag{8.183}$$

Proposition 7.16, (8.158), (8.173), (8.174), (8.180)–(8.183) imply that

$$\epsilon \le \|x(t_k) - x_f(t_k)\|$$

$$= \|e^{A(t_k-\tau)}(x(\tau) - x_f(\tau))\| + \|\Phi_{t_k-\tau}u(\tau + \cdot)\| + \|\Phi_{t_k-\tau}u_f(\tau + \cdot)\|$$

$$\le M_* e^{|\omega_*|\delta}\|x(\tau) - x_f(\tau)\|$$

$$+ \|\Phi_1\|(\int_\tau^{t_k} \|u(s)\|^2 ds)^{1/2} + \|\Phi_1\|(\int_\tau^{t_k} \|u_f(s)\|^2 ds)^{1/2}$$

$$< \epsilon/8 + \|\Phi_1\|(\int_\tau^{t_k} \|u(s)\|^2 ds)^{1/2} + \|\Phi_1\|(\int_\tau^{t_k} \|u_f(s)\|^2 ds)^{1/2}$$

and

$$(4^{-1}\epsilon(\|\Phi_1\| + 1)^{-1})^2 \le \max\{\int_\tau^{t_k} \|u(s)\|^2 ds, \int_\tau^{t_k} \|u_f(s)\|^2 ds\}. \tag{8.184}$$

Let

$$(y, v) \in \{(x, u), (x_f, u_f)\}, \tag{8.185}$$

$$\int_\tau^{t_k} \|v(s)\|^2 ds \ge \max\{\int_\tau^{t_k} \|u(s)\|^2 ds, \int_\tau^{t_k} \|u_f(s)\|^2 ds\}. \tag{8.186}$$

In view of (8.184)–(8.186),

$$\int_\tau^{t_k} \|v(s)\|^2 ds \ge (4^{-1}\epsilon)^2(\|\Phi_1\| + 1)^{-2}. \tag{8.187}$$

By (8.26) and (8.196),

$$\int_{\tau}^{t_k} f(s, y(s), v(s))ds + a_0(t_k - \tau) \geq K_0 \int_{\tau}^{t_k} \|v(s)\|^2 ds$$
$$\geq 4^{-1}\epsilon^2 K_0(\|\Phi_1\| + 1)^{-2}. \quad (8.188)$$

It follows from (8.180) and (8.188) that

$$\int_{\tau}^{t_k} f(s, y(s), v(s))ds \geq 4^{-1} K_0 \epsilon^2 (\|\Phi_1\| + 1)^{-2} - a_0\delta. \quad (8.189)$$

In view of (8.175) and (8.178), for each $S > t_k$,

$$\text{mes}(\{t \in [t_k, S] : \ \|y(t) - x_f(t)\| > \delta\}) \leq \delta/2. \quad (8.190)$$

Inequality (8.190) implies that there exist

$$S_1 \in [t_k, t_k + \delta] \quad (8.191)$$

such that

$$\|y(S_1) - x_f(S_1)\| \leq \delta \quad (8.192)$$

and

$$S_2 \in [S_1 + 2b_f, S_1 + 2b_f + \delta] \quad (8.193)$$

such that

$$\|y(S_2) - x_f(S_2)\| \leq \delta. \quad (8.194)$$

It follows from (8.180), (8.189), and (8.191) that

$$I^f(\tau, S_1, y, v) = I^f(\tau, t_k, y, v) + I^f(t_k, S_1, y, v)$$
$$\geq 4^{-1} K_0 \epsilon^2 (\|\Phi_1\| + 1)^{-2} - a_0\delta - a_0\delta. \quad (8.195)$$

Property (iii) and (8.178) and (8.191)–(8.194) imply that

$$I^f(S_1, S_2, y, v) \geq I^f(S_1, S_2, x_f, u_f) - 32^{-1} K_0 \epsilon^2 (\|\Phi_1\| + 1)^{-2}. \quad (8.196)$$

By (8.178)–(8.180), (8.191), (8.195), (8.196), and property (ii),

$$I^f(\tau, S_2, y, v) = I^f(\tau, S_1, y, v) + I^f(S_1, S_2, y, v)$$

$$\geq 4^{-1}K_0\epsilon^2(\|\Phi_1\|+1)^{-2} - 2a_0\delta$$

$$+ I^f(S_1, S_2, x_f, u_f) - 32^{-1}K_0\epsilon^2(\|\Phi_1\|+1)^{-2}$$

$$\geq K_0\epsilon^2(\|\Phi_1\|+1)^{-2}(4^{-1} - 32^{-1})$$

$$- 2a_0\delta + I^f(\tau, S_2, x_f, u_f) - I^f(\tau, S_1, x_f, u_f)$$

$$\geq I^f(\tau, S_2, x_f, u_f) + K_0\epsilon^2(\|\Phi_1\|+1)^{-2}(4^{-1} - 32^{-1})$$

$$- 2a_0\delta - 16^{-1}K_0\epsilon^2(\|\Phi_1\|+1)^{-2}. \tag{8.197}$$

Property (i), (8.180), (8.181), and (8.191)–(8.194) imply that there exists $(\tilde{x}, \tilde{u}) \in X(\tau, S_2)$ such that

$$\tilde{x}(\tau) = y(\tau),\ \tilde{x}(\tau + b_f) = x_f(\tau + b_f), \tag{8.198}$$

$$I^f(\tau, \tau+b_f, \tilde{x}, \tilde{u}) \leq I^f(\tau, \tau+b_f, x_f, u_f) + 32^{-1}\epsilon^2 K_0(\|\Phi_1\|+1)^{-2}, \tag{8.199}$$

$$\tilde{x}(S_2) = y(S_2),\ \tilde{x}(S_2 - b_f) = x_f(S_2 - b_f), \tag{8.200}$$

$$I^f(S_2 - b_f, S_2, \tilde{x}, \tilde{u}) \leq I^f(S_2 - b_f, S_2, x_f, u_f) + 32^{-1}\epsilon^2 K_0(\|\Phi_1\|+1)^{-2}, \tag{8.201}$$

$$(\tilde{x}(t), \tilde{u}(t)) = (x_f(t), u_f(t)),\ t \in [\tau + b_f, S_2 - b_f]. \tag{8.202}$$

In view of (8.199), (8.201), and (8.202),

$$I^f(\tau, S_2, \tilde{x}, \tilde{u}) \leq I^f(\tau, S_2, x_f, u_f) + 16^{-1}\epsilon^2 K_0(\|\Phi_1\|+1)^{-2}. \tag{8.203}$$

It follows from (8.174), (8.197), and (8.203) that

$$I^f(\tau, S_2, y, v) - I^f(\tau, S_2, \tilde{x}, \tilde{u})$$

$$\geq \epsilon^2 K_0(\|\Phi_1\|+1)^{-2}(4^{-1} - 32^{-1} - 8^{-1}) - 2\delta a_0$$

$$\geq \epsilon^2 K_0(\|\Phi_1\|+1)^{-2}16^{-1}. \tag{8.204}$$

By (8.198), (8.200), and (8.204),

$$I^f(\tau, S_2, y, v) \geq U^f(\tau, S_2, y(\tau), y(S_2)) + \epsilon^2 K_0(\|\Phi_1\|+1)^{-2}/16. \tag{8.205}$$

It follows from (8.176)–(8.178), (8.180), and (8.125) that $\tau > T_1$ and

$$I^f(\tau, S_2, y, v) \leq U^f(\tau, S_2, y(\tau), y(S_2)) + 32^{-1}\epsilon^2 K_0(\|\Phi_1\|+1)^{-2}.$$

This contradicts (8.205). The contradiction we have reached proves that (8.179) is not true and

$$\|x(t) - x_f(t)\| \geq \delta, \ t \in [t_k - \delta, t_k]$$

for each natural number k satisfying $t_k \geq T_0 + T_1 + 1 + \bar{L}$. Together with (8.157), this implies that

$$\text{mes}(\{t \in [0, T] : \ \|x(t) - x_f(t)\| \geq \delta\}) \to \infty \text{ as } T \to \infty.$$

This contradicts (8.175) holding for all $S > T_0$. The contradiction we have reached proves that $\lim_{t\to\infty}(x(t) - x_f(t) = 0$ and (P1) holds for the second class of problems too.

Consider the third class of problems. Fix

$$\epsilon_1 \in (0, 4^{-1}\epsilon). \tag{8.206}$$

By (8.3), (8.5), and (8.56), there exists $h_0 > 0$ such that

$$f(t, x, u) \geq 4\epsilon_1^{-1}(M_1 + M_2 + a_0 + 8)(G(t, x, u) - a_0\rho_E(x, \theta_0)_+) \tag{8.207}$$

for each $(t, x, u) \in \mathcal{M}$ satisfying $G(t, x, u) - a_0\rho_E(x, \theta_0) \geq h_0$.

There exists $\delta \in (0, 1)$ such that

$$\delta < \epsilon/8, \ \delta(a_0 M_1 + h_0) < \epsilon_1/4. \tag{8.208}$$

Let $k \geq 1$ be an integer. We show that for all $t \in [t_k - \delta, t_k]$,

$$\rho_E(|x(t), x_f(t)) \geq \delta. \tag{8.209}$$

Assume the contrary. Then there exists

$$\tau \in [t_k - \delta, t_k] \tag{8.210}$$

such that

$$\rho_E(x(\tau), x_f(\tau)) < \delta. \tag{8.211}$$

By (8.55) and (8.210),

$$\rho_E(x(t_k), x(\tau)) \leq \int_\tau^{t_k} G(s, x(s), u(s))ds, \tag{8.212}$$

$$\rho_E(x_f(t_k), x_f(\tau)) \leq \int_\tau^{t_k} G(s, x_f(s), u_f(s))ds. \tag{8.213}$$

Let

$$(y, v) \in \{(x, u),\ (x_f, u_f)\}. \tag{8.214}$$

Set

$$\Omega_1 = \{t \in [\tau, t_k] :\ G(t, y(t), v(t)) - a_0\rho_E(y(t), \theta_0) \ge h_0\},\ \Omega_2 = [\tau, t_k] \setminus \Omega_1. \tag{8.215}$$

By (8.5), (8.159), (8.161), (8.207), (8.208), (8.210), (8.214), and (8.215),

$$\int_\tau^{t_k} G(s, y(s), v(s))ds \le a_0 \int_\tau^{t_k} \rho_E(y(s), \theta_0)ds$$

$$+ \int_\tau^{t_k} (G(s, y(s), v(s)) - a_0\rho_E y(s), \theta_0))_+ ds$$

$$\le a_0\delta M_1 + h_0\delta + \int_{\Omega_1} (G(s, y(s), v(s)) - a_0\rho_E y(s), \theta_0))_+ ds$$

$$\le a_0\delta M_1 + h_0\delta + 4^{-1}\epsilon_1(M_1 + M_2 + a_0 + 8)^{-1} \int_{\Omega_1} f(s, y(s), v(s))ds$$

$$\le a_0\delta M_1 + h_0\delta + (4^{-1}\epsilon_1(M_1 + M_2 + a_0 + 8)^{-1})(I^f(\tau, t_k, y, v) + a_0\delta)$$

$$\le \delta(a_0M_1 + h_0) + (4^{-1}\epsilon_1(M_1 + M_2 + a_0 + 8)^{-1})(I^f(\tau, t_k, x_f, u_f) + 2M_1 + a_0\delta)$$

$$\le \delta(a_0M_1 + h_0) + (4^{-1}\epsilon_1(M_1 + M_2 + a_0 + 8)^{-1})(M_2 + 2M_1 + a_0\delta)$$
$$< \epsilon_1/4 + \epsilon_1/2. \tag{8.216}$$

In view of (8.158), (8.206), (8.208), (8.211)–(8.214), and (8.216),

$$\epsilon \le \rho_E(x(t_k), x_f(t_k))$$

$$\le \rho_E(x(t_k), x(\tau)) + \rho_E(x(\tau), x_f(\tau)) + \rho_E(x_f(\tau), x(t_k))$$

$$\le \delta + 3\epsilon_1/2 < \epsilon,$$

a contradiction. The contradiction we have reached proves that (8.209) holds for all $t \in [t_k - \delta, t_k]$. Together with (8.157), this implies that

$$\text{mes}(\{t \in [0, T] :\ \rho_E(x(t_k), x_f(t_k)) \ge \delta\}) \to \infty \text{ as } T \to \infty.$$

On the other hand, property (P3) and (8.160) imply that there exists $L > 0$ such that for all $T > 0$,

$$\text{mes}(\{t \in [0, T] : \ \rho_E(x(t_k), x_f(t_k)) \geq \delta\}) \leq L.$$

The contradiction we have reached proves that $\lim_{t\to\infty} \rho_E(x(t), x_f(t)) = 0$ and (P1) holds for the third class of problems. Proposition 8.28 is proved.

8.17 Auxiliary Results for Theorem 8.29

Recall that for each $z \in R^1$, $\lfloor z \rfloor = \max\{i \in R^1 : \ i \text{ is an integer}, i \leq z\}$.

Proposition 8.38 *Assume that f has (P1), (P2) and LSC property, $T_0 \geq 0$, $z_0 \in E$, $(T_0, z_0) \in \cup\{\mathcal{A}_L : \ L \in (0, \infty)\}$. Then there exists an (f)-good and (f)-minimal pair $(x_*, u_*) \in X(T_0, \infty)$ such that $x_*(T_0) = z_0$.*

Proof There exists $L_0 > 0$ such that

$$(T_0, z_0) \in \mathcal{A}_{L_0}. \tag{8.217}$$

It follows from Theorem 8.5 that there exists $S_0 > 0$ such that for each $T_2 > T_1 \geq 0$ and each $(x, u) \in X(T_1, T_2)$,

$$I^f(T_1, T_2, x, u) + S_0 \geq I^f(T_1, T_2, x_f, u_f). \tag{8.218}$$

Fix an integer $k_0 \geq L_0$. LSC property and (8.217) imply that for each integer $k \geq k_0$ there exists $(x_k, u_k) \in X(T_0, T_0 + k)$ satisfying

$$x_k(T_0) = z_0, \tag{8.219}$$

$$I^f(T_0, T_0 + k, x_k, u_k) = U^f(T_0, T_0 + k, z_0). \tag{8.220}$$

In view of (8.4), (8.49), and (8.217), for each integer $k \geq k_0$,

$$U^f(T_0, T_0 + k, z_0) \leq L_0 + I^f(T_0, T_0 + k, x_f, u_f) + a_0 L_0. \tag{8.221}$$

By (8.218), (8.220), and (8.221), for each integer $k \geq k_0$ and each pair of numbers $T_1, T_2 \in [T_0, T_0 + k]$ satisfying $T_1 < T_2$,

$$I^f(T_1, T_2, x_k, u_k) = I^f(T_0, T_0 + k, x_k, u_k)$$

$$- I^f(T_0, T_1, x_k, u_k) - I^f(T_0 + k, T_2, x_k, u_k)$$

$$\leq I^f(T_0, T_0 + k, x_f, u_f) + L_0(1 + a_0) - I^f(T_0, T_1, x_f, u_f)$$

$$+ S_0 - I^f(T_0 + k, T_2, x_k, u_k) + S_0$$

$$= I^f(T_1, T_2, x_f, u_f) + 2S_0 + L_0(1 + a_0). \qquad (8.222)$$

By (8.222) and LSC property, extracting subsequences, using the diagonalization process, and re-indexing, we obtain that there exists a strictly increasing sequence of natural numbers $\{k_p\}_{p=1}^{\infty}$ such that $k_1 \geq k_0$, and for each integer $i \geq 0$, there exists $\lim_{p\to\infty} I^f(T_0+i, T_0+i+1, x_{k_p}, u_{k_p})$ and there exists $(y_i, v_i) \in X(T_0+i, T_0+i+1)$ such that

$$x_{k_p}(t) \to y_i(t) \text{ as } p \to \infty \text{ for all } t \in [T_0 + i, T_0 + i + 1], \qquad (8.223)$$

$$I^f(T_0 + i, T_0 + i + 1, y_i, v_i) \leq \lim_{p\to\infty} I^f(T_0 + i, T_0 + i + 1, x_{k_p}, u_{k_p}). \qquad (8.224)$$

In view of (8.223), there exists $(x_*, u_*) \in X(T_0, \infty)$ such that for each integer $i \geq 0$,

$$(x_*(t), u_*(t)) = (y_i(t), v_i(t)), \ \ t \in [T_0 + i, T_0 + i + 1]. \qquad (8.225)$$

It follows from (8.222), (8.224), and (8.225) that for every integer $q \geq 1$,

$$I^f(T_0, T_0 + q, x_*, u_*) \leq \lim_{p\to\infty} I^f(T_0, T_0 + q, x_{k_p}, u_{k_p})$$

$$\leq I^f(T_0, T_0 + q, x_f, u_f) + 2S_0 + L_0(1 + a_0). \qquad (8.226)$$

Theorem 8.5, (8.219), (8.223), (8.225), and (8.226) imply that (x_*, u_*) is (f)-good and

$$x_*(T_0) = z_0. \qquad (8.227)$$

In order to complete the proof of the proposition, it is sufficient to show that (x_*, u_*) is (f)-minimal. Assume the contrary. Then there exist $\Delta > 0$, an integer $\tau_0 \geq 1$, and $(y, v) \in X(T_0, T_0 + \tau_0)$ such that

$$y(T_0) = x_*(T_0), \ \ y(T_0 + \tau_0) = x_*(T_0 + \tau_0), \qquad (8.228)$$

$$I^f(T_0, T_0 + \tau_0, x_*, u_*) > I^f(T_0, T_0 + \tau_0, y, v) + 2\Delta. \qquad (8.229)$$

By (A2) and Lemma 8.37, there exist $L_1, \delta > 0$ such that the following properties hold:

(i) For each $(T, \xi) \in \mathcal{A}$ satisfying $\rho_E(\xi, x_f(T)) \leq \delta$, there exist $\tau_1 \in (0, b_f]$ and $(\tilde{x}_1, \tilde{u}_1) \in X(T, T + \tau_1)$ such that

$$\tilde{x}_1(T) = \xi \ , \tilde{x}_1(T + \tau_1) = x_f(T + \tau_1),$$

$$I^f(T, T + \tau_1, \tilde{x}_1, \tilde{u}_1) \leq I^f(T, T + \tau_1, x_f, u_f) + \Delta/8,$$

and if $T \geq b_f$, then there exist $\tau_2 \in (0, b_f]$ and $(\tilde{x}_2, \tilde{u}_2) \in X(T - \tau_2, T)$ such that

$$\tilde{x}_2(T - \tau_2) = x_f(T - \tau_2) \ , \tilde{x}_2(T) = \xi,$$

$$I^f(T - \tau_2, T, \tilde{x}_2, \tilde{u}_2) \leq I^f(T - \tau_2, T, x_f, u_f) + \Delta/8.$$

(ii) If $T_2 > T_1 \geq L_1$, $(x, u) \in X(T_1, T_2)$, $\rho_E(x(T_i), x_f(T_i)) \leq \delta$, $i = 1, 2$, we have

$$I^f(T_1, T_2, x, u) \geq I^f(T_1, T_2, x_f, u_f) - \Delta/8.$$

Theorems 8.20 and 8.25 and (8.217), (8.219), and (8.220) imply that there exists an integer $L_2 > L_0 + L_1$ such that for each integer $k \geq k_0 + 2L_2$,

$$\rho_E(x_k(t), x_f(t)) \leq \delta, \ t \in [T_0 + L_2, T_0 + k - L_2]. \tag{8.230}$$

In view of (8.223), (8.225), and (8.230),

$$\rho_E(x_*(t), x_f(t)) \leq \delta \ \text{ for all } t \geq T_0 + L_2. \tag{8.231}$$

By (8.223)–(8.225), there exists a natural number p_0 such that

$$k_{p_0} > k_0 + 2L_1 + 2L_2 + 2\tau_0 + 2 + 2b_f + 2T_0, \tag{8.232}$$

$$I^f(T_0, T_0 + \tau_0 + L_2, x_*, u_*) \leq I^f(T_0, T_0 + \tau_0 + L_2, x_{k_{p_0}}, u_{k_{p_0}}) + \Delta/2. \tag{8.233}$$

Property (i) and (8.228), (8.230)–(8.234) imply that there is $(x, u) \in X(T_0, T_0 + k_{p_0})$ such that

$$(x(t), u(t)) = (y(t), v(t)), \ t \in [T_0, T_0 + \tau_0], \tag{8.234}$$

$$(x(t), u(t)) = (x_*(t), u_*(t)), \ t \in (T_0 + \tau_0, T_0 + \tau_0 + L_2], \tag{8.235}$$

$$x(T_0 + \tau_0 + L_2 + b_f) = x_f(T_0 + \tau_0 + L_2 + b_f), \tag{8.236}$$

$$I^f(T_0 + \tau_0 + L_2, T_0 + \tau_0 + L_2 + b_f, x, u)$$

$$\leq I^f(T_0 + \tau_0 + L_2, T_0 + \tau_0 + L_2 + b_f, x_f, u_f) + \Delta/8, \tag{8.237}$$

$$(x(t), u(t)) = (x_{k_{p_0}}(t), u_{k_{p_0}}(t)), \ t \in [T_0 + \tau_0 + L_2 + 2b_f, k_{p_0} + T_0], \tag{8.238}$$

$$I^f(T_0+\tau_0+L_2+b_f, T_0+\tau_0+L_2+2b_f, x, u)$$

$$\leq I^f(T_0+\tau_0+L_2+b_f, T_0+\tau_0+L_2+2b_f, x_f, u_f)+\Delta/8. \tag{8.239}$$

It follows from (8.219), (8.220), (8.227), (8.228), and (8.233) that

$$I^f(T_0, T_0+k_{p_0}, x, u) \geq I^f(T_0, T_0+k_{p_0}, x_{k_{p_0}}, u_{k_{p_0}}). \tag{8.240}$$

By (8.229), (8.230), (8.232), (8.236)–(8.240), and property (ii),

$$0 \leq I^f(T_0, T_0+k_{p_0}, x, u) - I^f(T_0, T_0+k_{p_0}, x_{k_{p_0}}, u_{k_{p_0}})$$

$$= I^f(T_0, T_0+\tau_0, y, v) + I^f(T_0+\tau_0, T_0+\tau_0+L_2, x_*, u_*)$$

$$+ I^f(T_0+\tau_0+L_2, T_0+\tau_0+L_2+b_f, x_f, u_f)+\Delta/8$$

$$+ I^f(T_0+\tau_0+L_2+b_f, T_0+\tau_0+L_2+2b_f, x_f, u_f)+\Delta/8$$

$$- I^f(T_0, T_0+\tau_0+L_2, x_{k_{p_0}}, u_{k_{p_0}})$$

$$- I^f(T_0+\tau_0+L_2, T_0+\tau_0+L_2+b_f, x_{k_{p_0}}, u_{k_{p_0}})$$

$$- I^f(T_0+\tau_0+L_2+b_f, T_0+\tau_0+L_2+2b_f, x_{k_{p_0}}, u_{k_{p_0}})$$

$$\leq I^f(T_0, T_0+\tau_0, y, v) + I^f(T_0+\tau_0, T_0+\tau_0+L_2, x_*, u_*)$$

$$+ I^f(T_0+\tau_0+L_2, T_0+\tau_0+L_2+b_f, x_f, u_f)+\Delta/8$$

$$+ I^f(T_0+\tau_0+L_2+b_f, T_0+\tau_0+L_2+2b_f, x_f, u_f)+\Delta/8$$

$$- I^f(T_0, T_0+\tau_0+L_2, x_{k_{p_0}}, u_{k_{p_0}})$$

$$- I^f(T_0+\tau_0+L_2, T_0+\tau_0+L_2+2b_f, x_{k_{p_0}}, u_{k_{p_0}})+\Delta/8$$

$$= I^f(T_0, T_0+\tau_0, y, v) + I^f(T_0+\tau_0, T_0+\tau_0+L_2, x_*, u_*)$$

$$- I^f(T_0, T_0+\tau_0+L_2, x_{k_{p_0}}, u_{k_{p_0}})+\Delta/2$$

$$< I^f(T_0, T_0+\tau_0, x_*, u_*) - 2\Delta + I^f(T_0+\tau_0, T_0+\tau_0+L_2, x_*, u_*)$$

$$- I^f(T_0, T_0+\tau_0+L_2, x_{k_{p_0}}, u_{k_{p_0}})+\Delta/2 \leq -2\Delta+\Delta/2+\Delta/2,$$

a contradiction. The contradiction we have reached completes the proof of Proposition 8.38.

8.18 Proof of Theorem 8.29

Clearly, (i) implies (ii). In view of Assertion 2 of Theorem 8.5, (ii) implies (iii). By (P1), (iii) implies (iv). Evidently, (iv) implies (v). We show that (v) implies (iii). Assume that (x_*, u_*) is (f)-minimal and satisfies

$$\liminf_{t\to\infty} \rho_E(x_*(t), x_f(t)) = 0. \tag{8.241}$$

Since $(\tilde{x}, \tilde{u})$ is (f)-good, there exists $S_0 > 0$ such that for all numbers $T > S$,

$$|I^f(S, T, \tilde{x}, \tilde{u}) - I^f(S, T, x_f, u_f)| \le S_0. \tag{8.242}$$

By (A2), there exists $\delta > 0$ such that the following property holds:

(a) For each $(T, z) \in \mathcal{A}$ satisfying $\rho_E(z, x_f(T)) \le \delta$, there exist $\tau_1 \in (0, b_f]$ and $(x_1, u_1) \in X(T, T + \tau_1)$ that satisfy

$$x_1(T) = z, \ x_1(T + \tau_1) = x_f(T + \tau_1),$$

$$I^f(T, T + \tau_1, x_1, u_1) \le I^f(T, T + \tau_1, x_f, u_f) + 1,$$

and if $T \ge b_f$, then there exist $\tau_2 \in (0, b_f]$ and $(x_2, u_2) \in X(T - \tau_2, T)$ such that

$$x_2(T - \tau_2) = x_f(T - \tau_2), \ x_2(T) = z,$$

$$I^f(T - \tau_2, T, x_2, u_2) \le I^f(T - \tau_2, T, x_f, u_f) + 1.$$

In view of (8.241) and (P1), there exists an increasing sequence $\{t_k\}_{k=1}^{\infty} \subset (S, \infty)$ such that

$$\lim_{k\to\infty} t_k = \infty, \tag{8.243}$$

$$\rho_E(x_*(t_k + 2b_f), x_f(t_k + 2b_f)) \le \delta \text{ for all integers } k \ge 1, \tag{8.244}$$

$$\rho_E(\tilde{x}(t), x_f(t)) \le \delta \text{ for all } t \ge t_0. \tag{8.245}$$

Let $k \ge 1$ be an integer. By property (a) and (8.244) and (8.245), there exists $(y, v) \in X(S, t_k + 2b_f)$ such that

$$(y(t), v(t)) = (\tilde{x}(t), \tilde{u}(t)), \ t \in [S, t_k],$$

$$y(t_k + b_f) = x_f(t_k + b_f),$$

$$I^f(t_k, t_k + b_f, y, v) \le I^f(t_k, t_k + b_f, x_f, u_f) + 1,$$

$$y(t_k + 2b_f) = x_*(t_k + 2b_f),$$

$$I^f(t_k + b_f, t_k + 2b_f, y, v) \le I^f(t_k + b_f, t_k + 2b_f, x_f, u_f) + 1.$$

The relations above and (8.242) imply that

$$I^f(S, t_k + 2b_f, x_*, u_*) \le I^f(S, t_k + 2b_f, y, v)$$

$$\le I^f(S, t_k, \tilde{x}, \tilde{u}) + I^f(t_k, t_k + 2b_f, x_f, u_f) + 2$$

$$\le I^f(S, t_k + 2b_f, x_f, u_f) + 2 + 2S_0.$$

Together with Theorem 8.5, this implies that (x_*, u_*) is (f)-good and (iii) holds.

We show that (iii) implies (i). Assume that (x_*, u_*) is (f)-minimal and (f)-good. (P1) implies that

$$\lim_{t \to \infty} \rho_E(x_*(t), x_f(t)) = 0. \tag{8.246}$$

Since (x_*, u_*) is (f)-good, there exists $S_0 > 0$ such that

$$|I^f(S, T, x_*, u_*) - I^f(S, T, x_f, u_f)| \le S_0 \text{ for all } T > S. \tag{8.247}$$

Let $(x, u) \in X(S, \infty)$ satisfy

$$x(S) = x_*(S). \tag{8.248}$$

We show that

$$\limsup_{T \to \infty} [I^f(S, T, x_*, u_*) - I^f(S, T, x, u)] \le 0.$$

In view of Theorem 8.5 and (8.247), we may assume that (x, u) is (f)-good. (P1) implies that

$$\lim_{t \to \infty} \rho_E(x(t), x_f(t)) = 0. \tag{8.249}$$

Let $\epsilon > 0$. By (A2) and Lemma 8.37, there exist $\delta \in (0, \epsilon)$ and $L_1 > 0$ such that the following properties hold:

(b) For each $(T, z) \in \mathcal{A}$ satisfying $\rho_E(z, x_f(T)) \le \delta$, there exist $\tau_1 \in (0, b_f]$ and $(x_1, u_1) \in X(T, T + \tau_1)$ satisfying

$$x_1(T) = z \, , x_1(T + \tau_1) = x_f(T + \tau_1),$$

$$I^f(T, T+\tau_1, x_1, u_1) \le I^f(T, T+\tau_1, x_f, u_f) + \epsilon/8,$$

and if $T \ge b_f$, then there exist $\tau_2 \in (0, b_f]$ and $(x_2, u_2) \in X(T - \tau_2, T)$ satisfying

$$x_2(T-\tau_2) = x_f(T-\tau_2)\ , x_2(T) = z,$$

$$I^f(T-\tau_2, T, x_2, u_2) \le I^f(T-\tau_2, T, x_f, u_f) + \epsilon/8.$$

(c) For each $T_1 \ge L_1$, each $T_2 > T_1$, and each $(y, v) \in X(T_1, T_2)$ that satisfy $\rho_E(y(T_i), x_f(T_i)) \le \delta$, $i = 1, 2$, we have

$$I^f(T_1, T_2, y, v) \ge I^f(T_1, T_2, x_f, u_f) - \epsilon/8.$$

It follows from (8.246) and (8.249) that there exists $\tau_0 > 0$ such that

$$\rho_E(x(t), x_f(t)) \le \delta,\ \ \rho_E(x_*(t), x_f(t)) \le \delta \text{ for all } t \ge \tau_0. \tag{8.250}$$

Let

$$T \ge \tau_0 + L_1. \tag{8.251}$$

Property (b) and (8.250) and (8.251) imply that there exists $(y, v) \in X(S, T+2b_f)$, which satisfies

$$(y(t), v(t)) = (x(t), u(t)),\ t \in [S, T],\ \ y(T+b_f) = x_f(T+b_f), \tag{8.252}$$

$$I^f(T, T+b_f, y, v) \le I^f(T, T+b_f, x_f, u_f) + \epsilon/8, \tag{8.253}$$

$$y(T+2b_f) = x_*(T+2b_f), \tag{8.254}$$

$$I^f(T+b_f, T+2b_f, y, v) \le I^f(T+b_f, T+2b_f, x_f, u_f) + \epsilon/8. \tag{8.255}$$

By property (c) and (8.250) and (8.251),

$$I^f(T, T+2b_f, x_*, u_*) \ge I^f(T, T+2b_f, x_f, u_f) - \epsilon/8. \tag{8.256}$$

It follows from (8.248), (8.252)–(8.256) that

$$I^f(S, T, x_*, u_*) + I^f(T, T+2b_f, x_f, u_f) - \epsilon/8$$

$$\le I^f(S, T+2b_f, x_*, u_*) \le I^f(S, T+2b_f, y, v)$$

$$= I^f(S, T, x, u) + I^f(T, T + 2b_f, x_f, u_f) + \epsilon/4,$$

$$I^f(S, T, x_*, u_*) \le I^f(S, T, x, u) + \epsilon \text{ for all } T \ge \tau_0 + L_1.$$

Since ϵ is any positive number, we conclude that

$$\limsup_{T\to\infty}[I^f(S, T, x_*, u_*) - I^f(S, T, x, u)] \le 0,$$

(x_*, u_*) is (f)-overtaking optimal, and (i) holds. Assertion 2 of Theorem 8.29 is proved. Assertion 2 of Theorem 8.29 and Proposition 8.38 imply Assertion 1 of Theorem 8.29.

8.19 Proof of Theorem 8.33

In view of Theorem 8.27, in order to show that (P1), (P2) hold, it is sufficient to show that f_r has (P3). Let $\epsilon, M > 0$. Theorem 8.5 implies that there exists $S_0 > 0$ such that for each $T_2 > T_1 \ge 0$ and each $(x, u) \in X(T_1, T_2)$,

$$I^f(T_1, T_2, x, u) + S_0 \ge I^f(T_1, T_2, x_f, u_f). \tag{8.257}$$

Property (i) of Sect. 8.9 implies there exists $\delta \in (0, \epsilon)$ such that

$$\text{if } y, z \in E \text{ satisfies } \phi(y, z) \le \delta, \text{ then } \rho_E(y, z) \le \epsilon. \tag{8.258}$$

Set

$$L = \delta^{-1}r^{-1}(S_0 + M). \tag{8.259}$$

Assume that $T_1 \ge 0$, $T_2 \ge T_1 + L$, $(x, u) \in X(T_1, T_2)$ satisfies

$$I^{f_r}(T_1, T_2, x, u) \le I^{f_r}(T_1, T_2, x_f, u_f) + M. \tag{8.260}$$

By (8.257), (8.258), the equality $\phi(0, 0) = 0$, and the definition of f_r,

$$I^f(T_1, T_2, x_f, u_f) + M = M + I^{f_r}(T_1, T_2, x_f, u_f)$$

$$\ge I^f(T_1, T_2, x, u) + r\int_{T_1}^{T_2} \phi(x(t), x_f(t))dt$$

$$\ge I^f(T_1, T_2, x_f, u_f) - S_0 + r\int_{T_1}^{T_2} \phi(x(t), x_f(t))dt,$$

$$r^{-1}(S_0 + M) \geq \int_{T_1}^{T_2} \phi(x(t), x_f(t))dt$$

$$\geq \delta \mathrm{mes}(\{t \in [T_1, T_2] : \ \phi(x(t), x_f(t)) \geq \delta\})$$

$$\geq \delta \mathrm{mes}(\{t \in [T_1, T_2] : \ \rho_E(x(t), x_f(t)) > \epsilon\})$$

and

$$\mathrm{mes}(\{t \in [T_1, T_2] : \ \rho_E(x(t), x_f(t)) > \epsilon\}) \leq \delta^{-1} r^{-1}(S_0 + M) \leq L.$$

Thus f_r has (P3) and (P1), (P2) too.

Assume that $(x_f, u_f) \in X(0, \infty)$ is (f)-minimal. It is not difficult to see that (x_f, u_f) is (f_r)-minimal too. Let $(x, u) \in X(0, \infty)$ be (f_r)-good,

$$x(0) = x_f(0). \tag{8.261}$$

Therefore,

$$\lim_{t \to \infty} \rho_E(x(t), x_f(t)) = 0. \tag{8.262}$$

Let $\epsilon > 0$. By (A4), there exists $\delta > 0$ such that the following property holds:

(i) For each $(T, z) \in \mathcal{A}$ satisfying $\rho_E(z, x_f(T)) \leq \delta$, there exist $\tau \in (0, b_f]$, $(y, v) \in X(T, T + \tau)$ such that $y(T) = z$, $y(T + \tau) = x_f(T + \tau)$,

$$I^{f_r}(T, T + \tau, y, v) \leq I^{f_r}(T, T + \tau, x_f, u_f) + \epsilon/8.$$

In view of (8.262), there exists $T_0 > 0$ such that

$$\rho_E(x(t), x_f(t)) \leq \delta \text{ for all } t \geq T_0. \tag{8.263}$$

Let $T \geq T_0$. Property (i) and (8.263) imply that there exist $\tau \in (0, b_f]$, $(y, v) \in X(T, T + \tau)$ such that

$$y(T) = x(T), \ \ y(T + \tau) = x_f(T + \tau),$$

$$I^{f_r}(T, T + \tau, y, v) \leq I^f(T, T + \tau, x_f, u_f) + \epsilon/8. \tag{8.264}$$

Set

$$y(t) = x(t), \ \ v(t) = u(t), \ \ t \in [0, T]. \tag{8.265}$$

Clearly, $(y, v) \in X(0, T + \tau)$. Since (x_f, u_f) is (f)-minimal, it follows from (8.261), (8.264), and (8.265) that

$$I^f(0, T + \tau, x_f, u_f) \le I^f(0, T + \tau, y, v). \tag{8.266}$$

By the definition of f_r and (8.264)–(8.266),

$$I^{f_r}(0, T, x_f, u_f) \le I^f(0, T, y, v) + I^f(T, T + \tau, y, v) - I^f(T, T + \tau, x_f, u_f)$$

$$= I^f(0, T, x, u) + I^{f_r}(T, T + \tau, y, v)$$

$$- r \int_T^{T+\tau} \phi(y(t), x_f(t))dt - I^f(T, T + \tau, x_f, u_f)$$

$$\le I^f(0, T, x, u) - r \int_T^{T+\tau} \phi(y(t), x_f(t))dt + \epsilon/8$$

$$= I^{f_r}(0, T, x, u) - r \int_0^T \phi(x(t), x_f(t))dt - r \int_T^{T+\tau} \phi(y(t), x_f(t))dt + \epsilon/8.$$

Since the relation above holds for any $T \ge T_0$, we obtain

$$\limsup_{T\to\infty}(I^{f_r}(0, T, x_f, u_f) - I^{f_r}(0, T, x, u)) \le -r \lim_{T\to\infty} \int_0^T \phi(x(t), x_f(t))dt + \epsilon/8.$$

Since ϵ is any positive number, we have

$$\limsup_{T\to\infty}(I^{f_r}(0, T, x_f, u_f) - I^f(0, T, x, u)) \le -r \lim_{T\to\infty} \int_0^T \phi(x(t), x_f(t))dt. \tag{8.267}$$

This implies that $(x_f, u_f) \in X(0, \infty)$ is (f_r)-overtaking optimal.

Assume that (x, u) is (f_r)-overtaking optimal. Then (8.267) implies that $x(t) = x_f(t)$ for all $t \ge 0$. Theorem 8.33 is proved.

8.20 The Strong Turnpike Property

We showed (see Proposition 8.3) that (A0)' holds for the third class of problems. By Lemma 6.44 of [148], (A0)' holds for the first class of problems with $A(t) = 0$, $t \ge 0$. Here we consider the first class of problems with $A(t) = 0$, $t \ge 0$, and the third class of problems and study them simultaneously.

We prove the following result.

Theorem 8.39 *Let f have LSC property and (x_f, u_f) be (f)-overtaking optimal. Then f has STP if and only if (P1) and (P2) hold and the following property holds:*

(a) *For each (f)-overtaking optimal pair of sequences $(y, v) \in X(0, \infty)$ satisfying $y(0) = x_f(0)$, the equality $y(t) = x_f(t)$ holds for all $t \geq 0$.*

The next result follows from (A2), (P1), and STP.

Theorem 8.40 *Assume that f has STP and $\epsilon > 0$. Then there exists $\delta > 0$ such that for every $T_1 \geq 0$ and every (f)-overtaking optimal pair $(x, u) \in X(T_1, \infty)$ satisfying $\rho_E(x(T_1), x_f(T_1)) \leq \delta$, the inequality $\rho_E(x(t), x_f(t)) \leq \epsilon$ holds for all $t \geq T_1$.*

8.21 Auxiliary Results for Theorem 8.39

For the first class, we assume that $\theta_0 = 0$ and $\rho_E(x, y) = \|x - y\|$, $x, y \in E$. For the third class, we assume that $\|G(t, x, u)\| = G(t, x, u)$, $(t, x, u) \in \mathcal{M}$.

Assumption (A0)' implies the following result.

Lemma 8.41 *Let $M_1 > 0$, $0 < \tau_0 < \tau_1$. Then there exists $M_2 > 0$ such that for each $T_1 \geq 0$, each $T_2 \in (T_1 + \tau_0, T_1 + \tau_1]$, and each $(x, u) \in X(T_1, T_2)$ satisfying*

$$I^f(T_1, T_2, x, u) \leq M_1$$

the following inequality holds:

$$\rho_E(x(t), \theta_0) \leq M_2 \text{ for all } t \in [T_1, T_2].$$

Lemma 8.42 *Let $M_1 > 0$, $\epsilon \in (0, 1)$, $0 < \tau_0 < \tau_1$. Then there exists $\delta > 0$ such that for each $T_1 \geq 0$, each $T_2 \in [T_1 + \tau_0, T_1 + \tau_1]$, each $(x, u) \in X(T_1, T_2)$ satisfying*

$$I^f(T_1, T_2, x, u) \leq M_0$$

and each $t_1, t_2 \in [T_1, T_2]$ satisfying $|t_1 - t_2| \leq \delta$ the inequality $\rho_E(x(t_1), x(t_2)) \leq \epsilon$ holds.

Proof Lemma 8.3 implies that there exists $M_2 > 0$ such that the following property holds:

(i) For each $T_1 \geq 0$, each $T_2 \in [T_1 + \tau_0, T_1 + \tau_1]$, and each $(x, u) \in X(T_1, T_2)$ satisfying

$$I^f(T_1, T_2, x, u) \leq M_1,$$

we have

$$\rho_E(x(t), \theta_0) \le M_2 \text{ for all } t \in [T_1, T_2].$$

In view of (8.3), (8.5), and (8.56), for the first class of problems, there exists $h_0 > 0$ such that for the first class of problems

$$f(t, x, u) \ge 4\epsilon^{-1}(M_1 + a_0\tau_1 + 8)(\|G(t, x, u)\| - a_0\|x\|)_+ \tag{8.268}$$

$$\text{for each } (t, x, u) \in \mathcal{M} \text{ satisfying } \|G(t, x, u)\| - a_0\|x\| \ge h_0$$

and that for the third class of problems

$$f(t, x, u) \ge 4\epsilon^{-1}(M_1 + a_0\tau_1 + 8)(G(t, x, u) - a_0\rho_E(x, \theta_0)_+ \tag{8.269}$$

$$\text{for each } (t, x, u) \in \mathcal{M} \text{ satisfying } G(t, x, u) - a_0\rho_E(x, \theta_0) \ge h_0.$$

Choose a number

$$\delta \in (0, \epsilon(4a_0M_2 + 4h_0 + 4)^{-1}). \tag{8.270}$$

Let

$$T_1 \ge 0,\ T_2 \in [T_1 + \tau_0, T_1 + \tau_1],\ (x, u) \in X(T_1, T_2),$$

$$I^f(T_1, T_2, x, u) \le M_1, \tag{8.271}$$

$$t_1, t_2 \in [T_1, T_2],\ 0 < t_2 - t_2 \le \delta. \tag{8.272}$$

Property (i) and (8.271) imply that

$$\rho_E((x(t), \theta_0) \le M_2 \text{ for all } t \in [T_1, T_2]. \tag{8.273}$$

Define

$$\Omega_1 = \{t \in [t_1, t_2] :\ \|G(t, x(t), u(t))\| - a_0\rho_E(x(t), \theta_0) \ge h_0\}, \tag{8.274}$$

$$\Omega_2 = [t_1, t_2] \setminus \Omega_1. \tag{8.275}$$

By (8.18) and (8.268)–(8.275),

$$\rho_E(x(t_2), x(t_1)) \le \int_{t_1}^{t_2} \|G(s, x(s), u(s))\| ds$$

$$\le a_0 \int_{t_1}^{t_2} \rho_E(x(t), \theta_0) dt + \int_{t_1}^{t_2} (\|G(t, x(t), u(t))\| - a_0\rho_E(x(t), \theta_0))_+ dt$$

$$\leq a_0\delta M_2 + \delta h_0 + \int_{\Omega_1} (\|G(t, x(t), u(t)) - a_0\rho_E(x(t), \theta_0))_+ dt$$

$$\leq a_0\delta M_2 + \delta h_0 + \epsilon(4(M_1 + a_0\tau_1 + 8))^{-1} \int_{\Omega_1} f(t, x(t), u(t))dt$$

$$\leq a_0\delta M_2 + \delta h_0 + 4^{-1}\epsilon < \epsilon.$$

Lemma 8.42 is proved.

Lemma 8.43 *Let $\Delta > 0$. Then there exists $\delta > 0$ such that for each $(T_1, z_1), (T_2, z_2) \in \mathcal{A}$ satisfying*

$$T_2 \geq T_1 + 2b_f,\ \rho_E(z_i, x_f(T_i)) \leq \delta,\ i = 1, 2, \tag{8.276}$$

the following inequality holds:

$$U^f(T_1, T_2, z_1, z_2) \leq I^f(T_1, T_2, x_f, u_f) + \Delta.$$

Proof Let $\delta > 0$ be as guaranteed by (A2) with $\epsilon = \Delta/4$. Let

$$(T_1, z_1), (T_2, z_2) \in \mathcal{A}$$

satisfy (8.276). By (8.276), the choice of δ, and (A2) with $\epsilon = \Delta/4$, there exists $(y, v) \in X(T_1, T_2)$ such that

$$y(T_1) = z_1,\ y(T_2) = z_2,$$

$$y(t) = x_f(t),\ v(t) = u_f(t),\ t \in [T_1 + b_f, T_2 - b_f],$$

$$I^f(T_1, T_1 + b_f, y, v) \leq I^f(T_1, T_1 + b_f, x_f, u_f) + \Delta/4,$$

$$I^f(T_2 - b_f, T_2, y, v) \leq I^f(T_2 - b_f, T_2, x_f, u_f) + \Delta/4.$$

By the relations above,

$$U^f(T_1, T_2, z_1, z_2) \leq I^f(T_1, T_2, y, v) \leq I^f(T_1, T_2, x_f, u_f) + \Delta/2.$$

Lemma 8.43 is proved.

Lemma 8.44 *Let $\Delta \in (0, 1)$. Then there exists $\delta > 0$ such that for each $T_1 \geq 0$, each $T_2 \geq T_1 + 3b_f$, each $(x, u) \in X(T_1, T_2)$ satisfying*

$$\rho_E(x(T_i), x_f(T_i)) \leq \delta,\ i = 1, 2,$$

$$I^f(T_1, T_2, x, u) \le U^f(T_1, T_2, x(T_1), x(T_2)) + \delta,$$

and each $\tau_1 \in [T_1, T_1 + \delta]$ *and each* $\tau_2 \in [T_2 - \delta, T_2]$, *the inequality*

$$I^f(\tau_1, \tau_2, x, u) \le I^f(\tau_1, \tau_2, x_f, u_f) + \Delta$$

holds.

Proof Lemma 8.43 implies that there exists

$$\delta_0 \in (0, \min\{4^{-1}\Delta, 2^{-1}b_f\})$$

such that the following property holds:

(ii) For each $(T_1, z_1), (T_2, z_2) \in \mathcal{A}$ satisfying

$$T_2 \ge T_1 + 2b_f, \ \rho_E(z_i, x_f(T_i)) \le \delta_0, \ i = 1, 2,$$

we have

$$U^f(T_1, T_2, z_1, z_2) \le I^f(T_1, T_2, x_f, u_f) + \Delta/4.$$

Theorem 8.5 implies that there exists $M_0 > 0$ such that the following property holds:

(iii) For each pair of numbers $T_2 > T_1 \ge 0$ and each $(x, u) \in X(T_1, T_2)$,

$$I^f(T_1, T_2, x, u) + M_0 \ge I^f(T_1, T_2, x_f, u_f).$$

Lemma 8.42 implies that there exists $\delta_1 \in (0, \delta_0)$ such that the following property holds:

(iv) For each $S \ge 0$, each $(y, v) \in X(S, S + b_f)$ satisfying

$$I^f(S, S + b_f, y, v) \le |\Delta_f|(b_f + 2) + 3a_0 + 2M_0 + 1,$$

and each $t_1, t_2 \in [S, S + b_f]$ satisfying $|t_1 - t_2| \le \delta_1$, we have $\rho_E(y(t_1), y(t_2)) \le \delta_0/8$.

Set

$$\delta = \delta_1/8. \tag{8.277}$$

Assume that

$$T_1 \ge 0, \ T_2 \ge T_1 + 3b_f, \tag{8.278}$$

$(x, u) \in X(T_1, T_2)$,

$$\rho_E(x(T_i), x_f(T_i)) \le \delta, \; i = 1, 2, \tag{8.279}$$

$$I^f(T_1, T_2, x, u) \le U^f(T_1, T_2, x(T_1), x(T_2)) + \delta, \tag{8.280}$$

$$\tau_1 \in [T_1, T_1 + \delta], \; \tau_2 \in [T_2 - \delta, T_2]. \tag{8.281}$$

Property (ii) and (8.278)–(8.280) imply that

$$I^f(T_1, T_2, x, u) \le I^f(T_1, T_2, x_f, u_f) + 1.$$

Property (iii) and the relation above imply that for each $s_1, s_2 \in [T_1, T_2]$ satisfying $s_1 < s_2$,

$$I^f(s_1, s_2, x, u) = I^f(T_1, T_2, x, u) - I^f(T_1, s_1, x, u) - I^f(s_2, T_2, x, u)$$

$$\le I^f(T_1, T_2, x_f, u_f) + 1 - I^f(T_1, s_1, x_f, u_f) + M_0 - I^f(s_2, T_2, x_f, u_f) + M_0$$

$$= I^f(T_1, T_2, x_f, u_f) + 2M_0 + 1. \tag{8.282}$$

In view of (8.5) and (8.74),

$$I^f(T_1, T_1 + b_f, x_f, u_f) \le I^f(\lfloor T_1 \rfloor, \lfloor T_1 \rfloor + \lfloor b_f \rfloor + 2, x_f, u_f) + 3a_0$$

$$\le (\lfloor b_f \rfloor + 2)\Delta_f + 3a_0, \tag{8.283}$$

$$I^f(T_2 - b_f, T_2, x_f, u_f) \le I^f(\lfloor T_2 \rfloor - \lfloor b_f \rfloor - 1, \lfloor T_2 \rfloor + 1, x_f, u_f) + 3a_0$$

$$\le (\lfloor b_f \rfloor + 2)\Delta_f + 3a_0. \tag{8.284}$$

It follows from (8.282)–(8.284) that

$$I^f(T_1, T_1 + b_f, x, u), \; I^f(T_2 - b_f, T_2, x, u) \le \Delta_f(\lfloor b_f \rfloor + 2) + 3a_0 + 2M_0 + 1. \tag{8.285}$$

Property (iv), (8.277), (8.281), and (8.285) imply that

$$\rho_E(x_f(T_i), x_f(\tau_i)) \le \delta_0/8, \; i = 1, 2, \tag{8.286}$$

$$\rho_E(x(T_i), x(\tau_i)) \le \delta_0/8, \; i = 1, 2. \tag{8.287}$$

By (8.277), (8.279), (8.286), and (8.287), for $i = 1, 2$,

$$\rho_E(x_f(\tau_i), x(\tau_i)) \le \rho_E(x_f(\tau_i), x_f(T_i)) + \rho_E(x_f(T_i), x(T_i)) + \rho_E(x(T_i), x(\tau_i))$$

$$\leq \delta_0/8 + \delta_0/8 + \delta \leq \delta_0/2. \tag{8.288}$$

Property (ii), (8.278), (8.279), (8.281), and (8.288) imply that

$$U^f(\tau_1, \tau_2, x(\tau_1), x(\tau_2)) \leq I^f(\tau_1, \tau_2, x_f, u_f) + \Delta/4. \tag{8.289}$$

By (8.280), (8.281), and (8.289),

$$I^f(\tau_1, \tau_2, x, u) \leq U^f(\tau_1, \tau_2, x(\tau_1), x(\tau_2)) + \delta$$

$$\leq I^f(\tau_1, \tau_2, x_f, u_f) + \Delta/4 + \delta \leq I^f(\tau_1, \tau_2, x_f, u_f) + \Delta.$$

Lemma 8.44 is proved.

8.22 Proof of Theorem 8.39

Here we again consider the first class of problems with $A(t) = 0$, $t = 0$, and the third class of problems and study them simultaneously. For the first class, $\theta_0 = 0$, $\rho_E(x, y) = \|x - y\|$, $x, y \in E$, and for the third class, $\|G(t, x, u)\| = G(t, x, u)$, $(t, x, u) \in \mathcal{M}$.

Assume that STP holds. Theorem 8.25 implies that (P1) and (P2) hold. Let $(\tilde{x}, \tilde{u}) \in X(0, \infty)$ be (f)-overtaking optimal and

$$\tilde{x}(0) = x_f(0). \tag{8.290}$$

(P1), STP, and (8.290) imply that $\tilde{x}(t) = x_f(t)$ for all $t \geq 0$ and (P4) holds.

Assume that (P1), (P2) and (a) of Theorem 8.32 hold.

Lemma 8.45 *Let $\epsilon > 0$. Then there exists $\delta > 0$ such that for each pair of numbers $T_1 \geq 0$, $T_2 \geq T_1 + 3b_f$ and each $(x, u) \in X(T_1, T_2)$ satisfying*

$$\rho_E(x(T_i), x_f(T_i)) \leq \delta, \ i = 1, 2,$$

$$I^f(T_1, T_2, x, u) \leq U^f(T_1, T_2, x(T_1), x(T_2)) + \delta,$$

the inequality $\rho_E(x(t), x_f(t)) \leq \epsilon$ for all $t \in [T_1, T_2]$.

Proof Lemma 8.36 and (A2) imply that there exist $\delta_0 \in (0, \min\{1, \epsilon/4\})$ and $L_0 > 0$ such that the following properties hold:

(i) (A2) holds with $\epsilon = 1$ and $\delta = \delta_0$.
(ii) For each pair of numbers $T_1 \geq L_0$, $T_2 \geq T_1 + 2b_f$ and each $(x, u) \in X(T_1, T_2)$ satisfying

$$\rho_E(x(T_i), x_f(T_i)) \le \delta_0,\ i = 1, 2$$

$$I^f(T_1, T_2, x, u) \le U^f(T_1, T_2, x(T_1), x(T_2)) + \delta_0,$$

we have

$$\rho_E(x(t), x_f(t)) \le \epsilon,\ t \in [T_1, T_2].$$

Theorem 8.5 implies that there exists $S_0 > 0$ such that for each $T_2 > T_1 \ge 0$ and each $(x, u) \in X(T_1, T_2)$,

$$I^f(T_1, T_2, x, u) + S_0 \ge I^f(T_1, T_2, x_f, u_f). \tag{8.291}$$

It follows from Theorem 8.26 that there exists $L_1 > 0$ such that the following property holds:

(iii) For each $T \ge 0$ and each $(x, u) \in X(T, T + L_1)$ satisfying

$$I^f(T_1, T_2, x, u) \le I^f(T_1, T_2, x_f, u_f) + 2S_0 + 3,$$

we have

$$\min\{\rho_E(x(t), x_f(t)) :\ t \in [T, T + L_1]\} \le \delta_0.$$

Consider a sequence $\{\delta_i\}_{i=1}^{\infty} \subset (0, 1)$ such that

$$\delta_i < 2^{-1}\delta_{i-1},\ i = 1, 2, \dots. \tag{8.292}$$

Assume that the lemma does not hold. Then for each integer $i \ge 1$, there exist

$$T_{i,1} \ge 0,\ T_{i,2} \ge T_{i,1} + 3b_f \tag{8.293}$$

and $(x_i, u_i) \in X(T_{i,1}, T_{i,2})$ such that

$$\rho_E(x_i(T_{i,1}), x_f(T_{i,1})) \le \delta_i,\ \rho_E(x_i(T_{i,2}), x_f(T_{i,2})) \le \delta_i, \tag{8.294}$$

$$I^f(T_{i,1}, T_{i,2}, x_i, u_i) \le U^f(T_{i,1}, T_{i,2}, x_i(T_{i,1}), x_i(T_{i,2})) + \delta_i \tag{8.295}$$

and $t_i \in [T_{i,1}, T_{i,2}]$ for which

$$\rho_E(x(t_i), x_f(t_i)) > \epsilon. \tag{8.296}$$

Let i be a natural number. Property (ii) and (8.292)–(8.296) imply that

$$T_{i,1} < L_0. \tag{8.297}$$

We show that

$$t_i \le 2b_f + 1 + L_1 + L_0. \tag{8.298}$$

Property (i), (A2) with $\epsilon = 1$, $\delta = \delta_0$, (8.292)–(8.294) imply that there exists $(y_i, v_i) \in X(T_{i,1}, T_{i,2})$ such that

$$y_i(T_{i,1}) = x_i(T_{i,1}),\ \ y_i(T_{i,2}) = x_i(T_{i,2}), \tag{8.299}$$

$$y_i(t) = x_f(t),\ \ v_i(t) = u_f(t),\ \ t \in [T_{i,1} + b_f, T_{i,2} - b_f], \tag{8.300}$$

$$I^f(T_{i,1}, T_{i,1} + b_f, y_i, v_i) \le I^f(T_{i,1}, T_{i,1} + b_f, x_f, u_f) + 1, \tag{8.301}$$

$$I^f(T_{i,2} - b_f, T_{i,2}, y_i, v_i) \le I^f(T_{i,2} - b_f, T_{i,2}, x_f, u_f) + 1. \tag{8.302}$$

It follows from (8.295) and (8.299)–(8.302) that

$$I^f(T_{i,1}, T_{i,2}, x_i, u_i) \le 1 + I^f(T_{i,1}, T_{i,2}, y_i, v_i) \le I^f(T_{i,1}, T_{i,2}, x_f, u_f) + 3. \tag{8.303}$$

In view of (8.291), (8.303), for each pair of integers $S_1, S_2 \in [T_{i,1}, T_{i,2}]$ satisfying $S_1 < S_2$,

$$\begin{aligned} I^f(S_1, S_2, x_i, u_i) &= I^f(T_{i,1}, T_{i,2}, x_i, u_i) - I^f(T_{i,1}, S_1, x_i, u_i) - I^f(S_2, T_{i,2}, x_i, u_i) \\ &\le I^f(T_{i,1}, T_{i,2}, x_f, u_f) + 3 - I^f(T_{i,1}, S_1, x_f, u_f) + S_0 - I^f(S_2, T_{i,2}, x_f, u_f) + S_0 \\ &= I^f(S_1, S_2, x_f, u_f) + 3 + 2S_0. \end{aligned} \tag{8.304}$$

Assume that (8.298) does not hold. Then

$$t_i > 2b_f + 1 + L_1 + L_0. \tag{8.305}$$

Consider the restriction of (x_i, u_i) to the interval

$$[t_i - 2b_f - 1 - L_1, t_i - 1 - 2b_f] \subset [L_0, \infty). \tag{8.306}$$

By property (iii) and (8.304), there exists

$$\tilde{t} \in [t_i - 2b_f - 1 - L_1, t_i - 1 - 2b_f] \tag{8.307}$$

such that

$$\rho_E(x_f(\tilde{t}), x(\tilde{t})) \le \delta_0. \tag{8.308}$$

Property (ii), (8.292), (8.294), (8.295), and (8.306)–(8.308) imply that

$$\rho_E(x(t), x_f(t)) \le \epsilon,\ t \in [\tilde{t}, T_{i,2}],$$

and in particular,

$$\rho_E(x(t_i), x_f(t_i)) \le \epsilon.$$

This contradicts (8.296). The contradiction we have reached proves (8.298).

In view of (8.74) and (8.304), there exists $M_1 > 0$ such that the following property holds:

(iv) For each integer $i \ge 1$ and each $S_2 > S_1 \ge 0$ satisfying $S_1, S_2 \in [T_{i,1}, T_{i,2}]$, $S_2 - S_1 \le 2b_f$,

$$I^f(S_1, S_2, x_i, u_i) \le M_1.$$

Lemma 8.41 and property (iv) imply that there exists $M_2 > 0$ such that

$$\rho_E(x_i(t), \theta_0) \le M_2,\ t \in [T_{i,1}, T_{i,2}],\ i = 1, 2, \dots. \tag{8.309}$$

In view of (8.207) and (8.298), extracting a subsequence and re-indexing, we may assume without loss of generality that there exists

$$\tilde{T}_1 = \lim_{i\to\infty} T_{i,1} \in [0, L_0], \tag{8.310}$$

$$\tilde{t} = \lim_{i\to\infty} t_i \in [\tilde{T}_1, L_0 + L_1 + 2b_f + 1], \tag{8.311}$$

$$\tilde{T}_2 = \lim_{i\to\infty} T_{i,2} \in [\tilde{t}, \infty]. \tag{8.312}$$

There exist a strictly decreasing sequence $\{T_1^{(k)}\}_{k=1}^{\infty}$ and a strictly increasing sequence $\{T_2^{(k)}\}_{k=1}^{\infty}$ such that

$$\tilde{T}_1 = \lim_{k\to\infty} T_1^{(k)},\ \tilde{T}_2 = \lim_{k\to\infty} T_2^{(k)}, \tag{8.313}$$

for all integers $k \ge 1$,

$$T_1^{(k)} < T_2^{(k)}. \tag{8.314}$$

If $\tilde{T}_2 = \infty$, then we may assume that

$$T_2^{(k)} = T_{k,2} \text{ for all integers } k \ge 1. \tag{8.315}$$

By LSC property and property (iv), extracting a subsequence and re-indexing, we may assume without loss of generality that there exist $\widehat{x} : (\tilde{T}_1, \tilde{T}_2) \to E$, $\widehat{u} : (\tilde{T}_1, \tilde{T}_2) \to F$ such that for each $t \in (\tilde{T}_1, \tilde{T}_2)$,

$$\widehat{x}(t) = \lim_{i \to \infty} x_i(t), \tag{8.316}$$

for each integer $k \geq 1$,

$$(\widehat{x}, \widehat{u}) \in X(T_1^{(k)}, T_2^{(k)}), \tag{8.317}$$

$$I^f(T_1^{(1)}, T_2^{(1)}, \widehat{x}, \widehat{u}) \leq \liminf_{i \to \infty} I^f(T_1^{(1)}, T_2^{(1)}, x_i, u_i), \tag{8.318}$$

and for each integer $k \geq 1$,

$$I^f(T_1^{(k+1)}, T_1^{(k)}, \widehat{x}, \widehat{u}) \leq \liminf_{i \to \infty} I^f(T_1^{(k+1)}, T_1^{(k)}, x_i, u_i), \tag{8.319}$$

$$I^f(T_2^{(k)}, T_2^{(k+1)}, \widehat{x}, \widehat{u}) \leq \liminf_{i \to \infty} I^f(T_2^{(k)}, T_2^{(k+1)}, x_i, u_i). \tag{8.320}$$

By (8.309) and (8.316),

$$\rho_E(\widehat{x}(t), \theta_0) \leq M_2, \ t \in (\tilde{T}_1, \tilde{T}_2). \tag{8.321}$$

We show that

$$x_f(\tilde{T}_1) = \lim_{t \to \tilde{T}_1^+} \widehat{x}(t). \tag{8.322}$$

Let $\gamma > 0$. Lemma 8.42, property (iv), and (8.74) imply that there exists $\delta > 0$ such that the following property holds:

(v) For each integer $i \geq 1$ and each $S_1, S_2 \in [T_{i,1}, T_{i,2}]$ satisfying $|S_2 - S_1| \leq 2\delta$, we have

$$\rho_E(x_i(S_1), x_i(S_2)) \leq \gamma/2.$$

(vi) For each $S_1, S_2 \in [0, \infty)$ satisfying $|S_2 - S_1| \leq 2\delta$, we have

$$\rho_E(x_f(S_1), x_f(S_2)) \leq \gamma/2.$$

Let

$$t \in (\tilde{T}_1, \tilde{T}_1 + 2\delta). \tag{8.323}$$

In view of property (v), (8.294), (8.313), and (8.323), for all sufficiently large natural numbers k,

$$t \in (T_{k,1}, T_{k,1} + 2\delta),$$

$$\rho_E(x_k(t), x_k(T_{k,1})) \le \gamma/2,$$

$$\rho_E(x_k(t), x_f(T_{k,1})) \le \rho_E(x_k(t), x_k(T_{k,1})) + \rho_E(x_k(T_{k,1}), x_f(T_{k,1})) \le \gamma/2 + \delta_k. \tag{8.324}$$

Property (vi) and (8.313) imply that for all sufficiently large natural numbers k,

$$\rho_E(x_f(T_{k,1}), x_f(\tilde{T}_1)) \le \gamma/2. \tag{8.325}$$

It follows from (8.292), (8.316), (8.324), (8.325) that for all t satisfying (8.323),

$$\rho_E(\widehat{x}(t), x_f(\tilde{T}_1)) = \lim_{k\to\infty} \rho_E(x_k(t), x_f(\tilde{T}_1))$$

$$\le \limsup_{k\to\infty} \rho_E(x_k(t), x_f(T_{k,1})) + \rho_E(x_f(T_{k,1}), x_f(\tilde{T}_1)) \le \gamma.$$

Since γ is an arbitrary positive number, we conclude that

$$\lim_{t\to\tilde{T}_1^+} \rho_E(\widehat{x}(t), x_f(\tilde{T}_1)) = 0$$

and (8.322) holds. Analogously, we can show that if $\tilde{T}_2 < \infty$, then

$$x_f(\tilde{T}_2) = \lim_{t\to\tilde{T}_2^-} \widehat{x}(t). \tag{8.326}$$

Set

$$\widehat{x}(\tilde{T}_1) = x_f(\tilde{T}_1),\ \widehat{u}(\tilde{T}_1) = u_f(\tilde{T}_1). \tag{8.327}$$

If $\tilde{T}_2 < \infty$, then we set

$$\widehat{x}(\tilde{T}_2) = x_f(\tilde{T}_2),\ \widehat{u}(\tilde{T}_2) = u_f(\tilde{T}_2). \tag{8.328}$$

We show that

$$\tilde{t} > \tilde{T}_1,\ \rho_E(\widehat{x}(\tilde{t}), x_f(\tilde{t})) \ge \epsilon. \tag{8.329}$$

Let $\gamma \in (0, \epsilon/4)$, and let $\delta > 0$ be such that properties (v) and (vi) hold. Let $k \ge 1$ be an integer. If $t_k - T_{k,1} \le \delta$, then in view of properties (v), (vi) and the choice of δ,

$$\rho_E(x_f(t_k), x_f(T_{k,1})) \le \gamma/2, \ \ \rho_E(x_k(t_k), x_k(T_{k,1})) \le \gamma/2,$$

and together with (8.294) and (8.296), these relations imply that

$$\epsilon < \rho_E(x_f(t_k), x_k(t_k)) \le \rho_E(x_f(t_k), x_f(T_{k,1})) + \rho_E(x_f(T_{k,1}), x_k(T_{k,1}))$$
$$+ \rho_E(x_k(T_{k,1}), x_k(t_k)) \le \gamma/2 + \delta_k + \gamma/2 \le \gamma + \epsilon/4 < \epsilon,$$

a contradiction. The contradiction we have reached proves that

$$t_k - T_{k,1} > \delta. \tag{8.330}$$

If $T_{k,2} - t_k \le \delta$, then in view of properties (v), (vi) and the choice of δ,

$$\rho_E(x_f(t_k), x_f(T_{k,2})) \le \gamma/2, \ \ \rho_E(x_k(t_k), x_k(T_{k,2})) \le \gamma/2,$$

and together with (8.294) and (8.296), these relations imply that

$$\epsilon < \rho_E(x_f(t_k), x_k(t_k)) \le \rho_E(x_f(t_k), x_f(T_{k,2})) + \rho_E(x_f(T_{k,2}), x_k(T_{k,2}))$$
$$+ \rho_E(x_k(T_{k,2}), x_k(t_k)) \le \gamma/2 + \delta_k + \gamma/2 \le \gamma + \epsilon/4 < \epsilon,$$

a contradiction. The contradiction we have reached proves that

$$T_{k,2} - t_k \ge \delta. \tag{8.331}$$

By (8.330) and (8.331), for all natural numbers k,

$$T_{k,2} - t_k \ge \delta, \ \ t_k - T_{k,1} \ge \delta. \tag{8.332}$$

It follows from (8.310), (8.312), and (8.332) that

$$\tilde{T}_2 - \tilde{t} \ge \delta, \ \tilde{t} - \tilde{T}_1 \ge \delta. \tag{8.333}$$

Properties (v) and (vi), (8.298), (8.311), and (8.316) imply that

$$\rho_E(\widehat{x}(\tilde{t}), x_f(\tilde{t})) = \lim_{k\to\infty} \rho_E(x_k(\tilde{t}), x_f(\tilde{t})) = \lim_{k\to\infty} \rho_E(x_k(t_k), x_f(t_k)) \ge \epsilon$$

and (8.329) holds. In view of (8.317), for every $\tau \in (\tilde{T}_1, \tilde{T}_2)$, the function $G(t, \widehat{x}(t), \widehat{u}(t))$, $t \in [\tilde{T}_1, \tau]$ is strongly measurable for the first class of problems and just measurable for the third class of problems.

Let k_0 be a natural number and

$$\tau = T_2^{(k_0)}. \tag{8.334}$$

We show that the function $G(t, \widehat{x}(t), \widehat{u}(t)),\ t \in [\tilde{T}_1, \tau]$, is (Bochner) integrable. By (8.5), (8.20), and (8.56), there exists $\gamma_0 > 0$ such that

$$f(t, x, u) \geq 8(\|G(t, x, u)\| - a_0 \rho_E(x, \theta_0))_+ \text{ for all } (t, x, u) \in \mathcal{M}$$

$$\text{satisfying } \|G(t, x, u)\| - a_0 \rho_E(x, \theta_0) \geq \gamma_0. \tag{8.335}$$

Let $k \geq 1$ be an integer and $T_1^{(k)} < \tau$. By (8.309),

$$\int_{T_1^{(k)}}^{\tau} \|G(t, \widehat{x}(t), \widehat{u}(t))\| dt \leq \int_{T_1^{(k)}}^{\tau} a_0 \rho_E(\widehat{x}(t), \theta_0) dt$$

$$+ \int_{T_1^{(k)}}^{\tau} (\|G(t, \widehat{x}(t), \widehat{u}(t))\| - a_0 \rho_E(\widehat{x}(t), \theta_0))_+ dt$$

$$\leq a_0 M_2(\tau - T_1^{(k)}) + \int_{T_1^{(k)}}^{\tau} (\|G(t, \widehat{x}(t), \widehat{u}(t))\| - a_0 \rho_E(\widehat{x}(t), \theta_0))_+ dt. \tag{8.336}$$

Set

$$E_1 = \{t \in [T_1^{(k)}, \tau] : \ \|G(t, \widehat{x}(t), \widehat{u}(t))\| \geq a_0 \rho_E(\widehat{x}(t), \theta_0) + \gamma_0\}, \tag{8.337}$$

$$E_2 = [T_1^{(k)}, \tau] \setminus E_1. \tag{8.338}$$

It follows from (8.5), (8.304), (8.318), (8.320), (8.334)–(8.338) that

$$\int_{T_1^{(k)}}^{\tau} \|G(t, \widehat{x}(t), \widehat{u}(t))\| dt \leq a_0 M_2(\tau - T_1^{(k)})$$

$$+ \int_{E_1} (\|G(t, \widehat{x}(t), \widehat{u}(t))\| - a_0 \rho_E(\widehat{x}(t), \theta_0))_+ dt$$

$$+ \int_{E_2} (\|G(t, \widehat{x}(t), \widehat{u}(t))\| - a_0 \rho_E(\widehat{x}(t), \theta_0))_+ dt$$

$$\leq a_0 M_2(\tau - T_1^{(k)}) + \gamma_0(\tau - T_1^{(k)})$$

$$+ \int_{E_1} (\|G(t, \widehat{x}(t), \widehat{u}(t))\| - a_0 \rho_E(\widehat{x}(t), \theta_0))_+ dt$$

$$\leq a_0 M_2(\tau - T_1^{(k)}) + \gamma_0(\tau - T_1^{(k)}) + \int_{E_1} f(t, \widehat{x}(t), \widehat{u}(t)) dt$$

$$\leq a_0 M_2(\tau - T_1^{(k)}) + \gamma_0(\tau - T_1^{(k)}) + I^f(T_1^{(k)}, \tau, \widehat{x}, \widehat{u}) + a_0(\tau - T_1^{(k)})$$

$$\leq (a_0 M_2 + \gamma_0 + a_0)(T_2^{(k_0)} - T_1^{(k)}) + I^f(T_1^{(k)}, T_2^{(k_0)}, \widehat{x}, \widehat{u})$$

$$\leq (T_2^{(k_0)} - T_1^{(k)})(a_0 M_2 + \gamma_0 + a_0) + \liminf_{i\to\infty} I^f(T_1^{(k)}, T_2^{(k_0)}, x_i, u_i)$$

$$\leq (T_2^{(k_0)} - T_1^{(k)})(a_0 M_2 + \gamma_0 + a_0) + I^f(T_1^{(k)}, T_2^{(k_0)}, x_f, u_f) + 3 + 2S_0. \tag{8.339}$$

By (8.313) and (8.339),

$$\lim_{k\to\infty} \int_{T_1^{(k)}}^{T_2^{(k_0)}} \|G(t, \widehat{x}(t), \widehat{u}(t))\| dt$$

$$\leq (T_2^{(k_0)} - \tilde{T}_1)(a_0 M_2 + \gamma_0 + a_0) + I^f(\tilde{T}_1, T_2^{(k_0)}, x_f, u_f) + 3 + 2S_0.$$

Fatou's lemma and (8.313) imply that

$$\int_{\tilde{T}_1}^{T_2^{(k_0)}} \|G(t, \widehat{x}(t), \widehat{u}(t))\| dt < \infty,$$

$G(t, \widehat{x}(t), \widehat{u}(t))$, $t \in [\tilde{T}_1, T_2^{(k_0)}]$ is Bochner integrable for all integers $k_0 \geq 1$, and

$$\int_{\tilde{T}_1}^{T_2^{(k_0)}} \|G(t, \widehat{x}(t), \widehat{u}(t))\| dt \leq I^f(\tilde{T}_1, T_2^{(k_0)}, x_f, u_f) + 3 + 2S_0$$

$$+ (T_2^{(k_0)} - \tilde{T}_1)(a_0 M_2 + \gamma_0 + a_0). \tag{8.340}$$

If $\tilde{T}_2 = \infty$, then by (8.340),

$$G(t, \widehat{x}(\cdot), \widehat{u}(\cdot)) \text{ is (Bochner) integrable on } [\tilde{T}_1, \tau] \text{ for any } \tau > \tilde{T}_1. \tag{8.341}$$

If $\tilde{T}_2 < \infty$, then by (8.313) and (8.340), $G(\cdot, \widehat{x}(\cdot), \widehat{u}(\cdot))$ is strongly measurable on $[\tilde{T}_1, \tilde{T}_2]$ and

$$\lim_{k\to\infty} \int_{\tilde{T}_1}^{T_2^{(k)}} \|G(t, \widehat{x}(t), \widehat{u}(t))\| dt \leq I^f(\tilde{T}_1, \tilde{T}_2, x_f, u_f) + 3 + 2S_0$$

$$+ (\tilde{T}_2 - \tilde{T}_1)(a_0 M_2 + \gamma_0 + a_0),$$

and in view of Fatou's lemma

$$G(\cdot, \widehat{x}(\cdot), \widehat{u}(\cdot)) \text{ is integrable on } [\tilde{T}_1, \tilde{T}_2]. \tag{8.342}$$

By (8.55), (8.322), (8.326), and (8.342),

$$(\widehat{x}, \widehat{u}) \in X(\tilde{T}_1, \tilde{T}_2). \tag{8.343}$$

Assume that $\tilde{T}_2 = \infty$. By (8.55), (8.322), and (8.341), for every $\tau > \tilde{T}_1$,

$$(\widehat{x}, \widehat{u}) \in X(\tilde{T}_1, \tau). \tag{8.344}$$

Assume that $\tilde{T}_2 < \infty$. It follows from (8.327) and (8.328) that

$$x_f(\tilde{T}_i) = \widehat{x}(\tilde{T}_i),\ i = 1, 2,\ \rho_E(\widehat{x}(\tilde{t}), x_f(\tilde{t})) \geq \epsilon. \tag{8.345}$$

Property (a) and (8.345) imply that

$$I^f(\tilde{T}_1, \tilde{T}_2, x_f, u_f) < I^f(\tilde{T}_1, \tilde{T}_2, \widehat{x}, \widehat{u}). \tag{8.346}$$

Let $\Delta > 0$. Lemma 8.44 implies that there exists $\delta > 0$ such that the following property holds:

(vii) For each $\tau_1 \geq 0$, each $\tau_2 \geq \tau_1 + 3b_f$, each $(x, u) \in X(\tau_1, \tau_2)$ satisfying

$$\rho_E(x(\tau_i), x_f(\tau_i)) \leq \delta,\ i = 1, 2,$$

$$I^f(\tau_1, \tau_2, x, u) \leq U^f(\tau_1, \tau_2, x(\tau_1), x(\tau_2)) + \delta,$$

each $t_1 \in [\tau_1, \tau_1 + \delta]$, each $t_2 \in [\tau_2 - \delta, \tau_2]$,

$$I^f(t_1, t_2, x, u) \leq I^f(t_1, t_2, x_f, u_f) + \Delta/2.$$

In view of (8.292), there exists an integer $k_0 \geq 1$ such that

$$\delta_k < \delta \text{ for all integers } k \geq k_0. \tag{8.347}$$

Let $q \geq 1$ be an integer such that for all integers $i \geq q$,

$$|\tilde{T}_1 - T_1^{(i)}| \leq \delta/2,\ |\tilde{T}_2 - T_2^{(i)}| \leq \delta/2. \tag{8.348}$$

By (8.310), (8.312), (8.313), and (8.348), there exists an integer $k_1 \geq k_0$ such that for each integer $k \geq k_1$,

$$T_{k,1} < T_1^{(q)} \leq T_{k,1} + \delta,\ T_{k,2} - \delta \leq T_2^{(q)} < T_{k,2}. \tag{8.349}$$

Property (vii), (8.293)–(8.295), (8.347), and (8.349) imply that for each integer $k \geq k_1$,

$$I^f(T_1^{(q)}, T_2^{(q)}, x_k, u_k) \leq I^f(T_1^{(q)}, T_2^{(q)}, x_f, u_f) + \Delta/2. \tag{8.350}$$

In view of (8.318)–(8.320) and (8.350),

$$I^f(T_1^{(q)}, T_2^{(q)}, \widehat{x}, \widehat{u}) \leq \liminf_{k\to\infty} I^f(T_1^{(q)}, T_2^{(q)}, x_k, u_k)$$

$$\leq I^f(T_1^{(q)}, T_2^{(q)}, x_f, u_f) + \Delta/2 \tag{8.351}$$

for every integer $q \geq 1$ such that (8.348) is true for all integers $i \geq q$. Fatou's lemma and (8.5) and (8.313) imply that

$$I^f(\tilde{T}_1, \tilde{T}_2, \widehat{x}, \widehat{u}) \leq I^f(\tilde{T}_1, \tilde{T}_2, x_f, u_f) + \Delta/2.$$

Since Δ is any positive number, we conclude that

$$I^f(\tilde{T}_1, \tilde{T}_2, \widehat{x}, \widehat{u}) \leq I^f(\tilde{T}_1, \tilde{T}_2, x_f, u_f).$$

This contradicts (8.346). The contradiction we have reached proves $\tilde{T}_2 = \infty$. Recall (see (8.327) and (8.329)) that

$$\widehat{x}(\tilde{T}_1) = x_f(\tilde{T}_1), \ \ \rho_E(\widehat{x}(\tilde{t}), x_f(\tilde{t})) \geq \epsilon. \tag{8.352}$$

Let $\Delta > 0$. Lemma 8.44 implies that there exists $\delta \in (0, b_f/4)$ such that the following property holds:

(viii) For each $\tau_1 \geq 0$, each $\tau_2 \geq \tau_1 + 3b_f$, each $(x, u) \in X(\tau_1, \tau_2)$ satisfying

$$\rho_E(x(\tau_i), x_f(\tau_i)) \leq \delta, \ i = 1, 2,$$

$$I^f(\tau_1, \tau_2, x, u) \leq U^f(\tau_1, \tau_2, x(\tau_1), x(\tau_2)) + \delta,$$

each $t_1 \in [\tau_1, \tau_1 + \delta]$, each $t_2 \in [\tau_2 - \delta, \tau_2]$,

$$I^f(t_1, t_2, x, u) \leq I^f(t_1, t_2, x_f, u_f) + \Delta/2.$$

Recall (see (8.315)) that for all integers $k \geq 1$,

$$T_2^{(k)} = T_{k,2}. \tag{8.353}$$

In view of (8.292), there exists an integer $k_0 \geq 1$ such that

$$\delta_k < \delta \text{ for all integers } k \geq k_0. \tag{8.354}$$

Property (viii), (8.293)–(8.295), (8.353), and (8.354) imply that the following property holds:

(ix) For each integer $k \geq k_0$, each $\tau_1 \in [T_{k,1}, T_{k,1} + \delta]$, each $\tau_2 \in [T_{k,2} - \delta, T_{k,2}]$,

$$I^f(\tau_1, \tau_2, x_k, u_k) \leq I^f(\tau_1, \tau_2, x_f, u_f) + \Delta/2.$$

By (8.318)–(8.320) and (8.353), for each integer $k \geq 1$ and each integer $s \geq k$,

$$I^f(T_1^{(s)}, T_{k,2}, \widehat{x}, \widehat{u})$$

$$= I^f(T_1^{(s)}, T_2^{(k)}, \widehat{x}, \widehat{u}) \leq \liminf_{i \to \infty} I^f(T_1^{(s)}, T_2^{(k)}, x_i, u_i)$$

$$= \liminf_{i \to \infty} I^f(T_1^{(s)}, T_{k,2}, x_i, u_i). \tag{8.355}$$

In view of (8.304) and (8.355), for each integer $k \geq 1$ and each integer $s \geq k$,

$$I^f(T_1^{(s)}, T_{k,2}, \widehat{x}, \widehat{u}) \leq I^f(T_1^{(s)}, T_{k,2}, x_f, u_f) + 3 + 2S_0. \tag{8.356}$$

Fatou's lemma, (8.5), (8.313), and (8.356) imply that for all integers $k \geq 1$,

$$I^f(\tilde{T}_1, T_{k,2}, \widehat{x}, \widehat{u}) \leq I^f(\tilde{T}_1, T_{k,2}, x_f, u_f) + 4 + 2S_0.$$

Together with Theorem 8.5 and (8.352), this implies that $(\widehat{x}, \widehat{u})$ is (f)-good. Property (P1) implies that

$$\lim_{t \to \infty} \rho_E(\widehat{x}(t), x_f(t)) = 0.$$

Therefore, there exists $\tau_0 > 1 + \tilde{T}_1 + 3b_f$ such that

$$\rho_E(\widehat{x}(t), x_f(t)) \leq \delta/4 \text{ for all } t \geq \tau_0. \tag{8.357}$$

By (8.310)–(8.313) and (8.315), there exists an integer $q \geq 1$ such that

$$T_{q,2} > \tau_0,\ |I^f(\tilde{T}_1, T_1^{(q)}, \widehat{x}, \widehat{u})| < \Delta/8,\ |I^f(\tilde{T}_1, T_1^{(q)}, x_f, u_f)| < \Delta/8, \tag{8.358}$$

for all integers $i \geq q$,

$$|\tilde{T}_1 - T_1^{(i)}| \leq \delta/2. \tag{8.359}$$

It follows from (8.310), (8.313), (8.316), and (8.359) that there exists an integer $k_1 \geq k_0 + q$ such that for each integer $k \geq k_1$,

$$T_{k,1} < T_1^{(q)} \leq T_{k,1} + \delta, \ \rho_E(\widehat{x}(T_{q,2}), x_k(T_{q,2})) \leq \delta/4. \tag{8.360}$$

Assume that an integer $k \geq k_1$. Then (8.360) holds. By (8.293)–(8.295), (8.315), (8.360) and the choice of k_0,

$$I^f(T_1^{(q)}, T_{q,2}, x_k, u_k) \leq U^f(T_1^{(q)}, T_{q,2}, x_k(T_1^{(q)}), x_k(T_{q,2})) + \delta, \tag{8.361}$$

$$\rho_E(x_k(T_{k,1}), x_f(T_{k,1})) \leq \delta_k \leq \delta. \tag{8.362}$$

In view of (8.357)–(8.360),

$$\rho_E(x_k(T_{q,2}), x_f(T_{q,2}))$$

$$\leq \rho_E(x_k(T_{q,2}), \widehat{x}(T_{q,2})) + \rho_E(\widehat{x}(T_{q,2}), x_f(T_{q,2})) \leq \delta/4 + \delta/4. \tag{8.363}$$

Property (x) applied with $\tau_1 = T_{k,1}$, $\tau_2 = T_{q,2}$, $(x, u) = (x_k, u_k)$, $t_1 = T_1^{(q)}$, $t_2 = T_{q,2}$, and (8.359)–(8.363) imply that

$$I^f(T_1^{(q)}, T_{q,2}, x_k, u_k) \leq I^f(T_1^{(q)}, T_{q,2}, x_f, u_f) + \Delta/2. \tag{8.364}$$

In view of (8.364),

$$I^f(T_1^{(q)}, T_{q,2}, \widehat{x}, \widehat{u}) \leq \liminf_{k\to\infty} I^f(T_1^{(q)}, T_{q,2}, x_k, u_k)$$

$$\leq I^f(T_1^{(q)}, T_{q,2}, x_f, u_f) + \Delta/2. \tag{8.365}$$

By (8.365) and (8.358),

$$I^f(\tilde{T}_1, T_{q,2}, \widehat{x}, \widehat{u}) \leq I^f(\tilde{T}_1, T_{q,2}, x_f, u_f) + \Delta. \tag{8.366}$$

Since the relation above holds for any integer $q \geq 1$ satisfying (8.358), we have

$$\liminf_{T\to\infty}(I^f(\tilde{T}_1, T, \widehat{x}, \widehat{u}) - I^f(\tilde{T}_1, T, x_f, u_f)) \leq \Delta.$$

Since Δ is any positive number, we conclude that

$$\liminf_{T\to\infty}(I^f(\tilde{T}_1, T, \widehat{x}, \widehat{u}) - I^f(\tilde{T}_1, T, x_f, u_f)) \leq 0. \tag{8.367}$$

For all $t \in [0, \tilde{T}_1] \setminus \{\tilde{T}_1\}$, set

$$\widehat{x}(t) = x_f(t), \ \widehat{u}(t) = u_f(t).$$

In view of (8.367), $(\widehat{x}, \widehat{u}) \in X(0, \infty)$ is (f)-weakly optimal. Theorem 8.29 and (a) imply that $(\widehat{x}, \widehat{u}) \in X(0, \infty)$ is (f)-overtaking optimal and that

$$\widehat{x}(t) = x_f(t), \ t \geq 0.$$

This contradicts (8.345). The contradiction we have reached completes the proof of Lemma 8.45.

Completion of Theorem 8.39 By Theorem 8.25, TP holds. Lemma 8.45 and TP imply STP.

8.23 Examples

In this section, we present a family of problems that belong to the second class of problems and for which the results of this chapter hold. We use the notation introduced in Sects. 8.1–8.3.

Let $(E, \langle\cdot,\cdot\rangle)_E$ be a Hilbert space equipped with an inner product $\langle\cdot,\cdot\rangle_E$ that induces the norm $\|\cdot\|_E$, and let $(F, \langle\cdot,\cdot\rangle_F)$ be a Hilbert space equipped with an inner product $\langle\cdot,\cdot\rangle_F$ that induces the norm $\|\cdot\|_F$. For simplicity, we set $\langle\cdot,\cdot\rangle_E = \langle\cdot,\cdot\rangle$, $\|\cdot\|_E = \|\cdot\|$, $\langle\cdot,\cdot\rangle_F = \langle\cdot,\cdot\rangle$, $\|\cdot\|_F = \|\cdot\|$, if E, F are understood. We suppose that $\mathcal{A}_0$ is a nonempty subset of E and $\mathcal{U}_0 : \mathcal{A}_0 \to 2^F$ is a point to set mapping with a graph

$$\mathcal{M}_0 = \{(x, u) : \ x \in \mathcal{A}_0, \ u \in \mathcal{U}_0(x)\}.$$

We suppose that $\mathcal{M}_0$ is a Borel measurable subset of $E \times F$. Define

$$\mathcal{A} = [0, \infty) \times \mathcal{A}_0,$$

$\mathcal{U} : \mathcal{A} \to 2^F$ by

$$\mathcal{U}(t, x) = \mathcal{U}_0(x), \ (t, x) \in \mathcal{A},$$

$$\mathcal{M} = [0, \infty) \times \mathcal{M}_0.$$

Let a linear operator $A : \mathcal{D}(A) \to E$ generate a C_0 semigroup $S(t) = e^{At}, t \geq 0$ on E, $E_1^d = \mathcal{D}(A^*)$, $E_{-1} = \mathcal{D}(A^*)'$, and let $B \in \mathcal{L}(F, E_{-1})$ be an admissible control operator for $e^{At}, t \geq 0$.

For $T_2 > T_1 \geq 0$, we consider the following control system:

$$x'(t) = Ax(t) + Bu(t), \ t \in [T_1, T_2] \text{ a.e.}$$

Assume that $(x_f, u_f) \in X(0, \infty)$ and

$$\sup\{\|x_f(t)\| : t \in [0, \infty)\} < \infty.$$

Recall that for every $T > 0$,

$$\Phi_T \in \mathcal{L}(L^2(0, \infty; F), E)$$

is defined by

$$\Phi_T u = \int_0^T S(T - s)Bu(s)ds, \ u \in L^2(0, T; F).$$

The following result is proved in Chapter 6 of [148].

Theorem 8.46 *Assume that* $T_0 > 0$, $Ran(\Phi_{T_0}) = E$. *Then there exists a constant* $c > 0$ *such that for each* $T \geq 0$ *and each* $z^0, z^1 \in E$, *there exist* $u \in L^2(T, T + T_0; F)$ *and* $z \in C^0([T, T + T_0]); E)$, *which is a solution of the initial value problem*

$$z'(t) = Az(t) + Bu(t), \ t \in [T, T + T_0] \ a.e. \ , z(T) = z^0$$

in E_{-1} *and satisfies* $z(T + T_0) = z^1$ *and such that the following inequalities hold:*

$$\|z(t) - x_f(t)\|, \|u(t) - u_f(t)\|$$

$$\leq c(\|z^1 - x_f(T + T_0)\| + \|z^0 - x_f(T)\|), \ t \in [T, T + T_0].$$

Assume that there exists $T_f > 0$ such that

$$\text{Ran}\Phi_{T_f} = E.$$

In other words, the pair (A, B) is exactly controllable. We suppose that there exists $r_* > 0$ such that for each $t \geq 0$,

$$\{(x, u) \in E \times F : \|x - x_f(t)\| \leq r_*, \ \|u - u_f(t)\| \leq r_*\} \subset \mathcal{M}_0.$$

Assume that $L : [0, \infty) \times E \times F \to [0, \infty)$ is a Borelian function,

$$L(t, x_f(t), u_f(t)) = 0, \ t \in [0, \infty),$$

$\psi_0 : [0, \infty) \to [0, \infty)$ is an increasing function, $K_1, a_1 > 0$, such that

$$\lim_{t\to\infty} \psi_0(t) = \infty,$$

$$L(t, x, u) \geq -a_1 + \max\{\psi_0(\|x\|)\|x\|, \ K_1\|u\|^2\}$$

for all $(t, x, u) \in \mathcal{M}$, $\mu \in R^1$, $\bar{p} \in \mathcal{D}(A^*)$. Let for all $(t, x, u) \in \mathcal{M}$.

It is not difficult to see that f is a Borelian function, and there exist $a_0, K_0 > 0$ and an increasing function $\psi : [0, \infty) \to [0, \infty)$ such that $\lim_{t\to\infty} \psi(t) = \infty$, and for all $(t, x, u) \in \mathcal{M}$,

$$f(t, x, u) \geq -a_0 + \max\{\psi(\|x\|), K_0\|u\|^2\}.$$

We suppose that the following property holds:

(a) For each $\epsilon > 0$, there exists $\delta > 0$ such that for each $(t, x, u) \in \mathcal{M}$ satisfying $\|x - x_f(t)\| + \|u - u_f(t)\| \leq \delta$ we have

$$f(t, x, u) \leq f(t, x_f(t), u_f(t)) + \epsilon.$$

It follows from property (a) and Theorem 8.46 that (A3) holds.

Let $0 \leq T_1 < T_2$, $(x, u) \in X(T_1, T_2)$. It is not difficult to see that

$$I^f(T_1, T_2, x, u) - I^f(T_1, T_2, x_f, u_f)$$

$$= \int_{T_1}^{T_2} L(t, x(t), u(t))dt + \int_{T_1}^{T_2} \langle x(t), A^*\bar{p}\rangle dt + \int_{T_1}^{T_2} \langle Bu(t), \bar{p}\rangle_{E_{-1},E_d} dt$$

$$- \int_{T_1}^{T_2} \langle x_f(t), A^*\bar{p}\rangle dt - \int_{T_1}^{T_2} \langle Bu_f(t), \bar{p}\rangle_{E_{-1},E_d} dt$$

$$\geq \langle x(T_2) - x(T_1), \bar{p}\rangle - \langle x_f(T_2) - x_f(T_1), \bar{p}\rangle.$$

This implies that (A1) holds.

We can easily obtain particular cases of this example with the pairs (A, B) considered in Sect. 7.6.

Chapter 9
Stability and Genericity Results

In this chapter we continue to study the turnpike phenomenon for the continuous-time optimal control problems on subintervals of half-axis in metric spaces which are discussed in Chap. 8. For these optimal control problems the turnpike is not a singleton. We show that the turnpike phenomenon is stable under small perturbations of an objective function. Using the Baire category approach we show that for some classes of problems a typical (generic) problem has a turnpike property.

9.1 Preliminaries

We continue to consider the class of problems introduced in Sect. 8.1. Let (E, ρ_E) be a complete metric space and (F, ρ_F) be a metric space. We suppose that $\mathcal{A}$ is a nonempty subset of $[0, \infty) \times E, \mathcal{U} : \mathcal{A} \to 2^F$ is a point to set mapping with a graph

$$\mathcal{M} = \{(t, x, u) : (t, x) \in \mathcal{A}, \ u \in \mathcal{U}(t, x)\}.$$

We suppose that $\mathcal{M}$ is a Borel measurable subset of $[0, \infty) \times E \times F$. Let $\theta_0 \in E$ and $\theta_1 \in F$.

Assume that for each pair of numbers T_1, T_2 satisfying $0 \le T_1 < T_2 < \infty$ we are given a set $X(T_1, T_2)$ of pairs (x, u) (called as trajectory–control pairs) such that $x : [T_1, T_2] \to E$ is a continuous function, $u : [T_1, T_2] \to F$ is a Lebesgue measurable function satisfying

$$(t, x(t)) \in \mathcal{A}, \ t \in [T_1, T_2],$$

$$u(t) \in \mathcal{U}(t, x(t)), \ t \in [T_1, T_2] \text{ almost everywhere (a.e.).}$$

A. J. Zaslavski, *Turnpike Phenomenon in Metric Spaces*, Springer Optimization and Its Applications 201, https://doi.org/10.1007/978-3-031-27208-0_9

We suppose that the following property holds:

if $0 \le T_1 \le S_1 < S_2 \le T_2 < \infty$, $(x, u) \in X(T_1, T_2)$, $\tilde{x}$ is the restriction of x to $[S_1, S_2]$ and $\tilde{u}$ is the restriction of u to $[S_1, S_2]$, then $(\tilde{x}, \tilde{u}) \in X(S_1, S_2)$;
for each triplet of nonnegative numbers $T_1 < T_2 < T_3$, each $(x_1, u_1) \in X(T_1, T_2)$ and each $(x_2, u_2) \in X(T_2, T_3]$ satisfying

$$x_1(T_2) = x_2(T_2)$$

a pair $(x_3, u_3) \in X(T_1, T_3)$, where

$$x_3(t) = x_1(t),\ t \in [T_1, T_2],\ x_3(t) = x_2(t),\ t \in (T_2, T_3],$$

$$u_3(t) = u_1(t),\ t \in [T_1, T_2],\ u_3(t) = u_2(t),\ t \in (T_2, T_3].$$

Let $a_0 > 0$ and let $\psi : [0, \infty) \to [0, \infty)$ be an increasing function such that

$$\psi(t) \to \infty \text{ as } t \to \infty \tag{9.1}$$

and let $\psi_1 : \mathcal{M} \to [0, \infty)$ satisfy

$$\psi_1(t, x, u) \ge \max\{\psi(\rho_E(x, \theta_0)),\ \psi(\rho_F(x, \theta_1))\},\ (t, x, u) \in \mathcal{M}. \tag{9.2}$$

Denote by $\mathfrak{M}_\psi$ the set of all borelian functions $g : \mathcal{M} \to R^1$ such that for each $(t, x, u) \in \mathcal{M}$,

$$g(t, x, u) \ge \psi_1(t, x, u) - a_0. \tag{9.3}$$

Recall that for each pair of numbers $T_2 > T_1 \ge 0$, each $(x, u) \in X(T_1, T_2)$ and each $g \in \mathfrak{M}_\psi$,

$$I^g(T_1, T_2, x, u) = \int_{T_1}^{T_2} g(t, x(t), u(t))dt \in (-\infty, \infty].$$

We consider functionals of the form $I^g(T_1, T_2, x, u)$, where $0 \le T_1 < T_2$, $(x, u) \in X(T_1, T_2)$ and $g \in \mathfrak{M}_\psi$. Let $g \in \mathfrak{M}_\psi$. Recall that for each pair of numbers $T_2 > T_1 \ge 0$ and each pair of points $(T_1, y), (T_2, z) \in \mathcal{A}$ we define

$$U^g(T_1, T_2, y, z) = \inf\{I^g(T_1, T_2, x, u) :$$

$$(x, u) \in X(T_1, T_2),\ x(T_1) = y,\ x(T_2) = z\},$$

$$U^g(T_1, T_2, y) = \inf\{I^g(T_1, T_2, x, u) : (x, u) \in X(T_1, T_2),\ x(T_1) = y\},$$

$$\widehat{U}^g(T_1, T_2, z) = \inf\{I^g(T_1, T_2, x, u) : (x, u) \in X(T_1, T_2),\ x(T_2) = z\},$$

$$U^g(T_1, T_2) = \inf\{I^g(T_1, T_2, x, u) : (x, u) \in X(T_1, T_2)\}.$$

We equip the space $\mathfrak{M}_\psi$ with the uniformity determine by the base (8.10) and assume that (A0) holds. Note that all the results of Chap. 8 hold.

Proposition 9.1 *Assume that $g \in \mathfrak{M}_\psi$, $D, \epsilon > 0$ and $c, M_0 > 0$. Then there exists a neighborhood V of g in $\mathfrak{M}_\psi$ such that for each $T_1 \geq 0$ each $T_2 \in (T_1, T_1 + c]$, each $h \in V$ and each $(x, u) \in X(T_1, T_2)$ which satisfies*

$$\rho_E(x(T_1), \theta_0) \leq M_0,$$

$$\min\{I^g(T_1, T_2, x, u),\ I^h(T_1, T_2, x, u)\} \leq D$$

the inequality

$$|I^g(T_1, T_2, x, u) - I^h(T_1, T_2, x, u)| \leq \epsilon$$

holds.

Proof By (A0), there exists $S > 0$ such that the following property holds:

(i) for each $h \in \mathfrak{M}_\psi$, each $T_1 \geq 0$ each $T_2 \in (T_1, T_1 + c]$ and each $(x, u) \in X(T_1, T_2)$ which satisfies

$$\rho_E(x(T_1), \theta_0) \leq M_0,\ I^h(T_1, T_2, x, u) \leq D + 1$$

we have

$$\rho_E(x(t), \theta_0) \leq S,\ t \in [T_1, T_2]. \tag{9.4}$$

By (9.1), there exist $\delta \in (0, 1)$, $N > S$ and $\Gamma > 1$ such that

$$\psi(N) > 4a_0,\ \delta c \leq 8^{-1}\epsilon, \tag{9.5}$$

$$(\Gamma - 1)(D + a_0 c + c) \leq \epsilon/8. \tag{9.6}$$

Define

$$V = \{h \in \mathfrak{M}_\psi :\ (g, h) \in E(N, \delta, \Gamma)\} \tag{9.7}$$

(see (8.10)).

Assume that

$$h \in V,\ T_1 \geq 0,\ T_2 \in (T_1, T_1 + c],\ (x, u) \in X(T_1, T_2) \tag{9.8}$$

$$\rho_E(x(T_1), \theta_0) \leq M_0, \tag{9.9}$$

$$\min\{I^g(T_1, T_2, x, u),\ I^h(T_1, T_2, x, u)\} \le D. \tag{9.10}$$

Property (i), (9.8), and (9.10) imply (9.4). Set

$$E_1 = \{t \in [T_1, T_2] :\ \rho_F(u(t), \theta_1) \le N\},\ E_2 = [T_1, T_2] \setminus E_1. \tag{9.11}$$

It follows from (8.10), (9.4), (9.7), (9.8), and (9.11) that for all $t \in E_1$,

$$|g(t, x(t), u(t)) - h(t, x(t), u(t))| \le \delta. \tag{9.12}$$

Define

$$\xi(t) = \min\{g(t, x(t), u(t)),\ h(t, x(t), u(t))\},\ t \in [T_1, T_2]. \tag{9.13}$$

It follows from (8.10), (9.2), (9.4), (9.5), (9.7), and (9.8) that for each $t \in E_2$,

$$(g(t, x(t), u(t)) + 1)(h(t, x(t), u(t)) + 1)^{-1} \in [\Gamma^{-1}, \Gamma],$$

$$|g(t, x(t), u(t)) - h(t, x(t), u(t))| \le (\Gamma - 1)(\xi(t) + 1). \tag{9.14}$$

By (9.5), (9.6), (9.8) and (9.10)–(9.14),

$$|I^g(T_1, T_2, x, u) - I^h(T_1, T_2, x, u)|$$

$$\le \delta c + (\Gamma - 1)\int_{E_2} \xi(t)dt + (\Gamma - 1)c$$

$$\le \delta c + (\Gamma - 1)(D + a_0 c + c) \le \epsilon.$$

Proposition 9.1 is proved.

9.2 Stability of TP

We continue to use the notation, definitions, and assumptions introduced in Chap. 8.

Assume that $f \in \mathfrak{M}_\psi$, $(x_f, u_f) \in X(0, \infty)$, $b_f > 0$, (8.73), (8.74) are true, (A1), (A2), and TP hold.

We prove the following result which shows the stability of the turnpike phenomenon.

Theorem 9.2 *Assume that $L > 0$ and $\epsilon \in (0, 1)$. Then there exist $L_1 > L$, $\delta \in (0, \epsilon)$ and a neighborhood V of f in $\mathfrak{M}_\psi$ such that for each $g \in V$ and each pair of integers $T_1 \ge 0$, $T_2 \ge T_1 + 4L_1$,*

$$I^g(S_1, S_2, x_f, u_f) \text{ is finite for each } S_2 > S_1 \geq 0,$$

$U^g(T_1, T_2)$ is finite, $U^g(T_1, T_2, z)$ is finite for each $(z, T_1) \in \mathcal{A}_L$ satisfying $\rho_E(z, \theta_0) \leq L$, $U^g(T_1, T_2, z_1, z_2)$ is finite for each $z_1, z_2 \in E$ satisfying $(T_1, z_1) \in \mathcal{A}_L$, $(T_2, z_2) \in \widehat{\mathcal{A}}_L$ and $\rho_E(z_1, \theta_0) \leq L$ and that for each $(x, u) \in X(T_1, T_2)$ which satisfies at least one of the following conditions:

(i)

$$I^g(T_1, T_2, x, u) \leq \min\{U^g(T_1, T_2) + L,\ U^g(T_1, T_2, x(T_1), x(T_2)) + \delta\};$$

(ii)

$$I^g(T_1, T_2, x, u) \leq U^g(T_1, T_2, x(T_1)) + \delta,\ (T_1, x(T_1)) \in \mathcal{A}_L,\ \rho_E(x(T_1), \theta_0) \leq L;$$

(iii)

$$I^g(T_1, T_2, x, u) \leq U^g(T_1, T_2, x(T_1), x(T_2)) + \delta,$$

$$(T_1, x(T_1)) \in \mathcal{A}_L,\ (T_2, x(T_2)) \in \widehat{\mathcal{A}}_L,\ \rho_E(x(T_1), \theta_0) \leq L$$

there exist $\tau_1 \in [T_1, T_1 + L_1]$, $\tau_2 \in [T_2 - L_1, T_2]$ such that

$$\rho_E(x(t), x_f(t)) \leq \epsilon,\ t \in [\tau_1, \tau_2].$$

Moreover, if $\rho_E(x(T_2), x_f(T_2)) \leq \delta$, then $\tau_2 = T_2$ and if $T_1 \geq L_1$ and $\rho_E(x(T_1), x_f(T_1) \leq \delta$, then $\tau_1 = T_1$.

9.3 Auxiliary Results for Theorem 9.2

Lemma 9.3 *Assume that $M \geq 1$, $\epsilon \in (0, 1)$ and $L > 0$. Then there exists an integer $L_0 \geq 4 + L$ such that for each $L_1 > L_0$ there exists a neighborhood V of f in $\mathfrak{M}_\psi$ such that for each $g \in V$ and each pair of numbers $T_1 \geq 0$, $T_2 \in [T_1 + L_0, T_1 + L_1]$,*

$$I^g(T_1, T_2, x_f, u_f)$$

is finite and for each $(x, u)) \in X(T_1, T_2)$ which satisfies

$$I^g(T_1, T_2, x, u) \leq I^f(T_1, T_2, x_f, u_f) + M$$

the inequality

$$mes(\{t \in [T_1, T_2] :\ \rho_E(x(t), x_f(t)) > \epsilon\}) \leq L_0/2$$

holds.

Proof Theorems 8.25 and 8.26 and WTP imply that there exists an integer $L_0 \geq 4 + L$ such that the following property holds:

(a) for each pair of numbers $T_1 \geq 0$, $T_2 \geq T_1 + L_0/2$, and for each $(x, u) \in X(T_1, T_2)$ which satisfies

$$I^f(T_1, T_2, x, u) \leq I^f(T_1, T_2, x_f, u_f) + M + 4 + 8a_0 + 4\Delta_f$$

the inequality

$$\text{mes}(\{t \in [T_1, T_2] : \rho_E(x(t), x_f(t)) > \epsilon\}) < L_0/4$$

holds.

Let $L_1 > L_0$. Choose a positive number M_0 such that

$$\psi(M_0) > M + (L_1 + 2)(a_0 + 2) + (L_1 + 2)(\Delta_f + 1). \tag{9.15}$$

Proposition 9.1 implies that there exists a neighborhood V of f in $\mathfrak{M}_\psi$ such that the following property holds:

(b) for each pair of numbers $T_1 \geq 0$, $T_2 \in (T_1, T_1 + L_1]$, each $g \in V$ and each $(x, u) \in X(T_1, T_2)$ which satisfies

$$\rho_E(x(T_1), \theta_0) \leq M_0,$$

$$\min\{I^g(T_1, T_2, x, u),\ I^f(T_1, T_2, x, u)\} \leq (L_1 + 2)\Delta_f + M + 4c_0 + 4$$

the inequality

$$|I^g(T_1, T_2, x, u) - I^f(T_1, T_2, x, u)| \leq 1$$

holds.

Assume that

$$g \in V,\ T_1 \geq 0,\ T_2 \in (T_1 + L_0, T_1 + L_1], \tag{9.16}$$

$(x, u) \in X(T_1, T_2)$ and that

$$I^g(T_1, T_2, x, u) \leq I^f(T_1, T_2, x_f, u_f) + M. \tag{9.17}$$

We show that there exist

$$S_1 \in [T_1, T_1 + 1],\ S_2 \in [T_2 - 1, T_2] \tag{9.18}$$

such that

$$\rho_E(x(S_i), \theta_0) \le M_0, \ i = 1, 2. \tag{9.19}$$

Assume the contrary. Then in view of (9.2), (9.3), (9.16), and (9.17),

$$I^f(T_1, T_2, x_f, u_f) + M \ge I^g(T_1, T_2, x, u)$$

$$\ge -a_0 + \psi(M_0) - a_0(T_2 - T_1 - 2) \ge \psi(M_0) - a_0 L_1.$$

Together with (8.74), (9.2), (9.3), and (9.16) this implies that

$$\psi(M_0) \le a_0 L_1 + M + (L_1 + 2)\Delta_f + 2a_0.$$

This contradicts (9.15). The contradiction we have reached proves that there exist numbers S_1, S_2 satisfying (9.18) and (9.19). In view of (9.16) and (9.18),

$$S_2 - S_1 \ge T_2 - T_1 - 2 \ge L_0 - 2 \ge L_0/2. \tag{9.20}$$

Equations (9.3), (9.16), and (9.17) imply that

$$I^g(S_1, S_2, x, u) = I^g(T_1, T_2, x, u) - I^g(T_1, S_1, x, u) - I^g(S_2, T_2, x, u)$$

$$\le M + 2a_0 + I^g(T_1, T_2, x_f, u_f) \tag{9.21}$$

and

$$I^g(S_1, S_2, x, u) \le M + 2a_0 + (L_1 + 2)\Delta_f + 2a_0.$$

Property (b), the relation above and Eqs. (9.16), (9.19) and (9.20) imply that

$$|I^f(S_1, S_2, x, u) - I^g(S_1, S_2, x, u)| \le 1. \tag{9.22}$$

By (8.74), (9.3), (9.18), (9.21), and (9.22),

$$I^f(S_1, S_2, x, u) \le I^g(S_1, S_2, x, u) + 1$$

$$\le M + 2a_0 + 1 + I^f(T_1, T_2, x_f, u_f)$$

$$\le M + 2a_0 + 1 + I^f(S_1, S_2, x_f, u_f) + 4(\Delta_f + a_0). \tag{9.23}$$

Property (a), (9.20), and (9.23) imply that

$$\text{mes}(\{t \in [S_1, S_2] : \ \rho_E(x(t), x_f(t)) > \epsilon\}) < L_0/4.$$

Together with (9.18) this implies that

$$\text{mes}(\{t \in [T_1, T_2] : \rho_E(x(t), x_f(t)) > \epsilon\}) < L_0/2.$$

Lemma 9.3 is proved.

Lemma 9.4 *Assume that* $M \geq 8$,

$$L > \sup\{\rho_E(\theta_0, x_f(t)) : t \in [0, \infty)\} \tag{9.24}$$

be an integer and $\epsilon > 0$. *Then there exist* $L_0 > L + 4$,

$$L_1 > 8L_0(2 + L(a_0 + 1))$$

and a neighborhood V *of* f *in* $\mathfrak{M}_\psi$ *such that for each* $g \in V$,

$$I^g(S_1, S_2, x_f, u_f) \text{ is finite for each } S_2 > S_1 \geq 0$$

and for each pair of numbers $T_1 \geq 0$, $T_2 \geq T_1 + L_1$ *and each* $(x, u) \in X(T_1, T_2)$ *which satisfies at least one of the following conditions:*

(i)

$$I^g(T_1, T_2, x, u) \leq U^g(T_1, T_2) + M;$$

(ii)

$$I^g(T_1, T_2, x, u) \leq U^g(T_1, T_2, x(T_1)) + M, \ (T_1, x(T_1)) \in \mathcal{A}_L, \ \rho_E(x(T_1), \theta_0) \leq L;$$

(iii)

$$I^g(T_1, T_2, x, u) \leq U^g(T_1, T_2, x(T_1), x(T_2)) + M,$$

$$(T_1, x(T_1)) \in \mathcal{A}_L, \ (T_2, x(T_2)) \in \widehat{\mathcal{A}}_L, \ \rho_E(x(T_1), \theta_0) \leq L;$$

(iv)

$$I^g(T_1, T_2, x, u) \leq \widehat{U}^g(T_1, T_2, x(T_2)) + M, \ (T_2, x(T_2)) \in \widehat{\mathcal{A}}_L$$

the value $I^g(T_1, T_2, x, u)$ *is finite and the inequality*

$$mes(\{t \in [T_1, T_2] : \rho_E(x(t), x_f(t)) \leq \epsilon\}) \geq L_0/2 \tag{9.25}$$

holds.

Proof Lemma 9.3 implies that there exist an integer $L_0 \geq 4+L$ and a neighborhood V_0 of f in $\mathfrak{M}_\psi$ such that the following property holds:

(a) for each $g \in V_0$ and each pair of numbers $T_1 \geq 0$, $T_2 \in [T_1+L_0, T_1+8(L_0+1)]$

$$I^g(T_1, T_2, x_f, u_f)$$

is finite and for each $(x, u) \in X(T_1, T_2)$ which satisfies

$$I^g(T_1, T_2, x, u) \leq I^f(T_1, T_2, x_f, u_f) + M$$

the inequality

$$\text{mes}(\{t \in [T_1, T_2] : \ \rho_E(x(t), x_f(t)) > \epsilon\}) \leq L_0/2$$

holds.

Proposition 9.1 implies that there exists a neighborhood $V \subset V_0$ of f in $\mathfrak{M}_\psi$ such that the following property holds:

(b) for each pair of numbers $T_1 \geq 0$, $T_2 \in (T_1, T_1 + 8L_0]$, each $g \in V$ and each $(x, u) \in X(T_1, T_2)$ which satisfies

$$\rho_E(x(T_1), \theta_0) \leq L,$$

$$\min\{I^g(T_1, T_2, x, u), \ I^f(T_1, T_2, x, u)\} \leq 2L_0(\Delta_f + 2) + 2L_0(1 + a_0)$$

the inequality

$$|I^g(T_1, T_2, x, u) - I^f(T_1, T_2, x, u)| \leq 1/8$$

holds.

Choose

$$L_1 > 8L_0(2 + L(a_0 + 1)). \tag{9.26}$$

Property (b), (8.74), and (9.24) imply that for each $g \in V$, each $S_1 \geq 0$ and each $S_2 > S_1$

$$I^g(S_1, S_2, x_f, u_f)$$

is finite.

Assume that

$$g \in V, \ T_1 \geq 0, \ T_2 \geq T_1 + L_1, \ (x, u) \in X(T_1, T_2) \tag{9.27}$$

and at least one of the following conditions (i)–(iii) holds. Define $(\widehat{x}, \widehat{u}) \in X(T_1, T_2)$. If condition (i) holds, then we set

$$\widehat{x}(t) = x_f(t), \ \widehat{u}(t) = u_f(t), \ t \in [T_1, T_2]. \tag{9.28}$$

If condition (ii) holds, then there exist $(\widehat{x}_t, \widehat{u}) \in X(T_1, T_2)$ and $\tau_1 \in (0, L]$ such that

$$\widehat{x}(T_1) = x(T_1), \widehat{x}_t = x_f(t), \ \widehat{u}(t) = u_f(t), \ t \in [T_1 + \tau_1, T_2], \tag{9.29}$$

$$I^f(T_1, T_1 + \tau_1, \widehat{x}, \widehat{u}) \le L. \tag{9.30}$$

If condition (iii) holds, then there exist $(\widehat{x}_t, \widehat{u}) \in X(T_1, T_2)$ and $\tau_1, \tau_2 \in (0, L]$ such that

$$\widehat{x}(T_1) = x(T_1), \ \widehat{x}(T_2) = x(T_2), \tag{9.31}$$

$$\widehat{x}_t = x_f(t), \ \widehat{u}(t) = u_f(t), \ t \in [T_1 + \tau_1, T_2 - \tau_2], \tag{9.32}$$

$$I^f(T_1, T_1 + \tau_1, \widehat{x}, \widehat{u}) \le L, \ I^f(T_2 - \tau_2, T_2, \widehat{x}, \widehat{u}) \le L. \tag{9.33}$$

If condition (iv) holds, then there exist $(\widehat{x}, \widehat{u}) \in X(T_1, T_2)$ and $\tau_1 \in (0, L]$ such that

$$\widehat{x}(T_2) = x(T_2), \ I^f(T_2 - \tau_1, T_2, \widehat{x}, \widehat{u}) \le L, \tag{9.34}$$

$$\widehat{x}_t = x_f(t), \ \widehat{u}(t) = u_f(t), \ t \in [T_1, T_2 - \tau_1]. \tag{9.35}$$

There exists a natural number q such that

$$qL_0 \le T_2 - T_1 < (q+1)L_0. \tag{9.36}$$

By (9.27) and (9.36),

$$q \ge 8 \text{ and } q \ge (T_2 - T_1)L_0^{-1} - 1 \ge L_1 L_0^{-1} - 1. \tag{9.37}$$

It follows from property (b), the relation $L_0 \ge L + 4$, (8.74), (9.24), (9.27)–(9.29), (9.32), and (9.35)–(9.37) that for all $i = 1, \dots, q-2$,

$$|I^g(T_1 + iL_0, T_1 + (i+1)L_0, \widehat{x}, \widehat{u}) - I^f(T_1 + iL_0, T_1 + (i+1)L_0, x_f, u_f)| \le 8^{-1}. \tag{9.38}$$

By (9.3) and (9.28)–(9.36),

$$I^f(T_1 + (q-1)L_0, T_2, \widehat{x}, \widehat{u}) \le I^f(T_1 + (q-1)L_0, T_2, \widehat{x}, \widehat{u}) + L(1 + a_0). \tag{9.39}$$

Equations (9.24), (9.27), (9.28)–(9.36), and (9.39) imply that

$$I^g(T_1 + (q-1)L_0, T_2, \widehat{x}, \widehat{u}) \le I^f(T_1 + (q-1)L_0, T_2, x_f, u_f) + L(a_0 + 1) + 1. \tag{9.40}$$

In view of (9.3) and (9.28)–(9.35) and the inequality $L_0 > L + 4$,

$$I^f(T_1, T_1 + L_0, \widehat{x}, \widehat{u}) \le I^f(T_1, T_1 + L_0, x_f, u_f) + L(1 + a_0). \tag{9.41}$$

Property (b) and Eqs. (8.74), (9.27) and (9.41) imply that

$$I^g(T_1, T_1 + L_0, \widehat{x}, \widehat{u}) \le I^f(T_1, T_1 + L_0, \widehat{x}, \widehat{u}) + 1$$

$$\le I^f(T_1, T_1 + L_0, x_f, u_f) + L(a_0 + 1) + 1. \tag{9.42}$$

By (9.38), (9.40), and (9.42),

$$I^g(T_1, T_2, \widehat{x}, \widehat{u}) \le I^f(T_1, T_2, x_f, u_f) + (q-2) + 2L(a_0 + 1) + 2. \tag{9.43}$$

It follows from the construction of $(\widehat{x}, \widehat{u})$ that in all the cases (i)–(iv),

$$I^g(T_1, T_2, x, u) \le I^g(T_1, T_2, x_f, u_f) + M \tag{9.44}$$

and it is finite. In order to complete the proof of our lemma we need only to show that (9.24) holds. Assume the contrary. Then

$$\mathrm{mes}(\{t \in [T_1, T_2] : \rho_E(x(t), x_f(t)) \le \epsilon\}) < L_0/2. \tag{9.45}$$

Property (a), (9.27), and (9.45) imply that for all $i = 0, \dots, q-2$,

$$I^g(T_1 + iL_0, T_1 + (i+1)L_0, x, u) > I^f(T_1 + iL_0, T_1 + (i+1)L_0, x_f, u_f) + M,$$

$$I^g(T_1 + (q-1)L_0, T_2, x, u) \ge I^f(T_1 + (q-1)L_0, T_2, x_f, u_f) + M.$$

By the equations above

$$I^g(T_1, T_2, x, u) \ge I^f(T_1, T_2, x_f, u_f) + Mq. \tag{9.46}$$

In view of (9.43), (9.44), and (9.46),

$$I^f(T_1, T_2, x_f, u_f) \le I^g(T_1, T_2, x, u) - qM$$

$$\le I^g(T_1, T_2, \widehat{x}, \widehat{u}) - qM + M \le M(1-q) + I^f(T_1, T_2, x_f, u_f) + q + 1 + 2L(a_0 + 1),$$

$$M(q-1) \le q + 1 + 2(a_0 + 1)L, \quad (M-1)(q-1) \le 2(1 + a_0)L + 2$$

and

$$q \leq 1 + L(a_0 + 1). \tag{9.47}$$

Equations (9.27), (9.36), and (9.47) imply that

$$L_1 \leq L_0(q+1) \leq L_0(2 + L(a_0 + 1)).$$

This contradicts (9.26). The contradiction we have reached proves (9.24) completes the proof of Lemma 9.4.

9.4 Proof of Theorem 9.2

By (A2), we may assume without loss of generality that

$$L > 8 + 4b_f + \sup\{\rho_E(\theta_0, x_f(t)) : t \in [0, \infty)\} \tag{9.48}$$

and that the following property holds:

(a) if $(T, z) \in \mathcal{A}$ and $\rho_E(z, x_f(t)) \leq \epsilon$, then $z \in \mathcal{A}_L$ and if in addition $T \geq L$, then $z \in \widehat{\mathcal{A}}_L$.

Lemma 8.36 and (A2) imply that there exist $\delta_0 \in (0, \epsilon/4)$ and $\bar{L}_0 > 0$ such that the following properties hold:

(b) for each $T_1 \geq \bar{L}_0$, each $T_2 \geq T_1 + 2b_f$ and each $(x, u) \in X(T_1, T_2)$ which satisfies

$$\rho_E(x_{T_i}, x_f(T_i)) \leq 4\delta_0, \ i = 1, 2,$$

$$I^f(T_1, T_2, x, u) \leq U^f(T_1, T_2, x(T_1), x(T_2)) + 4\delta_0$$

the inequality $\rho_E(x(t), x_f(t)) \leq \epsilon$ holds for all $t \in [T_1, T_2]$;

(c) for each $(T_i, z_i) \in \mathcal{A}$, $i = 1, 2$ satisfying $\rho_E(z_i, x_f(T_i)) \leq \delta_0$, $i = 1, 2$ and $T_2 \geq b_f$ there exist $\tau_1, \tau_2 \in (0, b_f]$ and $(x_1, u_1) \in X(T_1, T_1 + \tau_1)$, $(x_2, u_2) \in X(T_2 - \tau_2, T_2)$ which satisfies

$$x_1(T_1) = z_1, \ x_1(T_1 + \tau_1) = x_f(T_1 + \tau_1),$$

$$I^f(T_1, T_1 + \tau_1, x_1, u_1) \leq I^f(T_1, T_1 + \tau_1, x_f, u_f) + 1,$$

$$x_2(T_2) = z_2, \ x_2(T_2 - \tau_2) = x_f(T_2 - \tau_2),$$

$$I^f(T_2 - \tau_2, T_2, x_2, u_2) \leq I^f(T_2 - \tau_2, T_2, x_f, u_f) + 1.$$

Lemma 9.4 implies that there exists an integer $L_0 > L + 4 + 2\bar{L}_0$,

$$L_1 > 8L_0(2 + L(a_0 + 1))$$

and a neighborhood V_0 of f in $\mathfrak{M}_\psi$ such that the following property holds:

(d) for each $g \in V_0$,

$$I^g(S_1, S_2, x_f, u_f) \text{ is finite for each } S_2 > S_1 \geq 0$$

and for each pair of numbers $T_1 \geq 0$, $T_2 \geq T_1 + L_1$, $U^g(T_1, T_2)$ is finite, $U^g(T_1, T_2, z)$ is finite for each $(T_1, z) \in \mathcal{A}_L$ satisfying $\rho_E(z, \theta_0) \leq L$, $\widehat{U}^g(T_1, T_2, z)$ is finite for each $(T_2, z) \in \widehat{\mathcal{A}}_L$, $U^g(T_1, T_2, z_1, z_2)$ is finite for each $z_1, z_2 \in E$ satisfying $(T_1, z_1) \in \mathcal{A}_L$, $(T_2, z_2) \in \widehat{\mathcal{A}}_L$ and $\rho_E(z_1, \theta_0) \leq L$ and for each $(x, u) \in X(T_1, T_2)$ which satisfies at least one of the following conditions:

$$I^g(T_1, T_2, x, u) \leq U^g(T_1, T_2) + L;$$

$$I^g(T_1, T_2, x, u) \leq U^g(T_1, T_2, x(T_1)) + L, \ (T_1, x(T_1)) \in \mathcal{A}_L, \ \rho_E(x(T_1), \theta_0) \leq L;$$

$$I^g(T_1, T_2, x, u) \leq U^g(T_1, T_2, x(T_1), x(T_2)) + L,$$

$$(T_1, x(T_1)) \in \mathcal{A}_L, \ (T_2, x(T_2)) \in \widehat{\mathcal{A}}_L, \ \rho_E(x(T_1), \theta_0) \leq L;$$

(iv)

$$I^g(T_1, T_2, x, u) \leq \widehat{U}^g(T_1, T_2, x(T_2)) + L, \ (T_2, x(T_2)) \in \widehat{\mathcal{A}}_L$$

the value $I^g(T_1, T_2, x, u)$ is finite and the inequality

$$\text{mes}(\{t \in [T_1, T_2] : \ \rho_E(x(t), x_f(t)) \leq \delta_0\}) \geq L_0/2$$

holds.

Proposition 9.1 implies that there exists a neighborhood $V \subset V_0$ of f in $\mathfrak{M}_\psi$ such that the following property holds:

(e) for each $S_1 \geq 0$ each $S_2 \in (S_1, S_1 + L_1]$, each $g \in V$ and each $(x, u) \in X(S_1, S_2)$ which satisfies

$$\rho_E(x(S_1), \theta_0) \leq L,$$

$$\min\{I^g(S_1, S_2, x, u), \ I^f(S_1, S_2, x, u)\} \leq (L_1 + 2)\Delta_f + 4 + 2a_0$$

the inequality

$$|I^g(S_1, S_2, x, u) - I^f(S_1, S_2, x, u)| \leq \delta_0$$

is valid.

Fix

$$\delta \in (0, \delta_0/2). \tag{9.49}$$

Assume that

$$g \in V,\ T_1 \geq 0,\ T_2 \geq T_1 + 4L_1. \tag{9.50}$$

Property (d) and (9.50) imply that

$$I^g(S_1, S_2, x_f, u_f) \text{ is finite for each } S_2 > S_1 \geq 0,$$

$U^g(T_1, T_2)$ is finite, $U^g(T_1, T_2, z)$ is finite for each $(T_1, z) \in \mathcal{A}_L$ satisfying $\rho_E(z, \theta_0) \leq L$, $\widehat{U}^g(T_1, T_2, z)$ is finite for each $(T_2, z) \in \widehat{\mathcal{A}}_L$, $U^g(T_1, T_2, z_1, z_2)$ is finite for each $z_1, z_2 \in E$ satisfying $(T_1, z_1) \in \mathcal{A}_L$, $(T_2, z_2) \in \widehat{\mathcal{A}}_L$ and $\rho_E(z_1, \theta_0) \leq L$. Assume that $(x, u) \in X(T_1, T_2)$ satisfies at least one of the following conditions:

(i)

$$I^g(T_1, T_2, x, u) \leq \min\{U^g(T_1, T_2) + L,\ U^g(T_1, T_2, x(T_1), x(T_2)) + \delta\};$$

(ii)

$$I^g(T_1, T_2, x, u) \leq U^g(T_1, T_2, x(T_1)) + \delta,\ (T_1, x(T_1)) \in \mathcal{A}_L,\ \rho_E(x(T_1), \theta_0) \leq L;$$

(iii)

$$I^g(T_1, T_2, x, u) \leq U^g(T_1, T_2, x(T_1), x(T_2)) + \delta,$$

$$(T_1, x(T_1)) \in \mathcal{A}_L,\ (T_2, x(T_2)) \in \widehat{\mathcal{A}}_L,\ \rho_E(x(T_1), \theta_0) \leq L.$$

In order to complete the proof of the theorem it is sufficient to show that there exist $\tau_1 \in [T_1, T_1 + L_1]$, $\tau_2 \in [T_2 - L_1, T_2]$ such that

$$\rho_E(x(t), x_f(t)) \leq \epsilon,\ t \in [\tau_1, \tau_2];$$

if $\rho_E(x(T_2), x_f(T_2)) \leq \delta$, then $\tau_2 = T_2$ and if $T_1 \geq L_1$ and $\rho_E(x(T_1), x_f(T_1)) \leq \delta$, then $\tau_1 = T_1$.

Property (d) and (9.50) imply that

$$\text{mes}(\{t \in [T_1, T_2] :\ \rho_E(x(t), x_f(t)) \leq \delta_0\}) \geq L_0/2. \tag{9.51}$$

Set

$$S_1 = \min\{t \in [\max\{T_1, \bar{L}_0\}, T_2] : \ \rho_E(x(t), x_f(t)) \le \delta_0\}. \tag{9.52}$$

(Note that if $T_1 \ge \bar{L}_0$ and $\rho_E(x(T_1), x_f(T_1)) \le \delta$, then $S_1 = T_1$.) In view of (9.48) and (9.51), S_1 is well-defined. If $S_1 \ge T_1 + L_1$, then it follows from properties (a), (d) applied to $(x, u) \in X(T_1, S_1)$, (i)–(iii), (9.50) and (9.52) that

$$\mathrm{mes}(\{t \in [T_1, S_1] : \ \rho_E(x(t), x_f(t)) \le \delta_0\}) \ge L_0/2.$$

On the other hand, in view of (9.52),

$$\mathrm{mes}(\{t \in [T_1, S_1] : \ \rho_E(x(t), x_f(t)) \le \delta_0\}) \le \bar{L}_0$$

and $\bar{L}_0 \ge L_0/2$. This contradicts (9.48). The contradiction we have reached proves that

$$S_1 \le T_1 + L_1. \tag{9.53}$$

Set

$$S_2 = \max\{t \in [S_1, T_2] : \ \rho_E(x(t), x_f(t)) \le \delta_0\}. \tag{9.54}$$

If $S_2 \le T_2 - L_1$, then it follows from properties (a), (d) applied to $(x, u) \in X(S_2, T_2)$, and (9.50) that

$$\mathrm{mes}(\{t \in [S_2, T_2] : \ \rho_E(x(t), x_f(t)) \le \delta_0\}) \ge L_0/2$$

and this contradicts (9.54). Thus

$$S_2 \ge T_2 - L_1. \tag{9.55}$$

In order to complete the proof of the theorem it is sufficient to show that

$$\rho_E(x(t), x_f(t)) \le \epsilon, \ t \in [S_1, S_2].$$

Assume the contrary. Then there exists

$$\tau_0 \in [S_1, S_2] \tag{9.56}$$

such that

$$\rho_E(x(\tau_0), x_f(\tau_0)) > \epsilon. \tag{9.57}$$

In view of (9.50), (9.53), and (9.55),

$$S_2 - S_1 \ge T_2 - T_1 - 2L_1 \ge 2L_1. \tag{9.58}$$

By (9.58) and (9.59), there exist numbers S_3, S_4 such that

$$S_4 = S_3 + L_0/4, \ \ S_1 \le S_3 \le \tau_0 \le S_4 \le S_2. \tag{9.59}$$

Set

$$S_5 = \max\{t \in [S_1, S_3] : \ \rho_E(x(t), x_f(t)) \le \delta_0\}, \tag{9.60}$$

$$S_6 = \min\{t \in [S_4, S_2] : \ \rho_E(x(t), x_f(t)) \le \delta_0\}. \tag{9.61}$$

It follows from (9.52), (9.54) and (9.59)–(9.61) that

$$S_6 - S_5 \ge S_4 - S_3 = L_0/4. \tag{9.62}$$

Assume that

$$S_6 - S_5 \ge L_1.$$

Properties (a) and (d) applied to $(x, u) \in X(S_5, S_6)$, conditions (i)–(iii) and Eqs. (9.50), (9.60) and (9.61) imply that

$$\mathrm{mes}(\{t \in [S_5, S_6] : \ \rho_E(x(t), x_f(t)) \le \delta\}) \ge L_0/2.$$

On the other hand, by (9.59)–(9.61),

$$\mathrm{mes}(\{t \in [S_5, S_6] : \ \rho_E(x(t), x_f(t)) \le \delta_0\}) \le S_4 - S_3 = L_0/4.$$

The contradiction we have reached proves that

$$S_6 - S_5 \le L_1.$$

Together with (9.62) this implies that

$$L_0/4 \le S_6 - S_5 \le L_1. \tag{9.63}$$

Property (c) and Eqs. (9.48), (9.60), (9.61), and (9.63) imply that there exist $(y, v) \in X(S_5, S_6)$ and $t_1, t_2 \in (0, b_f]$ such that

$$y(S_5) = x(S_5), \ \ y(S_6) = x(S_6),$$

$$y(t) = x_f(t), \ \ v(t) = u_f(t), \ \ t \in [S_5 + t_1, S_6 - t_2],$$

$$I^f(S_5, S_5 + t_1, y, v) \le I^f(S_5, S_5 + t_1, x_f, u_f) + 1,$$

$$I^f(S_6 - t_2, S_6, y, v) \le I^f(S_6 - t_2, S_6, x_f, u_f) + 1. \tag{9.64}$$

By (8.74), (9.3), (9.63), and (9.64),

$$I^f(S_5, S_6, y, v) \le I^f(S_5, S_6, x_f, u_f) + 2 \le 2 + (L_1 + 2)\Delta_f + 2a_0. \tag{9.65}$$

In view of conditions (i)–(iii),

$$I^f(S_5, S_6, x, u) \le U^g(S_5, S_6, x(S_5), x(S_6)) + \delta_0. \tag{9.66}$$

It follows from (9.64) and (9.65) that

$$U^g(S_5, S_6, x(S_5), x(S_6)) = \inf\{I^f(S_5, S_6, \xi, \eta) :$$

$$(\xi, \eta) \in X(S_5, S_6),\ I^f(S_5, S_6, \xi, \eta) \le 2 + (L_1 + 2)\Delta_f + 2a_0\}. \tag{9.67}$$

Property (e) and Eqs. (9.50), (9.60), (9.63)–(9.65), and (9.67) imply that

$$U^g(S_5, S_6, x(S_5), x(S_6)) \le \inf\{I^g(S_5, S_6, \xi, \eta) :\ (\xi, \eta) \in X(S_5, S_6),$$

$$I^f(S_5, S_6, \xi, \eta) \le 2 + (L_1 + 2)\Delta_f + 2a_0\}$$

$$\le \delta_0 + \inf\{I^g(S_5, S_6, \xi, \eta) :\ (\xi, \eta) \in X(S_5, S_6),$$

$$I^f(S_5, S_6, \xi, \eta) \le 2 + (L_1 + 2)\Delta_f + 2a_0\}$$

$$\le \delta_0 + U^f(S_5, S_6, x(S_5), x(S_6)). \tag{9.68}$$

By (9.66)–(9.68),

$$I^g(S_5, S_6, x, u) \le (L_1 + 2)\Delta_f + 2a_0 + 3.$$

Together with property (e) and Eqs. (9.50), (9.60) and (9.63) this implies that

$$|I^g(S_5, S_6, x, u) - I^f(S_5, S_6, x, u)| \le \delta_0.$$

Combined with (9.65) and (9.68) this implies that

$$I^f(S_5, S_6, x, u) \le I^g(S_5, S_6, x, u) + \delta_0$$

$$\le U^g(S_5, S_6, x(S_5), x(S_6)) + 2\delta_0 \le U^f(S_5, S_6, x(S_5), x(S_6)) + 3\delta_0.$$

Property (b), (9.60), (9.61), (9.63) and the equation above imply that

$$\rho_E(x(t), x_f(t)) \le \epsilon,\ t \in \{S_5, S_6].$$

Together with (9.59)–(9.61) this implies that

$$\rho_E(x(\tau_0), x_f(\tau_0)) \le \epsilon.$$

This contradicts (9.57). The contradiction we have reached competes the proof of Theorem 9.2.

9.5 Problems with Discount

We use the notation, definitions, and assumptions introduced in Sects. 9.1 and 9.2.

Let $g \in \mathfrak{M}_\psi$ and $\alpha : [0, \infty) \to [0, \infty)$ be a Borelian bounded function. For each pair of numbers $T_2 > T_1 \ge 0$ and each $(x, u) \in X(T_1, T_2)$ define

$$I^{\alpha g}(T_1, T_2, x, u) = \int_{T_1}^{T_2} \alpha(t) g(t, x(t), u(t)) dt \in (-\infty, \infty]$$

and for each pair of numbers $T_2 > T_1 \ge 0$ and each pair of points $(T_1, y), (T_2, z) \in \mathcal{A}$ we define

$$U^{\alpha g}(T_1, T_2, y, z) = \inf\{I^{\alpha g}(T_1, T_2, x, u) :$$

$$(x, u) \in X(T_1, T_2),\ x(T_1) = y,\ x(T_2) = z\},$$

$$U^{\alpha g}(T_1, T_2, y) = \inf\{I^{\alpha g}(T_1, T_2, x, u) : (x, u) \in X(T_1, T_2),\ x(T_1) = y\},$$

$$\widehat{U}^{\alpha g}(T_1, T_2, z) = \inf\{I^{\alpha g}(T_1, T_2, x, u) : (x, u) \in X(T_1, T_2),\ x(T_2) = z\},$$

$$U^{\alpha g}(T_1, T_2) = \inf\{I^{\alpha g}(T_1, T_2, x, u) : (x, u) \in X(T_1, T_2)\}. \tag{9.69}$$

We prove the following stability result.

Theorem 9.5 *Assume that $L > 0$ and $\epsilon \in (0, 1)$. Then there exist $L_1 > L$, $\delta \in (0, \epsilon)$ $\lambda \in (0, 1)$ and a neighborhood V of f in $\mathfrak{M}_\psi$ such that for each $g \in V$, each pair of numbers $T_1 \ge 0$, $T_2 \ge T_1 + 4L_1$, each borelian function $\alpha : [0, \infty) \to [0, 1]$ which satisfies*

$$\alpha(t_1)\alpha(t_2)^{-1} \ge \lambda \textit{ for each } t_1, t_2 \in [T_1, T_2] \textit{ satisfying } |t_2 - t_1| \le L_1$$

$U^{\alpha g}(T_1, T_2)$ is finite, $U^{\alpha g}(T_1, T_2, z)$ is finite for each $(z, T_1) \in \mathcal{A}_L$ satisfying $\rho_E(z, \theta_0) \le L$, $U^{\alpha g}(T_1, T_2, z_1, z_2)$ is finite for each $z_1, z_2 \in E$ satisfying $(T_1, z_1) \in$

$\mathcal{A}_L$, $(T_2, z_2) \in \widehat{\mathcal{A}}_L$ and $\rho_E(z_1, \theta_0) \leq L$ and that for each $(x, u) \in X(T_1, T_2)$ which satisfies at least one of the following conditions:

(i)

$$I^{\alpha g}(T_1, T_2, x, u) \leq U^{\alpha g}(T_1, T_2) + \delta \inf\{\alpha(t) : t \in [T_1, T_2]\};$$

(ii)

$$I^{\alpha g}(T_1, T_2, x, u) \leq U^{\alpha g}(T_1, T_2, x(T_1))$$

$$+ \delta \inf\{\alpha(t) : t \in [T_1, T_2]\}, \ (T_1, x(T_1)) \in \mathcal{A}_L, \ \rho_E(x(T_1), \theta_0) \leq L;$$

(iii)

$$I^{\alpha g}(T_1, T_2, x, u) \leq U^{\alpha g}(T_1, T_2, x(T_1), x(T_2)) + \delta \inf\{\alpha(t) : t \in [T_1, T_2]\},$$

$$(T_1, x(T_1)) \in \mathcal{A}_L, \ (T_2, x(T_2)) \in \widehat{\mathcal{A}}_L, \ \rho_E(x(T_1), \theta_0) \leq L$$

there exist $\tau_1 \in [T_1, T_1 + L_1]$, $\tau_2 \in [T_2 - L_1, T_2]$ such that

$$\rho_E(x(t), x_f(t)) \leq \epsilon, \ t \in [\tau_1, \tau_2].$$

Moreover, if $\rho_E(x(T_2), x_f(T_2)) \leq \delta$, then $\tau_2 = T_2$ and if $T_1 \geq L_1$ and $\rho_E(x(T_1), x_f(T_1)) \leq \delta$, then $\tau_1 = T_1$.

9.6 Auxiliary Results for Theorem 9.5

Lemma 9.6 *Assume that $M \geq 2$, $\epsilon \in (0, 1)$ and $L > 0$. Then there exists an integer $L_0 \geq 4 + L$ such that for each $L_1 > L_0$ there exist $\lambda \in (0, 1)$ and a neighborhood V of f in $\mathfrak{M}_\psi$ such that for each $g \in V$ and each pair of numbers $T_1 \geq 0$, $T_2 \in [T_1 + L_0, T_1 + L_1]$, each borelian function $\alpha : [0, \infty) \to [0, 1]$ which satisfies*

$$\alpha(t_1)\alpha(t_2)^{-1} \geq \lambda \text{ for each } t_1, t_2 \in [T_1, T_2] \text{ satisfying } |t_2 - t_1| \leq L_1 \tag{9.70}$$

and for each $(x, u) \in X(T_1, T_2)$ which satisfies

$$I^{\alpha g}(T_1, T_2, x, u) \leq I^{\alpha f}(T_1, T_2, x_f, u_f) + 4^{-1} M \inf\{\alpha(t) : t \in [T_1, T_2]\} \tag{9.71}$$

the inequality

$$mes(\{t \in [T_1, T_2] : \rho_E(x(t), x_f(t)) > \epsilon\}) \leq L_0/2 \tag{9.72}$$

holds.

Proof Lemma 9.3 implies that there exists an integer $L_0 \geq 4 + L$ such that for each $L_1 > L_0$ the following property holds:

(a) there exists a neighborhood V of f in $\mathfrak{M}_\psi$ such that for each $g \in V$ and each pair of numbers $T_1 \geq 0$, $T_2 \in [T_1 + L_0, T_1 + L_1]$, $I^g(T_1, T_2, x_f, u_f)$ is finite and for each $(x, u) \in X(T_1, T_2)$ which satisfies

$$I^g(T_1, T_2, x, u) \leq I^f(T_1, T_2, x_f, u_f) + M$$

inequality (9.72) holds.

Assume that $L_1 > L_0$ and a neighborhood V of f in $\mathfrak{M}_\psi$ be as guaranteed by property (a). Choose $\lambda \in (0, 1)$ such that

$$(1 - \lambda)(\Delta_f + 1)(L_1 + 2)(a_0 + 1) < 1. \tag{9.73}$$

Assume that

$$g \in V,\ T_1 \geq 0,\ T_2 \in [T_1 + L_0, T_1 + L_1], \tag{9.74}$$

a borelian function $\alpha : [0, \infty) \to [0, 1]$ satisfies (9.70) and that $(x, u) \in X(T_1, T_2)$ satisfies (9.71). In order to complete the proof of the lemma it is sufficient to show that (9.72) holds.

Assume then contrary. Then by property (a) and (9.74),

$$I^g(T_1, T_2, x, u) > M + I^f(T_1, T_2, x_f, u_f). \tag{9.75}$$

By (9.3) and (9.75),

$$\begin{aligned}
I^{\alpha g}(T_1, T_2, x, u) &= -a_0 \int_{T_1}^{T_2} \alpha(t)dt + (I^{\alpha g}(T_1, T_2, x, u) + a_0 \int_{T_1}^{T_2} \alpha(t)dt) \\
&= -a_0 \int_{T_1}^{T_2} \alpha(t)dt + \int_{T_1}^{T_2} (\alpha(t)g(t, x(t), u(t)) + a_0)dt \\
&\geq -a_0 \int_{T_1}^{T_2} \alpha(t)dt + \inf\{\alpha(t) :\ t \in [T_1, T_2]\} \int_{T_1}^{T_2} (g(t, x(t), u(t)) + a_0)dt \\
&\geq -a_0 \int_{T_1}^{T_2} \alpha(t)dt + \inf\{\alpha(t) :\ t \in [T_1, T_2]\} \int_{T_1}^{T_2} (f(t, x_f(t), u_f(t)) + a_0)dt \\
&\quad + M \inf\{\alpha(t) :\ t \in [T_1, T_2]\}.
\end{aligned} \tag{9.76}$$

By (8.74), (9.3), (9.70), (9.73), and (9.74),

$$\sup\{\alpha(t) : t \in [T_1, T_2]\} \int_{T_1}^{T_2} (f(t, x_f(t), u_f(t)) + a_0) dt$$

$$- \inf\{\alpha(t) : t \in [T_1, T_2]\} \int_{T_1}^{T_2} (f(t, x_f(t), u_f(t)) + a_0) dt$$

$$\leq (1 - \lambda) \sup\{\alpha(t) : t \in [T_1, T_2]\} \int_{T_1}^{T_2} (f(t, x_f(t), u_f(t)) + a_0) dt$$

$$\leq (1 - \lambda) \sup\{\alpha(t) : t \in [T_1, T_2]\}(\Delta_f(L_1 + 2) + 2a_0 + a_0 L_1)$$

$$\leq \sup\{\alpha(t) : t \in [T_1, T_2]\}. \tag{9.77}$$

By (9.3), (9.76), and (9.77),

$$I^{\alpha g}(T_1, T_2, x, u) \geq -a_0 \int_{T_1}^{T_2} \alpha(t) dt$$

$$+ \sup\{\alpha(t) : t \in [T_1, T_2]\} \int_{T_1}^{T_2} (f(t, x_f(t), u_f(t)) + a_0) dt$$

$$- \sup\{\alpha(t) : t \in [T_1, T_2]\} + M \inf\{\alpha(t) : t \in [T_1, T_2]\}$$

$$\geq I^{\alpha f}(T_1, T_2, x_f, u_f) - \sup\{\alpha(t) : t \in [T_1, T_2]\} + M \inf\{\alpha(t) : t \in [T_1, T_2]\}$$

$$\geq I^{\alpha f}(T_1, T_2, x_f, u_f) + 2^{-1} M \inf\{\alpha(t) : t \in [T_1, T_2]\}.$$

This contradicts (9.71). The contradiction we have reached proves Lemma 9.6.

Lemma 9.7 *Assume that* $\epsilon > 0$ *and*

$$L > \sup\{\rho_E(\theta_0, x_f(t)) : t \in [0, \infty)\}. \tag{9.78}$$

Then there exists an integer $L_0 \geq L + 4$, $L_1 > 8L_0 + 8$, $\lambda \in (0, 1)$ *and a neighborhood* V *of* f *in* $\mathfrak{M}_\psi$ *such that for each* $g \in V$ *and each pair of numbers* $T_1 \geq 0$, $T_2 \geq T_1 + L_1$, *each borelian function* $\alpha : [0, \infty) \to [0, 1]$ *which satisfies*

$$\alpha(t_1)\alpha(t_2)^{-1} \geq \lambda \text{ for each } t_1, t_2 \in [T_1, T_2] \text{ satisfying } |t_2 - t_1| \leq L_1 \tag{9.79}$$

the following assertion holds.

Assume that $(x, u) \in X(T_1, T_2)$ *satisfies at least one of the following conditions:*

(i)

$$I^{\alpha g}(T_1, T_2, x, u) \le U^{\alpha g}(T_1, T_2) + 4\inf\{\alpha(t) : t \in [T_1, T_2]\};$$

(ii)

$$(T_1, x(T_1)) \in \mathcal{A}_L, \ \rho_E(x(T_1), \theta_0) \le L,$$

$$I^{\alpha g}(T_1, T_2, x, u) \le U^{\alpha g}(T_1, T_2, x(T_1)) + 4\inf\{\alpha(t) : t \in [T_1, T_2]\};$$

(iii)

$$(T_2, x(T_2)) \in \widehat{\mathcal{A}}_L,$$

$$I^{\alpha g}(T_1, T_2, x, u) \le \widehat{U}^{\alpha g}(T_1, T_2, x(T_2)) + 4\inf\{\alpha(t) : t \in [T_1, T_2]\};$$

(iv)

$$(T_1, x(T_1)) \in \mathcal{A}_L, \ (T_2, x(T_2)) \in \widehat{\mathcal{A}}_L, \ \rho_E(x(T_1), \theta_0) \le L,$$

$$I^{\alpha g}(T_1, T_2, x, u) \le U^{\alpha g}(T_1, T_2, x(T_1), x(T_2)) + 4\inf\{\alpha(t) : t \in [T_1, T_2]\}.$$

Then $I^{\alpha g}(T_1, T_2, x, u)$ is finite and

$$mes(\{t \in [T_1, T_2] : \rho_E(x(t), x_f(t)) \le \epsilon\}) \ge L_0/2. \tag{9.80}$$

Proof Lemma 9.6 implies that there exist an integer $L_0 \ge 4 + L$, $\lambda_0 \in (0, 1)$ and a neighborhood V_0 of f in $\mathfrak{M}_\psi$ such that the following property holds:

(a) for each $g \in V_0$, each pair of numbers $T_1 \ge 0$, $T_2 \in [T_1 + L_0, T_1 + 8(L_0 + 1)]$, each borelian function $\alpha : [0, \infty) \to [0, 1]$ which satisfies

$$\alpha(t_1)\alpha(t_2)^{-1} \ge \lambda_0 \text{ for each } t_1, t_2 \in [T_1, T_2]$$

and for each $(x, u) \in X(T_1, T_2)$ which satisfies

$$I^{\alpha g}(T_1, T_2, x, u) \le I^{\alpha f}(T_1, T_2, x_f, u_f) + 8\inf\{\alpha(t) : t \in [T_1, T_2]\}$$

the inequality

$$\text{mes}(\{t \in [T_1, T_2] : \rho_E(x(t), x_f(t)) > \epsilon\}) \le L_0/2 \tag{9.81}$$

holds.

We may assume without loss of generality that

$$\lambda_0 > 1/2, \ (1 - \lambda_0)((2L_0\Delta_f + 2a_0 + 2L_0 a_0) < 8^{-1}. \tag{9.82}$$

Proposition 9.1 implies that there exists a neighborhood $V \subset V_0$ of f in $\mathfrak{M}_\psi$ such that the following property holds:

(b) for each pair of numbers $T_1 \geq 0$, $T_2 \in (T_1, T_1 + 8(L_0 + 1)]$, each $g \in V$ and each $(x, u) \in X(T_1, T_2)$ which satisfies

$$\rho_E(x(T_1), \theta_0) \leq L + 4,$$

$$\min\{I^g(T_1, T_2, x, u), \ I^f(T_1, T_2, x, u)\} \leq 2L_0(\Delta_f + 1) + 2L(1 + a_0)$$

the inequality

$$|I^g(T_1, T_2, x, u) - I^f(T_1, T_2, x, u)| \leq 1/8$$

holds.

Choose integers

$$q_0 \geq 4 + 16((L_0 + 1)(a_0 + 1) + 2),$$

$$L_1 > L_0(8 + L(a_0 + 1)) + L_0(2q_0 + 4) \tag{9.83}$$

and

$$\lambda \in (\lambda_0, 1)$$

such that

$$\lambda^{q_0} > 2^{-1}. \tag{9.84}$$

Assume that

$$g \in V, \ T_1 \geq 0, \ T_2 \geq T_1 + L_1,$$

a borelian function $\alpha : [0, \infty) \to [0, 1]$ satisfies (9.79), $(x, u) \in X(T_1, T_2)$ and at least one of cases (i)–(iv) holds. Define $(\widehat{x}, \widehat{u}) \in X(T_1, T_2)$. If case (i) holds, then we set

$$\widehat{x}(t) = x_f(t), \ \widehat{u}(t) = u_f(t), \ t \in [T_1, T_2], \tag{9.85}$$

If case (ii) holds, then there exist $(\widehat{x}, \widehat{u}) \in X(T_1, T_2)$ and $\tau_1 \in (0, L]$ such that

$$\widehat{x}(T_1) = x(T_1), \widehat{x}(t) = x_f(t), \ \widehat{u}(t) = u_f(t), \ t \in [T_1 + \tau_1, T_2],$$

$$I^f(T_1, T_1 + \tau_1, \widehat{x}, \widehat{u}) \le L. \tag{9.86}$$

If case (iv) holds, then there exist $(\widehat{x}, \widehat{u}) \in X(T_1, T_2)$ and $\tau_1, \tau_2 \in (0, L]$ such that

$$\widehat{x}(T_1) = x(T_1),\ \widehat{x}(T_2) = x(T_2),$$

$$\widehat{x}(t) = x_f(t),\ \widehat{u}(t) = u_f(t),\ t \in [T_1 + \tau_1, T_2 - \tau_2],$$

$$I^f(T_1, T_1 + \tau_1, \widehat{x}, \widehat{u}) \le L,\ I^f(T_2 - \tau_2, T_2, \widehat{x}, \widehat{u}) \le L. \tag{9.87}$$

If case (iii) holds, then there exist $(\widehat{x}, \widehat{u}) \in X(T_1, T_2)$ and $\tau_1 \in (0, L]$ such that

$$\widehat{x}(T_2) = x(T_2),\ I^f(T_2 - \tau_1, T_2, \widehat{x}, \widehat{u}) \le L,$$

$$\widehat{x}(t) = x_f(t),\ \widehat{u}(t) = u_f(t),\ t \in [T_1, T_2 - \tau_1]. \tag{9.88}$$

There exists a natural number q such that

$$qL_0 \le T_2 - T_1 < (q + 1)L_0. \tag{9.89}$$

By (9.83), (9.87), and (9.89),

$$q \ge (T_2 - T_1)L_0^{-1} - 1 \ge L_1 L_0^{-1} - 1 \ge 8. \tag{9.90}$$

It follows from property (b), (8.74), (9.3), (9.76), (9.78), (9.84)–(9.90) and the relation $L_0 \ge L + 4$ that for all $i = 1, \dots, q - 2$,

$$|I^g(T_1+iL_0, T_1+(i+1)L_0, \widehat{x}, \widehat{u}) - I^f(T_1+iL_0, T_1+(i+1)L_0, x_f, u_f)| \le 8^{-1}. \tag{9.91}$$

By (9.3), (9.78) and (9.85)–(9.89),

$$I^f(T_1 + (q - 1)L_0, T_2, \widehat{x}, \widehat{u}) \le I^f(T_1 + (q - 1)L_2, T_2, \widehat{x}, \widehat{u}) + L(1 + a_0). \tag{9.92}$$

Property (b) and Eqs. (8.74), (9.3), (9.84), and (9.92) imply that

$$I^g(T_1+(q-1)L_0, T_2, \widehat{x}, \widehat{u}) \le I^f(T_1+(q-1)L_0, T_2, x_f, u_f) + L(a_0+1) + 1. \tag{9.93}$$

In view of (9.3), (9.78), (9.83) and (9.85)–(9.88),

$$I^f(T_1, T_1 + L_0, \widehat{x}, \widehat{u}) \le I^f(T_1, T_1 + L_0, x_f, u_f) + L(1 + a_0). \tag{9.94}$$

Property (b) and Eqs. (8.74), (9.3), (9.76), (9.78), (9.84), and (9.94) imply that

$$I^g(T_1, T_1 + L_0, \widehat{x}, \widehat{u}) \le I^f(T_1, T_1 + L_0, \widehat{x}, \widehat{u}) + 1$$

$$\le I^f(T_1, T_1 + L_0, x_f, u_f) + L(1 + a_0) + 1. \tag{9.95}$$

Let

$$\{S_1, S_2\} \in \{(T_1 + iL_0, T_1 + (i+1)L_0 : i = 1, \dots, q-2\}. \tag{9.96}$$

In view of (9.3), (9.79), (9.82)–(9.84), and (9.91),

$$I^{\alpha g}(S_1, S_2, \widehat{x}, \widehat{u}) = \int_{T_1}^{T_2} (\alpha(t) g(t, \widehat{x}(t), \widehat{u}(t)) + a_0) dt - a_0 \int_{S_1}^{S_2} \alpha(t) dt$$

$$\le \sup\{\alpha(t) : t \in [S_1, S_2]\} \int_{S_1}^{S_2} (g(t, \widehat{x}(t), \widehat{u}(t)) + a_0) dt - a_0 \int_{S_1}^{S_2} \alpha(t) dt$$

$$\le \sup\{\alpha(t) : t \in [S_1, S_2]\} \int_{S_1}^{S_2} (f(t, x_f(t), u_f(t)) + a_0) dt$$

$$+ 8^{-1} \sup\{\alpha(t) : t \in [S_1, S_2]\} - a_0 \int_{S_1}^{S_2} \alpha(t) dt$$

$$\le \int_{S_1}^{S_2} (f(t, x_f(t), u_f(t)) + a_0) \alpha(t) dt$$

$$+ (1 - \lambda) \sup\{\alpha(t) : t \in [S_1, S_2]\} \int_{S_1}^{S_2} (f(t, x_f(t), u_f(t)) + a_0) dt$$

$$+ 8^{-1} \sup\{\alpha(t) : t \in [S_1, S_2]\} - a_0 \int_{S_1}^{S_2} \alpha(t) dt$$

$$\le I^{\alpha f}(S_1, S_2, x_f, u_f) + (1 - \lambda) \sup\{\alpha(t) : t \in [S_1, S_2]\}((L_0 + 2)\Delta_f + 2a_0)$$

$$+ 8^{-1} \sup\{\alpha(t) : t \in [S_1, S_2]\}$$

$$\le I^{\alpha f}(S_1, S_2, x_f, u_f) + (1 - \lambda)\lambda^{-1} \inf\{\alpha(t) : t \in [S_1, S_2]\}((L_0 + 2)\Delta_f + 2a_0)$$

$$+ 8^{-1} \sup\{\alpha(t) : t \in [S_1, S_2]\}$$

$$\le I^{\alpha f}(S_1, S_2, x_f, u_f) + 2^{-1} \inf\{\alpha(t) : t \in [S_1, S_2]\}.$$

Thus for all $i = 1, \dots, q - 2$,

$$I^{\alpha g}(T_1 + iL_0, T_1 + (i+1)L_0, \widehat{x}, \widehat{u})$$

$$\leq I^{\alpha f}(T_1 + iL_0, T_1 + (i+1)L_0, x_f, u_f)$$

$$+ 2^{-1} \inf\{\alpha(t) : t \in [T_1 + iL_0, T_1 + (i+1)L_0]\}. \tag{9.97}$$

By (8.74), (9.3), (9.79), (9.82), (9.84), (9.93), and (9.95),

$$I^{\alpha g}(S_1, S_2, \widehat{x}, \widehat{u}) = \int_{S_1}^{S_2} (\alpha(t) g(t, \widehat{x}(t), \widehat{u}(t)) + a_0) dt - a_0 \int_{S_1}^{S_2} \alpha(t) dt$$

$$\leq \sup\{\alpha(t) : t \in [S_1, S_2]\} \int_{S_1}^{S_2} (g(t, \widehat{x}(t), \widehat{u}(t)) + a_0) dt - a_0 \int_{S_1}^{S_2} \alpha(t) dt$$

$$\leq \sup\{\alpha(t) : t \in [S_1, S_2]\} (\int_{S_1}^{S_2} (f(t, x_f(t), u_f(t)) + a_0) dt$$

$$+ L(a_0 + 1) + 1) - a_0 \int_{S_1}^{S_2} \alpha(t) dt$$

$$\leq \sup\{\alpha(t) : t \in [S_1, S_2]\} (1 + a(L_0 + 1)) - a_0 \int_{S_1}^{S_2} \alpha(t) dt$$

$$+ I^{\alpha f}(S_1, S_2, x_f, u_f) + a_0 \int_{S_1}^{S_2} \alpha(t) dt$$

$$+ (\int_{S_1}^{S_2} (f(t, x_f(t), u_f(t)) + a_0) dt) \sup\{\alpha(t) : t \in [S_1, S_2]\} (1 - \lambda)$$

$$\leq I^{\alpha f}(S_1, S_2, x_f, u_f) + \lambda^{-1} \sup\{\alpha(t) : t \in [S_1, S_2]\} (1 + a(L_0 + 1))$$

$$+ (\Delta_f (L_0 + 2) + 2a_0) \lambda^{-1} (1 - \lambda) \inf\{\alpha(t) : t \in [S_1, S_2]\}$$

$$\leq I^{\alpha f}(S_1, S_2, x_f, u_f) + 2^{-1} \inf\{\alpha(t) : t \in [S_1, S_2]\}$$

$$+ 2 \inf\{\alpha(t) : t \in [S_1, S_2]\} (1 + a(L_0 + 2)). \tag{9.98}$$

It follows from (9.89) and (9.98) that

$$I^{\alpha g}(T_1, T_2, \widehat{x}, \widehat{u}) \leq I^{\alpha f}(T_1, T_2, x_f, u_f)$$

$$+2^{-1}\sum_{i=0}^{q-2}\inf\{\alpha(t):\ t\in[T_1+iL_0,T_1+(i+1)L_0]\}$$

$$+2^{-1}\inf\{\alpha(t):\ t\in[T_1+(q-1)L_0,T_2]\}$$

$$+(1+a_0(L_0+2))(2\inf\{\alpha(t):\ t\in[T_1,T_1+L_0]\}$$

$$+2\inf\{\alpha(t):\ t\in[T_1+(q-1)L_0,T_2]\}). \tag{9.99}$$

In view of (9.99), $I^{\alpha g}(T_1,T_2,\widehat{x},\widehat{u})$ is finite. Equations (9.86) and (9.88) imply that $I^{\alpha g}(T_1,T_2,x,u)$ is finite too. In order to complete the proof of our lemma we need only to show that (9.80) holds. Assume the contrary. Then

$$\text{mes}(\{t\in[T_1,T_2]:\ \rho_E(x(t),x_f(t))\le\epsilon\})<L_0/2. \tag{9.100}$$

Property (a), (9.79), (9.83), (9.84), (9.89), and (9.100) imply that for all $i=0,\dots,q-2$,

$$I^{\alpha g}(T_1+iL_0,T_1+(i+1)L_0,x,u)>I^{\alpha f}(T_1+iL_0,T_1+(i+1)L_0,x_f,u_f)$$

$$+8\inf\{\alpha(t):\ t\in[T_1+iL_0,T_1+(i+1)L_0]\}, \tag{9.101}$$

$$I^{\alpha g}(T_1+(q-1)L_0,T_2,x,u)>I^{\alpha f}(T_1+(q-1)L_0,T_2,x_f,u_f)$$

$$+8\inf\{\alpha(t):\ t\in[T_1+(q-1)L_0,T_2]\}. \tag{9.102}$$

By (9.98), (9.101), and (9.102),

$$I^{\alpha g}(T_1,T_2,x,u)>I^{\alpha f}(T_1,T_2,x_f,u_f)$$

$$+8(\sum_{i=0}^{q-2}\inf\{\alpha(t):\ t\in[T_1+iL_0,T_1+(i+1)L_0]\}$$

$$+\inf\{\alpha(t):\ t\in[T_1+(q-1)L_0,T_2]\})$$

$$\ge I^{\alpha g}(T_1,T_2,\widehat{x},\widehat{u})+7(\sum_{i=0}^{q-2}\inf\{\alpha(t):\ t\in[T_1+iL_0,T_1+(i+1)L_0]\}$$

$$+\inf\{\alpha(t):\ t\in[T_1+(q-1)L_0,T_2]\})$$

$$-(2\inf\{\alpha(t):\ t\in[T_1,T_1+L_0]\}$$

$$+ 2\inf\{\alpha(t) : t \in [T_1 + (q-1)L_0, T_2]\})(2 + (a_0+1)(L_0+2)). \tag{9.103}$$

Properties (a)–(d) and Eqs. (9.85), (9.88) imply that

$$I^{\alpha g}(T_1, T_2, x, u) \le I^{\alpha g}(T_1, T_2, \widehat{x}, \widehat{u}) + 4\inf\{\alpha(t) : t \in [T_1, T_2]\}. \tag{9.104}$$

It follows from (9.103) and (9.104) that

$$4(2 + (a_0+1)(L_0+2))(\inf\{\alpha(t) : t \in [T_1, T_1 + L_0]\}$$

$$+ \inf\{\alpha(t) : t \in [T_1 + (q-1)L_0, T_2]\})$$

$$\ge \sum_{i=0}^{q-2} \inf\{\alpha(t) : t \in [T_1 + iL_0, T_1 + (i+1)L_0]\}$$

$$+ \inf\{\alpha(t) : t \in [T_1 + (q-1)L_0, T_2]\}. \tag{9.105}$$

There exists

$$j \in \{0, \dots, q-2\} \tag{9.106}$$

such that

$$\inf\{\alpha(t) : t \in [T_1 + iL_0, T_1 + (i+1)L_0]\}$$

$$\le \inf\{\alpha(t) : t \in [T_1 + jL_0, T_1 + (j+1)L_0]\},\ i = 0, \dots, q-2. \tag{9.107}$$

In view of (9.83) and (9.90),

$$q - 2 \ge L_1 L_0^{-1} - 3 \ge 2q_0 + 2. \tag{9.108}$$

By (9.106) and (9.108), there exists an integer $j_0 \ge 0$ such that

$$j_0 + q_0 \le q - 2,\ j_0 \le j \le j_0 + q_0. \tag{9.109}$$

It follows from (9.79), (9.83), (9.107), and (9.109) that

$$\sum_{i=0}^{q-2} \inf\{\alpha(t) : t \in [T_1 + iL_0, T_1 + (i+1)L_0]\}$$

$$\ge \sum_{i=j_0}^{j_0+q_0-1} \inf\{\alpha(t) : t \in [T_1 + iL_0, T_1 + (i+1)L_0]\}$$

$$\geq q_0\lambda^{q_0}\inf\{\alpha(t):\ t\in[T_1+jL_0,T_1+(j+1)L_0]\}. \tag{9.110}$$

Equations (9.83), (9.84), (9.105), (9.107), and (9.110) imply that

$$8(2+((a_0+1)(L_0+1))\min\{\alpha_t:\ t\in[T_1+jL_0,T_1+(j+1)L_0]\}\lambda^{-1}$$

$$\geq \lambda^{q_0}q_0\min\{\alpha_t:\ t\in[T_1+jL_0,T_i+(j+1)L_0]\}.$$

Together with (9.84) this implies that

$$16(2+(a_0+1)(L_0+1))\geq q_0.$$

This contradicts (9.82). The contradiction we have reached proves Lemma 9.7.

9.7 Proof of Theorem 9.5

We may assume without loss of generality that

$$L>8b_f+\sup\{\rho_E(\theta_0,x_f(t)):\ t\in[0,\infty)\}. \tag{9.111}$$

Lemma 8.36 and (A2) imply that there exist $\delta_0\in(0,\epsilon)$ and $\bar{L}_0>0$ such that the following properties hold:

(a) for each $T_1\geq\bar{L}_0$, each $T_2\geq T_1+2b_f$ and each $(x,u)\in X(T_1,T_2)$ which satisfies

$$\rho_E(x(T_i),x_f(T_i))\leq 4\delta_0,\ i=1,2,$$

$$I^f(T_1,T_2,x,u)\leq U^f(T_1,T_2,x(T_1),x(T_2))+8\delta_0$$

the inequality $\rho_E(x(t),x_f(t))\leq\epsilon$ holds for all $t\in[T_1,T_2]$;

(b) for each $(T_i,z_i)\in\mathcal{A}$, $i=1,2$ satisfying $\rho_E(z_i,x_f(T_i))\leq\delta_0$, $i=1,2$ and $T_2\geq b_f$ there exist $\tau_1,\tau_2\in(0,b_f]$ and $(x_1,u_1)\in X(T_1,T_1+\tau_1)$, $(x_2,u_2)\in X(T_2-\tau_2,T_2)$ which satisfies

$$x_1(T_1)=z_1,\ x_1(T_1+\tau_1)=x_f(T_1+\tau_1),$$

$$I^f(T_1,T_1+\tau_1,x_1,u_1)\leq I^f(T_1,T_1+\tau_1,x_f,u_f)+1,$$

$$x_2(T_2)=z_2,\ x_2(T_2-\tau_2)=x_f(T_2-\tau_2),$$

$$I^f(T_2-\tau_2,T_2,x_2,u_2)\leq I^f(T_2-\tau_2,T_2,x_f,u_f)+1.$$

Let

$$\delta \in (0, \delta_0/2). \tag{9.112}$$

Lemma 9.7 implies that there exist an integer

$$L_0 \geq L + 4 + 2\bar{L}_0, \ L_1 > 8L_0 + 8, \tag{9.113}$$

$\lambda \in (1/2, 1)$ and a neighborhood V_0 of f in $\mathfrak{M}_\psi$ such that

$$4\lambda^{-1}(1-\lambda)a_0 L_1 < \delta_0, \ 4\lambda^{-1}(1-\lambda)(2+(L_1+2)\Delta_f+2a_0) < \delta_0 \tag{9.114}$$

and that the following property holds:

(c) for each $g \in V_0$ and each pair of numbers $T_1 \geq 0$, $T_2 \geq T_1 + L_1$, each borelian function $\alpha : [0, \infty) \to (0, 1]$ which satisfies

$$\alpha(t_1)\alpha(t_2)^{-1} \geq \lambda \text{ for each } t_1, t_2 \in [T_1, T_2] \text{ satisfying } |t_2 - t_1| \leq L_1$$

$U^{\alpha g}(T_1, T_2)$ is finite, $U^{\alpha g}(T_1, T_2, \xi)$ is finite for each $(T_1, \xi) \in \mathcal{A}_L$ satisfying $\rho_E(\xi, \theta_0) \leq L$, $\widehat{U}^{\alpha g}(T_1, T_2, \xi)$ is finite for each $(T_2, \xi) \in \widehat{\mathcal{A}}_L$, $U^{\alpha g}(T_1, T_2, \xi_1, \xi_2)$ is finite for each $(T_1, \xi_1) \in \mathcal{A}_L$, $(T_2, \xi_2) \in \widehat{\mathcal{A}}_L$ satisfying $\rho_E(z_1, \theta_0) \leq L$ and if $(x, u) \in X(T_1, T_2)$ satisfies at least one of the following conditions:

$$I^{\alpha g}(T_1, T_2, x, u) \leq U^{\alpha g}(T_1, T_2) + 4\inf\{\alpha(t) : t \in [T_1, T_2]\};$$

$$(T_1, x(T_1)) \in \mathcal{A}_L, \ \rho_E(x(T_1), \theta_0) \leq L,$$

$$I^{\alpha g}(T_1, T_2, x, u) \leq U^{\alpha g}(T_1, T_2, x(T_1)) + 4\inf\{\alpha(t) : t \in [T_1, T_2]\};$$

$$(T_2, x(T_2)) \in \widehat{\mathcal{A}}_L,$$

$$I^{\alpha g}(T_1, T_2, x, u) \leq \widehat{U}^{\alpha g}(T_1, T_2, x(T_2)) + 4\inf\{\alpha(t) : t \in [T_1, T_2]\};$$

$$(T_1, x(T_1)) \in \mathcal{A}_L, \ (T_2, x(T_2)) \in \widehat{\mathcal{A}}_L, \ \rho_E(x(T_1), \theta_0) \leq L,$$

$$I^{\alpha g}(T_1, T_2, x, u) \leq U^{\alpha g}(T_1, T_2, x(T_1), x(T_2)) + 4\inf\{\alpha(t) : t \in [T_1, T_2]\},$$

then $I^{\alpha g}(T_1, T_2, x, u)$ is finite and

$$\text{mes}(\{t \in [T_1, T_2] : \rho_E(x(t), x_f(t)) \le \delta_0\}) \ge L_0/2.$$

Proposition 9.1 implies that there exists a neighborhood $V \subset V_0$ of f in $\mathfrak{M}_\psi$ such that the following property holds:

(d) for each pair of numbers $S_1 \ge 0$, $S_2 \in (S_1, S_1 + L_1]$, each $g \in V$ and each $(x, u) \in X(S_1, S_2)$ which satisfies

$$\rho_E(x(T_1), \theta_0) \le L,$$

$$\min\{I^g(S_1, S_2, x, u),\ I^f(S_1, S_2, x, u)\} \le (L_1 + 2)(\Delta_f + 4 + 2a_0)$$

the inequality

$$|I^g(S_1, S_2, x, u) - I^f(S_1, S_2, x, u)| \le \delta_0$$

is valid.

Assume that

$$g \in V,\ T_1 \ge 0, T_2 \ge T_1 + 4L_1, \tag{9.115}$$

a borelian function $\alpha : [0, \infty) \to [0, 1]$ satisfies

$$\alpha(t_1)\alpha(t_2)^{-1} \ge \lambda \text{ for each } t_1, t_2 \in [T_1, T_2] \text{ satisfying } |t_2 - t_1| \le L_1. \tag{9.116}$$

Property (c), (9.115), and (9.116) imply that $U^{\alpha g}(T_1, T_2)$ is finite, for each $(T_1, \xi_1) \in \mathcal{A}_L$ satisfying $\rho_E(\xi_1, \theta_0) \le L$ and each $(T_2, \xi_2) \in \widehat{\mathcal{A}}_L$, $U^{\alpha g}(T_1, T_2, \xi_1)$, $\widehat{U}^{\alpha g}(T_1, T_2, \xi_2)$ and $U^{\alpha g}(T_1, T_2, \xi_1, \xi_2)$ are finite.

Assume that $(x, u) \in X(T_1, T_2)$ satisfies at least one of the following conditions:

(i)

$$I^{\alpha g}(T_1, T_2, x, u) \le U^{\alpha g}(T_1, T_2) + \delta \inf\{\alpha(t) : t \in [T_1, T_2]\};$$

(ii)

$$(T_1, x(T_1)) \in \mathcal{A}_L,\ \rho_E(x(T_1), \theta_0) \le L,$$

$$I^{\alpha g}(T_1, T_2, x, u) \le U^{\alpha g}(T_1, T_2, x(T_1)) + \delta \inf\{\alpha(t) : t \in [T_1, T_2]\};$$

(iii)

$$(T_1, x(T_1)) \in \mathcal{A}_L,\ (T_2, x(T_2)) \in \widehat{\mathcal{A}}_L,\ \rho_E(x(T_1), \theta_0) \le L,$$

$$I^{\alpha g}(T_1, T_2, x, u) \le U^{\alpha g}(T_1, T_2, x(T_1), x(T_2)) + \delta \inf\{\alpha(t) : t \in [T_1, T_2]\}.$$

Property (c), conditions (i)–(iii), (9.115), and (9.116) imply that

$$\text{mes}(\{t \in [T_1, T_2] : \rho_E(x(t), x_f(t)) \le \delta_0\}) \ge L_0/2. \tag{9.117}$$

Set

$$S_1 = \min\{t \in [\max\{T_1, \bar{L}_0\}, T_2] : \rho_E(x(t), x_f(t)) \le \delta_0\}. \tag{9.118}$$

(Note that if $T_1 \ge \bar{L}_0$ and $\rho_E(x(T_1), x_f(T_1)) \le \delta$, then $S_1 = T_1$.) Clearly, in view of (9.112) and (9.117), S_1 is well-defined.

Assume that $S_1 \ge T_1 + L_1$. It follows from property (c) applied to $(x, u) \in X(T_1, S_1)$, conditions (i)–(iii) and Eqs. (9.115), (9.116) that

$$\text{mes}(\{t \in [T_1, S_1] : \rho_E(x(t), x_f(t)) \le \delta_0\}) \ge L_0/2.$$

On the other hand in view of (9.112) and (9.118),

$$\text{mes}(\{t \in [T_1, S_1] : \rho_E(x(t), x_f(t)) \le \delta_0\}) \le \bar{L}_0 < L_0/2.$$

The contradiction we have reached proves that

$$S_1 \le T_1 + L_1. \tag{9.119}$$

Set

$$S_2 = \max\{t \in [S_1, T_2] : \rho_E(x(t), x_f(t)) \le \delta_0\}. \tag{9.120}$$

If $S_2 \le T_2 - L_1$, then it follows from property (c) applied to $(x, u) \in X(S_2, T_2)$, conditions (i)–(iii) and Eqs. (9.115)–(9.117) that

$$\text{mes}(\{t \in [S_2, T_2] : \rho_E(x(t), x_f(t)) \le \delta_0\}) \ge L_0/2.$$

This contradicts (9.120). Thus

$$S_2 \ge T_2 - L_1. \tag{9.121}$$

In order to complete the proof of Theorem 9.5 it is sufficient to show that

$$\rho_E(x(t), x_f(t)) \le \epsilon, \; t \in [S_1, S_2].$$

Assume the contrary. Then there exists an integer

$$\tau_0 \in [S_1, S_2] \tag{9.122}$$

such that

$$\rho_E(x(\tau_0), x_f(\tau_0)) > \epsilon. \tag{9.123}$$

In view of (9.115) and (9.121),

$$S_2 - S_1 \geq T_2 - T_1 - 2L_1 \geq 2L_1. \tag{9.124}$$

By (9.112), (9.122), and (9.124), there exist numbers S_3, S_4 such that

$$S_1 \leq S_3 \leq \tau_0 \leq S_4 \leq S_2, \tag{9.125}$$

$$S_4 = S_3 + L_0/4. \tag{9.126}$$

Set

$$S_5 = \max\{t \in [S_1, S_3] : \ \rho_E(x(t), x_f(t)) \leq \delta_0\}, \tag{9.127}$$

$$S_6 = \min\{t \in [S_4, S_2] : \ \rho_E(x(t), x_f(t)) \leq \delta_0\}. \tag{9.128}$$

It follows from (9.126)–(9.128) that

$$S_6 - S_5 \geq S_4 - S_3 = L_0/4. \tag{9.129}$$

Assume that

$$S_6 - S_5 \geq L_1. \tag{9.130}$$

Property (c) applied to $(x, u) \in X(S_5, S_6)$, conditions (i)–(iii) and Eqs. (9.101), (9.115) and (9.116) imply that

$$\text{mes}(\{t \in [S_5, S_6] : \ \rho_E(x(t), x_f(t)) \leq \delta_0\}) \geq L_0/2.$$

On the other hand it follows from (9.126)–(9.128) that

$$\text{mes}(\{t \in [S_5, S_6] : \ \rho_E(x(t), x_f(t)) \leq \delta_0\}) \leq S_4 - S_3 = L_0/4.$$

The contradiction we have reached proves that

$$S_6 - S_5 \leq L_1.$$

Together with (9.126)–(9.128) this implies that

$$L_0/4 \leq S_6 - S_5 \leq L_1. \tag{9.131}$$

Property (b) and Eqs. (9.111), (9.112), (9.127), (9.128), and (9.131) imply that there exist $(y, v) \in X(S_5, S_6)$ and $t_1, t_2 \in (0, b_f]$ such that

$$y(S_5) = x(S_5),\ y(S_6) = x(S_6),$$

$$y(t) = x_f(t),\ v(t) = u_f(t),\ t \in [S_5 + t_1, S_6 - t_2],$$

$$I^f(S_5, S_5 + t_1, y, v) \le I^f(S_5, S + t_1, x_f, u_f) + 1,$$

$$I^f(S_6 - t_2, S_6, y, v) \le I^f(S_6 - t_2, S_6, x_f, u_f) + 1. \tag{9.132}$$

Equations (8.74), (9.3), (9.131), and (9.132) imply that

$$I^f(S_5, S_6, y, v) \le I^f(S_5, S_6, x_f, u_f) + 2 \le 2 + (L_1 + 2)\Delta_f + 2a_0. \tag{9.133}$$

Property (d) and (9.111), (9.115), (9.127), (9.131), and (9.133) imply that

$$I^f(S_5, S_6, y, v) \le 3 + (L_1 + 2)\Delta_f + 2a_0. \tag{9.134}$$

By (9.3), (9.116), (9.131), and (9.134),

$$I^{\alpha g}(S_5, S_6, x, u) = \int_{S_5}^{S_6} (\alpha(t)(g(t, x(t), u(t)) + a_0)dt - a_0 \int_{S_5}^{S_6} \alpha(t)dt$$

$$\ge \inf\{\alpha(t) : t \in [S_5, S_6]\} \int_{S_5}^{S_6} \alpha(t)(g(t, x(t), u(t)) + a_0)dt - a_0 \int_{S_5}^{S_6} \alpha(t)dt$$

$$\ge I^g(S_5, S_6, x, u) \inf\{\alpha(t) : t \in [S_5, S_6]\} - a_0 \int_{S_5}^{S_6} \alpha(t)dt. \tag{9.135}$$

Property (d), conditions (i)–(iii) and Eqs. (9.111), (9.115), (9.127), (9.133), and (9.135) imply that

$$I^g(S_5, S_6, x, u) \inf\{\alpha(t) : t \in [S_5, S_6]\}$$

$$\le I^{\alpha g}(S_5, S_6, x, u) + (1 - \lambda)a_0 \int_{S_5}^{S_6} \alpha(t)dt$$

$$\le U^{\alpha g}(S_5, S_6, x(S_5), x(S_6)) + \delta_0 \inf\{\alpha(t) : t \in [S_5, S_6]\}$$

$$+ (1 - \lambda)a_0 \int_{S_5}^{S_6} \alpha(t)dt. \tag{9.136}$$

Conditions (i)–(iii) imply that

$$I^{\alpha g}(S_5, S_6, x, u) \le U^{\alpha g}(S_5, S_6, x(S_5), x(S_6)) + \delta_0 \inf\{\alpha(t) : t \in [T_1, T_2]\}. \tag{9.137}$$

By (9.3), (9.116), (9.131), and (9.132),

$$U^{\alpha g}(S_5, S_6, x(S_5), x(S_6)) \le \inf\{I^{\alpha g}(S_5, S_6, \xi, \eta) : (\xi, \eta) \in X(S_5, S_6),$$

$$I^f(S_5, S_6, \xi, \eta) \le I^f(S_5, S_6, y, v),\ \xi(S_5) = x(S_5),\ \xi(S_6) = x(S_6)\}$$

$$= \inf\{\int_{S_5}^{S_6} \alpha(t)(g(t, x(t), u(t)) + a_0)dt - a_0 \int_{S_5}^{S_6} \alpha(t)dt :$$

$$(\xi, \eta) \in X(S_5, S_6),\ I^f(S_5, S_6, \xi, \eta) \le I^f(S_5, S_6, y, v),$$

$$\xi(S_5) = x(S_5),\ \xi(S_6) = x(S_6)\}$$

$$\le \inf\{\lambda^{-1} \inf\{\alpha(t) :\ t \in [S_5, S_6]\} \int_{S_5}^{S_6} (g(t, x(t), u(t)) + a_0)dt$$

$$- a_0 \int_{S_5}^{S_6} \alpha(t)dt :$$

$$(\xi, \eta) \in X(S_5, S_6),\ I^f(S_5, S_6, \xi, \eta) \le I^f(S_5, S_6, y, v),$$

$$\xi(S_5) = x(S_5),\ \xi(S_6) = x(S_6)\}$$

$$\le \lambda^{-1} \inf\{\alpha(t) :\ t \in [S_5, S_6]\} \inf\{I^g(S_5, S_6, \xi, \eta) : (\xi, \eta) \in X(S_5, S_6),$$

$$I^f(S_5, S_6, \xi, \eta) \le I^f(S_5, S_6, y, v),\ \xi(S_5) = x(S_5),\ \xi(S_6) = x(S_6)\}$$

$$+ a_0 \inf\{\alpha(t) :\ t \in [S_5, S_6]\}(\lambda^{-1} - 1)(S_6 - S_5)$$

$$\le \lambda^{-1} \inf\{\alpha(t) :\ t \in [S_5, S_6]\} \inf\{I^f(S_5, S_6, \xi, \eta) + \delta_0 : (\xi, \eta) \in X(S_5, S_6),$$

$$\xi(S_5) = x(S_5),\ \xi(S_6) = x(S_6),$$

$$I^f(S_5, S_6, \xi, \eta) \le I^f(S_5, S_6, y, v)\}$$

$$+ a_0 \inf\{\alpha(t) :\ t \in [S_5, S_6]\}(\lambda^{-1} - 1)(S_6 - S_5)$$

$$\le \lambda^{-1} \inf\{\alpha(t) :\ t \in [S_5, S_6]\} I^f(S_5, S_6, y, v)$$

$$+ a_0 \inf\{\alpha(t) :\ t \in [S_5, S_6]\}(\lambda^{-1} - 1)L_1 + \delta_0 \lambda^{-1} \inf\{\alpha(t) :\ t \in [S_5, S_6]\}. \tag{9.138}$$

By (9.132) and (9.138),

$$U^{\alpha g}(S_5, S_6, x(S_5), x(S_6))$$

$$\leq \lambda^{-1} \inf\{\alpha(t) : t \in [S_5, S_6]\} \inf\{I^f(S_5, S_6, \xi, \eta) + \delta_0 : (\xi, \eta) \in X(S_5, S_6),$$

$$I^f(S_5, S_6, \xi, \eta) \leq I^f(S_5, S_6, y, v), \ \xi(S_5) = x(S_5), \ \xi(S_6) = x(S_6)\}$$

$$+ a_0 \inf\{\alpha(t) : t \in [S_5, S_6]\}(\lambda^{-1} - 1)L_1$$

$$\leq \lambda^{-1} \inf\{\alpha(t) : t \in [S_5, S_6]\} U^f(S_5, S_6, x(S_5), x(S_6))$$

$$+ \delta_0 \lambda^{-1} \inf\{\alpha(t) : t \in [S_5, S_6]\}$$

$$+ a_0 \inf\{\alpha(t) : t \in [S_5, S_6]\} L_1(\lambda^{-1} - 1). \tag{9.139}$$

In view of (9.132) and (9.133),

$$U^{\alpha g}(S_5, S_6, x(S_5), x(S_6)) \leq 2 + 2a_0 + (L_1 + 2)\Delta_f. \tag{9.140}$$

It follows from (9.137) and (9.139) that

$$I^{\alpha g}(S_5, S_6, x, u) \leq U^{\alpha g}(S_5, S_6, x(S_5), x(S_6)) + \delta \inf\{\alpha(t) : t \in [T_1, T_2]\}$$

$$\leq \delta \inf\{\alpha(t) : t \in [S_5, S_6]\}$$

$$+ \lambda^{-1} \inf\{\alpha(t) : t \in [S_5, S_6]\} U^f(S_5, S_6, x(S_5), x(S_6)$$

$$+ \delta_0 \lambda^{-1} \inf\{\alpha(t) : t \in [S_5, S_6]\}$$

$$+ a_0 L_1 \lambda^{-1}(1 - \lambda) \inf\{\alpha(t) : t \in [S_5, S_6]\}. \tag{9.141}$$

By (9.13), (9.135), and (9.141),

$$I^g(S_5, S_6, x, u) \inf\{\alpha(t) : t \in [S_5, S_6]\}$$

$$\leq a_0(1 - \lambda) \int_{S_5}^{S_6} \alpha(t) dt + I^{\alpha g}(S_5, S_6, x, u)$$

$$\leq a_0 L_1 \lambda^{-1}(1 - \lambda) \inf\{\alpha(t) : t \in [S_5, S_6]\}$$

$$+ \inf\{\alpha(t) : t \in [S_5, S_6]\}(\delta + \delta_0 \lambda^{-1} + a_0 L_1 \lambda^{-1}(1 - \lambda)$$

$$+ U^f(S_5, S_6, x(S_5), x(S_6)\lambda^{-1}).$$

Together with (9.113), (9.114), and (9.140) this implies that

$$I^g(S_5, S_6, x, u) \leq 2a_0 L_1 \lambda^{-1}(1 - \lambda) + \delta + \delta_0 \lambda^{-1}$$

$$+ U^f(S_5, S_6, x(S_5), x(S_6)) + \lambda^{-1}(1-\lambda)(2 + 2a_0 + \Delta_f(L_1+2))$$

$$\leq U^f(S_5, S_6, x(S_5), x(S_6)) + \delta + 2\delta_0 + 2\delta_0(2+\delta_0)$$

$$\leq U^f(S_5, S_6, x(S_5), x(S_6)) + 4\delta_0. \tag{9.142}$$

Property (d), (9.111), (9.115), (9.127), (9.131), (9.140), and (9.142) imply that

$$|I^f(S_5, S_6, x, u) - I^g(S_5, S_6, x, u)| \leq \delta_0.$$

Together with (9.142) this implies that

$$I^f(S_5, S_6, x, u) \leq U^f(S_5, S_6, x(S_5), x(S_6)) + 5\delta_0.$$

The equation above and (9.111), (9.118), (9.127), (9.128) and (9.131) this implies that

$$\rho_E(x(t), x_f(t)) \leq \epsilon, \ t \in [S_5, S_6].$$

Combined with (9.125)–(9.128) this implies that

$$\rho_E(x(\tau_0), x_f(\tau_0)) \leq \epsilon.$$

This contradicts (9.123). The contradiction we have reached completes the proof of Theorem 9.5.

9.8 Genericity

Assume that $\mathcal{M}_0$ is a borelian measurable subset of $[0, \infty) \times E \times F$ such that $\mathcal{M} \subset \mathcal{M}_0$. Let $\mathfrak{M}_0$ be a set of borelian functions $g : \mathcal{M}_0 \to R^1$ such that for every $(t, x, u) \in \mathcal{M}_0$,

$$g(t, x, u) \geq \psi_1(t, x, u) - a_0. \tag{9.143}$$

For each $g \in \mathfrak{M}_0$ denote by $\mathcal{R}(g)$ the restriction of g to $\mathcal{M}$ which belongs to $\mathfrak{M}_\psi$.

For each pair of numbers $T_2 > T_1 \geq 0$, each $(x, u) \in X(T_1, T_2)$, each $g \in \mathfrak{M}_0$ and each $(T_1, y), (T_2, z) \in \mathcal{A}$ define

$$I^g(T_1, T_2, x, u) = I^{\mathcal{R}(g)}(T_1, T_2, x, u), \ U^g(T_1, T_2, y, z) = U^{\mathcal{R}(g)}(T_1, T_2, y, z),$$

$$U^g(T_1, T_2) = U^{\mathcal{R}(g)}(T_1, T_2), \ U^g(T_1, T_2, y) = U^{\mathcal{R}(g)}(T_1, T_2, y),$$

$$\widehat{U}^g(T_1, T_2, z) = \widehat{U}^{\mathcal{R}(g)}(T_1, T_2, z).$$

Recall that the set $\mathfrak{M}_\psi$ is equipped with the uniformity which is defined by the base (8.10) and that assumption (A0) holds. This uniformity induces the topology in $\mathfrak{M}_\psi$ denoted by τ_w.

A set $V \subset \mathfrak{M}_0$ is open in a weak topology if $V = \mathcal{R}^{-1}(W)$, where W is an open set in $\mathfrak{M}_\psi$ equipped with the topology τ_w.

Assume that $\mathfrak{M}_0$ is equipped with the metric $d_{\mathfrak{M}}$ which induces a topology which is stronger than the weak topology and is called a strong topology. Denote by $\mathfrak{M}_{reg}$ the set of all $f \in \mathfrak{M}_0$ such that there exist $(x_f, u_f) \in X(0, \infty)$ and $b_f > 0$ satisfying (8.73) and (8.74) and such that (A1), (A4), (TP) hold for $\mathcal{R}(f)$.

Denote by $\bar{\mathfrak{M}}_{reg}$ the closure of $\mathfrak{M}_{reg}$ in $\mathfrak{M}_0$ equipped with the strong topology.

We prove the following result.

Theorem 9.8 *Let $M > 0$ and $\epsilon \in (0, 1)$. Then there exists an open (in the weak topology) and everywhere dense (in the strong topology) set $\mathcal{F}$ satisfying*

$$\mathfrak{M}_{reg} \subset \mathcal{F} \subset \bar{\mathfrak{M}}_{reg}$$

and such that for each $g \in \mathcal{F}$ there exists an open set in the weak topology $V \subset \mathfrak{M}_0$ containing g, $\delta \in (0, 1)$, $L > 0$ and $(x_*, u_*) \in X(0, \infty)$ such that for each $h \in V$, each pair of numbers $T_1 \geq 0$, $T_2 \geq T_1 + 2L$, $U^h(T_1, T_2)$ is finite and for each $(x, u) \in X(T_1, T_2)$ which satisfies

$$I^h(T_1, T_2, x, u) \leq \min\{U^h(T_1, T_2) + M, U^h(T_1, T_2, x(T_1), x(T_2)) + \delta\}$$

there exist $\tau_1 \in [T_1, T_1 + L]$, $\tau_2 \in [T_2 - L, T_2]$ such that

$$\rho_E(x(t), x_f(t)) \leq \epsilon, \ t \in [\tau_1, \tau_2].$$

Moreover, if $\rho_E(x(T_2), x_f(T_2)) \leq \delta$, then $\tau_2 = T_2$ and if $\rho_E(x(T_1), x_f(T_1)) \leq \delta$, $T_1 \geq L$, then $\tau_1 = T_1$.

Proof Let $f \in \mathfrak{M}_{reg}$. Theorem 9.2 implies that there exists an open neighborhood $V(f)$ of f in $\mathfrak{M}_0$ in the weak topology and $L(f) > 0, \delta(f) \in (0, 1)$ such that the following property holds:

for each $h \in V(f)$, each pair of numbers $T_1 \geq 0$, $T_2 \geq T_1 + 2L(f)$, $U^h(T_1, T_2)$ is finite and for each $(x, u) \in X(T_1, T_2)$ which satisfies

$$I^h(T_1, T_2, x, u) \leq \min\{U^h(T_1, T_2) + M, U^h(T_1, T_2, x(T_1), x(T_2)) + \delta(f)\}$$

there exist $\tau_1 \in [T_1, T_1 + L(f)]$, $\tau_2 \in [T_2 - L(f), T_2]$ such that

$$\rho_E(x(t), x_f(t)) \leq \epsilon, \ t \in [\tau_1, \tau_2];$$

moreover, if $\rho_E(x(T_2), x_f(T_2)) \le \delta(f)$, then $\tau_2 = T_2$ and if $\rho_E(x(T_1), x_f(T_1)) \le \delta(f)$, $T_1 \ge L$, then $\tau_1 = T_1$.

Define

$$\mathcal{F} = \cup\{V(f) : f \in \mathfrak{M}_{reg}\} \cap \bar{\mathfrak{M}}_{reg}.$$

It is not difficult to see that the set $\mathcal{F}$ is open in the weak topology, everywhere dense in the strong topology of $\bar{\mathfrak{M}}_{reg}$ and that our assertion holds. This completes the proof of Theorem 9.8.

The next generic result is proved in Sect. 9.9.

Theorem 9.9 *There exists an everywhere dense (in the strong topology) set* $\mathcal{F} \subset \bar{\mathfrak{M}}_{reg}$ *which is a countable intersection of open (in the weak topology) subsets of* $\bar{\mathfrak{M}}_{reg}$ *such that*

$$\mathfrak{M}_{reg} \subset \mathcal{F}$$

and that for each $f \in \mathcal{F}$ *there exist* $(x_*, u_*) \in X(0, \infty)$ *such that the following assertion holds.*

For each $M, \epsilon > 0$ *there exists a neighborhood* V *of* f *in* $\mathfrak{M}_0$ *in the weak topology and* $\delta, L > 0$ *such that for each* $g \in V$*, each pair of integers* $T_1 \ge 0$*,* $T_2 \ge T_1 + 2L$*,* $U^g(T_1, T_2)$ *is finite and for each* $(x, u) \in X(T_1, T_2)$ *which satisfies*

$$I^g(T_1, T_2, x, u) \le \min\{U^g(T_1, T_2) + M, U^g(T_1, T_2, x(T_1), x(T_2)) + \delta\}$$

there exist $\tau_1 \in [T_1, T_1 + L]$*,* $\tau_2 \in [T_2 - L, T_2]$ *such that*

$$\rho_E(x(t), x_f(t)) \le \epsilon, \ t \in [\tau_1, \tau_2].$$

9.9 Proof of Theorem 9.9

Let $f \in \mathfrak{M}_{reg}$ and $n \ge 1$ be an integer. By (A4), there exists

$$\Delta(f, n) \in (0, 2^{-n}) \tag{9.144}$$

such that the following property holds:

(a) for each $(T_i, z_i) \in \mathcal{A}$, $i = 1, 2$ satisfying $\rho_E(z_i, x_f(T_i)) \le \Delta(f, n)$, $i = 1, 2$ and $T_2 \ge b_f$ there exist $(x_1, u_1) \in X(T_1, T_1 + b_f)$, $(x_2, u_2) \in X(T_2 - b_f, T_2)$ such that

$$x_1(T_1) = z_1, \ x_1(T_1 + b_f) = x_f(T_1 + b_f),$$

$$x_2(T_2) = z_2, \ x_2(T_2 - b_f) = x_f(T_2 - b_f),$$

$$I^f(T_1, T_1 + b_f, x_1, u_1) \le I^f(T_1, T_1 + b_f, x_f, u_f) + 2^{-n},$$

$$I^f(T_2 - b_f, T_2, x_2, u_2) \le I^f(T_2 - b_f, T_2, x_f, u_f) + 2^{-n},$$

$$\rho_E(x_1(t), x_f(t)) \le 2^{-n}, \ t \in [T_1, T_1 + b_f],$$

$$\rho(x_2(t), x_f(t)) \le 2^{-n}, \ t \in [T_2 - b_f, T_2].$$

Theorem 9.2, Proposition 9.1 and Eqs. (8.73), (8.74) imply that there exists an open neighborhood $\mathcal{U}(f, n)$ of f in $\mathfrak{M}_0$ in the weak topology and

$$L(f, n) \ge n, \ \delta(f, n) \in (0, 2^{-n}) \tag{9.145}$$

such that the following properties hold:

(b) for each $h \in \mathcal{U}(f, n)$, each pair of numbers $T_1 \ge 0$, $T_2 \ge T_1 + 2L(f, n)$,

$$I^h(T_1, T_2, x_f, u_f)$$

is finite and for each $(x, u) \in X(T_1, T_2)$ which satisfies

$$I^h(T_1, T_2, x, u) \le \min\{U^h(T_1, T_2) + n, U^h(T_1, T_2, x(T_1), x(T_2)) + \delta(f, n)\}$$

there exist $\tau_1 \in [T_1, T_1 + L(f, n)]$, $\tau_2 \in [T_2 - L(f, n), T_2]$ such that

$$\rho_E(x(t), x_f(t)) \le 2^{-n}\Delta(f, n), \ t \in [\tau_1, \tau_2];$$

moreover, if

$$\rho_E(x(T_2), x_f(T_2)) \le \delta(f, n),$$

then $\tau_2 = T_2$ and if $\rho_E(x(T_1), x_f(T_1)) \le \delta(f, n)$, $T_1 \ge L(f, n)$, then $\tau_1 = T_1$;

(c) for each $h \in \mathcal{U}(f, n)$, each $T \ge 0$ and each $(x, u) \in X(T, T + b_f)$ which satisfies

$$x(T) = x_f(T),$$

$$I^f(T, T + b_f, x, u) \le I^f(T, T + b_f, x_f, u_f) + 4$$

we have

$$|I^h(T, T + b_f, x, u) - I^f(T, T + b_f, x, u)| \le 2^{-n}\Delta(f, n)$$

is true.

Define

$$\mathcal{F} = (\cap_{n=1}^{\infty} \cup \{\mathcal{U}(f, n) : f \in \mathfrak{M}_{reg}\}) \cap \bar{\mathfrak{M}}_{reg}. \tag{9.146}$$

It is not difficult to see that the set $\mathcal{F}$ is a countable intersection of open in the weak topology, everywhere dense in the strong topology subsets of $\bar{\mathfrak{M}}_{reg}$. Assume that

$$f \in \mathcal{F}. \tag{9.147}$$

By (9.146) and (9.147), for each integer $n \geq 1$ there exists

$$f^{(n)} \in \mathfrak{M}_{reg}$$

such that

$$f \in \mathcal{U}(f^{(n)}, n). \tag{9.148}$$

Property (b) and (9.148) imply that the following property holds:

(d) for each integer $n \geq 1$, each pair of numbers $T_1 \geq 0$, $T_2 \geq T_1 + 2L(f^{(n)}, n)$, $U^f(T_1, T_2)$ is finite and for each $(x, u) \in X(T_1, T_2)$ which satisfies

$$I^f(T_1, T_2, x, u) \leq \min\{U^f(T_1, T_2) + n, U^f(T_1, T_2, x(T_1), x(T_2)) + \delta(f, n)\}$$

we have

$$\rho_E(x(t), x_{f^{(n)}}(t)) \leq 2^{-n}\Delta(f, n), \ t \in [T_1 + L(f^{(n)}, n), T_2 - L(f^{(n)}, n)].$$

Property (d) implies that the following property holds:

(e) for each pair of integers $n, m \geq 1$, each pair of numbers $T_1 \geq 0$,

$$T_2 \geq T_1 + 2\max\{L(f^{(n)}, n), \ L(f^{(m)}, m)\}$$

and each $(x, u) \in X(T_1, T_2)$ which satisfies

$$I^f(T_1, T_2, x, u)$$

$$\leq \min\{U^f(T_1, T_2) + 1, U^f(T_1, T_2, x(T_1), x(T_2)) + \min\{\delta(f^{(i)}, i) : i = n, m\}$$

we have for $i = n, m$,

$$\rho_E(x(t), x_{f^{(i)}}(t)) \leq 2^{-i}\Delta(f^{(i)}, i), \ t \in [T_1 + L(f^{(i)}, i), T_2 - L(f^{(i)}, i)]$$

and for every $T \geq 2\max\{L(f^{(n)}, n),\ L(f^{(m)}, m)]\}$ and every

$$t \in [\max\{L(f^{(n)}, n),\ L(f^{(m)}, m)\}, T - \max\{L(f^{(n)}, n),\ L(f^{(m)}, m)\}]$$

we have

$$\rho_E(x_{f^{(m)}}(t), x_{f^{(n)}}(t)) \leq 2^{-n}\Delta(f^{(n)}, n) + 2^{-m}\Delta(f^{(m)}, m).$$

Property (e) implies that the following property holds:

(f) for each pair of integers $n, m \geq 1$ and each integer

$$t \geq \max\{L(f^{(n)}, n),\ L(f^{(m)}, m)\}$$

we have

$$\rho_E(x_{f^{(m)}}(t), x_{f^{(n)}}(t)) \leq 2^{-n}\Delta(f^{(n)}, n) + 2^{-m}\Delta(f^{(m)}, m).$$

Consider a strictly increasing sequence of natural numbers $\{n_k\}_{k=1}^{\infty}$ such that

$$n_1 = 1 \tag{9.149}$$

and that for each integer $k \geq 1$, we have

$$n_{k+1} > L(f^{(n_k)}, n_k) + n_k(b_{f^{(n_k)}} + 1), \tag{9.150}$$

$$2^{-n_{k+1}}\Delta(f^{(n_{k+1})}, n_{k+1}) < 2^{-n_k}\Delta(f^{(n_k)}, n_k). \tag{9.151}$$

Set

$$x_*(t) = x_{f^{(n_1)}}(t),\ u_*(t) = u_{f^{(n_1)}}(t),\ t \in [0, L(f^{(n_2)}, n_2)]. \tag{9.152}$$

In view of property (b) and (9.152),

$$I^f(0, L(f^{(n_2)}, n_2), x_*, u_*)$$

is finite.

Assume that $k \geq 1$ is an integer and that we defined

$$(x_*, u_*) \in X(0, L(f^{(n_{k+1})}), n_{k+1}))$$

such that

$$x_*(L(f^{(n_{k+1})}, n_{k+1})) = x_{f^{(n_k)}}(L(f^{(n_{k+1})}, n_{k+1})). \tag{9.153}$$

(Clearly, our assumption holds for $k = 1$.) Property (f), Eqs. (9.145), (9.150), and (9.151) imply that

$$\rho_E(x_{f^{(n_k)}}(L(f^{(n_{k+1})}, n_{k+1}) + b_{f^{(n_k)}}), x_{f^{(n_{k+1})}}(L(f^{(n_{k+1})}, n_{k+1}) + b_{f^{(n_k)}}))$$

$$\le 2^{-n_k} \Delta(f^{(n_k)}, n_k) + 2^{-n_{k+1}} \Delta(f^{(n_{k+1})}, n_{k+1})$$

$$\le 2^{-n_k+1} \Delta(f^{(n_k)}, n_k) \le \Delta(f^{(n_k)}, n_k). \tag{9.154}$$

Property (a) and Eqs. (9.153) and (9.154) imply that there exists

$$(x_*, u_*) \in X(L(f^{(n_{k+1})}, n_{k+1}), L(f^{(n_{k+1})}, n_{k+1}) + b_{f^{(n_k)}})$$

such that

$$x_*(L(f^{(n_{k+1})}, n_{k+1})) = x_{f^{(n_k)}}(L(f^{(n_{k+1})}, n_{k+1})), \tag{9.155}$$

$$x_*(L(f^{(n_{k+1})}, n_{k+1}) + b_{f^{(n_k)}}) = x_{f^{(n_{k+1})}}(L(f^{(n_{k+1})}, n_{k+1}) + b_{f^{(n_k)}}), \tag{9.156}$$

$$I^{f^{(n_k)}}(L(f^{(n_{k+1})}, n_{k+1}), L(f^{(n_{k+1})}, n_{k+1}) + b_{f^{(n_k)}}), x_*, u_*)$$

$$\le I^{f^{(n_k)}}(L(f^{(n_{k+1})}, n_{k+1}), L(f^{(n_{k+1})}, n_{k+1}) + b_{f^{(n_k)}}), x_{f^{(n_k)}}, u_{f^{(n_k)}}) + 2^{-n_k} \tag{9.157}$$

and that for every $t \in [L(f^{(n_{k+1})}, n_{k+1}), L(f^{(n_{k+1})}, n_{k+1}) + b_{f^{(n_k)}})]$,

$$\rho_E(x_*(t), x_{f^{(n_k)}}(t)) \le 2^{-n_k}. \tag{9.158}$$

Set

$$x_*(t) = x_{f^{(n_{k+1})}}(t), \ \ u_*(t) = u_{f^{(n_{k+1})}}(t),$$

$$t \in [L(f^{(n_{k+1})}, n_{k+1}) + b_{f^{(n_k)}}, L(f^{(n_{k+2})}, n_{k+2})]. \tag{9.159}$$

It is not difficult to see that

$$(x_*, u_*) \in X(0, L(f^{(n_{k+2})}, n_{k+2})).$$

Properties (b), (c) and Eqs. (9.157), (9.159) imply that

$$I^f(0, L(f^{(n_{k+2})}, n_{k+2}), x_*, u_*)$$

is finite and the assumption made for k also holds for $k + 1$. Therefore by induction we defined $(x_*, u_*) \in X(0, \infty)$ such that $I^f(0, T, x_*, u_*)$ is finite for every $T > 0$, (9.152) holds and for each integer $k \geq 1$ Eqs. (9.153) and (9.156)–(9.159) are true.

Assume that $k \geq 2$ is an integer and that

$$t \geq L(f^{(n_k)}, n_k). \tag{9.160}$$

By (9.60), (9.145) and (9.150), there exists an integer

$$q \geq k \tag{9.161}$$

such that

$$t \in [L(f^{(n_q)}, n_q), L(f^{(n_{q+1})}, n_{q+1})]. \tag{9.162}$$

It follows from property (f) and Eqs. (9.145), (9.150), (9.161), and (9.162) that

$$\rho_E(x_{f^{(n_k)}}(t), x_{f^{(n_q)}}(t)) \leq 2^{-n_k+1}\Delta(f^{(n_k)}, n_k). \tag{9.163}$$

If

$$t \in [L(f^{(n_q)}, n_q), L(f^{(n_q)}, n_q) + b_{f^{(n_{q-1})}}],$$

then it follows from property (f), (9.144) and (9.157) that

$$\rho_E(x_*(t), x_{f^{(n_q)}}(t)) \leq \rho_E(x_*(t), x_{f^{(n_{q-1})}}(t)) + \rho_E(x_{f^{(n_{q-1})}}(t), x_{f^{(n_q)}}(t))$$

$$\leq 2^{-n_{q-1}} + 2^{-n_{q-1}+1}\Delta(f^{(n_{q-1})}, n_{q-1}) \leq 2^{-n_{q-1}+1}.$$

By (9.159), for each

$$t \in [L(f^{(n_q)}, n_q) + b_{f^{(n_{q-1})}}, L(f^{(n_{q+1})}, n_{q+1})]$$

we have

$$x_*(t) = x_{f^{(n_q)}}(t).$$

Thus in all the cases

$$\rho_E(x_*(t), x_{f^{(n_q)}}(t)) \leq 2^{-n_{q-1}+1}.$$

In view of (9.161), (9.163) and the relation above,

$$\rho_E(x_*(t), x_{f^{(n_k)}}(t)) \leq \rho_E(x_{f^{(n_k)}}(t), x_{f^{(n_q)}}(t)) + \rho_E(x_{f^{(n_q)}}(t), x_*(t))$$

$$\leq 2^{-n_k+1}\Delta(f^{(n_k)}, n_k) + 2^{-n_{q-1}} \leq 2^{-n_k+1}\Delta(f^{(n_k)}, n_k) + 2^{-n_{k-1}+1} \leq 3 \cdot 2^{-n_{k-1}}.$$

Thus we have shown that the following property holds:

(g) for each integer $k \geq 2$ and each $t \geq L(f^{(n_k)}, n_k)$ we have

$$\rho_E(x_*(t), x_{f^{(n_k)}}(t)) \leq 3 \cdot 2^{-n_{k-1}}.$$

Assume that $M > 1$ and $\epsilon \in (0, 1)$. Choose an integer $k > 4$ such that

$$n_k > M, \ 4 \cdot 2^{-n_{k-1}} < \epsilon \tag{9.164}$$

and let

$$\delta = \delta(f^{(n_k)}, n_k). \tag{9.165}$$

Assume that

$$g \in \mathcal{U}(f^{(n_k)}, n_k), \tag{9.166}$$

$$T_1 \geq 0, \ T_2 \geq T_1 + 2L(f^{(n_k)}, n_k) \tag{9.167}$$

and that $(x, u) \in X(T_1, T_2)$ satisfies

$$I^g(T_1, T_2, x, u) \leq \min\{U^g(T_1, T_2)+M, U^g(T_1, T_2, x(T_1), x(T_2))+\delta\}. \tag{9.168}$$

By (9.164), (9.165), (9.167), and (9.168),

$$I^g(T_1, T_2, x, u) \leq \min\{U^g(T_1, T_2) + n_k, U^g(T_1, T_2, x(T_1), x(T_2)) + \delta(f^{(n_k)}, n_k)\}. \tag{9.169}$$

Property (b) and Eqs. (9.166) and (9.169) imply that there exist

$$\tau_1 \in [T_1, T_1 + L(f^{(n_k)}, n_k)], \ \tau_2 \in [T_2 - L(f^{(n_k)}, n_k), T_2]$$

such that

$$\rho_E(x(t), x_{f^{(n_k)}}(t)) \leq 2^{-n_k} \Delta(f^{(n_k)}, n_k), \ t \in [\tau_1, \tau_2]. \tag{9.170}$$

Property (g) and Eqs. (9.164) and (9.170) imply that for each $t \in [\max\{\tau_1, L(f^{(n_k)}, n_k)\}, \tau_2]$,

$$\rho_E(x(t), x_*(t)) \leq \rho_E(x(t,)x_{f^{(n_k)}}(t)) + \rho_E(x_{f^{(n_k)}}(t), x_*(t))$$

$$\leq 2^{-n_k} \Delta(f^{(n_k)}, n_k) + 3 \cdot 2^{-n_{k-1}} \leq 4 \cdot 2^{-n_{k-1}} < \epsilon.$$

This completes the proof of Theorem 9.9.

9.10 A Turnpike Result

We continue to use the notation, definitions, and assumptions introduced in the previous sections of this chapter. We also assume that the following assumption holds:

(A5) For each $f \in \mathfrak{M}_\psi$, each $M, \epsilon > 0$ there exists $\delta \in (0, \epsilon)$ such that for each $T_1 \geq 0$, each $T_2 \in [T_1 + 2^{-1}, T_1 + 1]$ and each $(x_i, u_i) \in X(T_1, T_2)$, $i = 1, 2$ satisfying

$$I^f(T_1, T_2, x_i, u_i) \leq M, \ i = 1, 2$$

and each $\tau \in [T_1, T_2]$ satisfying

$$0 < T_2 - \tau \leq \delta, \ \rho_E(x_1(\tau), x_2(\tau)) \leq \delta$$

the inequality $\rho_E(x_1(T_2), x_2(T_2)) \leq \epsilon$ holds.

In the sequel we show that it holds for our three classes of optimal control problems. Denote by $\mathfrak{M}_{0,reg}$ the set of all $f \in \mathfrak{M}_0$ such that there exist $(x_f, u_f) \in X(0, \infty)$ and $b_f > 0$ satisfying (8.73) and (8.74) and such that (A1), (A4) hold. Clearly, $\mathfrak{M}_{reg} \subset \mathfrak{M}_{0,reg}$.

We prove the following turnpike result.

Theorem 9.10 *Assume that $f \in \mathfrak{M}_\psi$, $(x_f, u_f) \in X(0, \infty)$ and $b_f > 0$ and that (8.73) and (8.74), (A1), (A4) hold. Then the integrand f has (TP) if and only if the following property holds:*

(P3) for each $\epsilon > 0$ and each $M > 0$ there exists $L > 0$ such that for each $T_1 \geq 0$, each $T_2 \geq T_1 + L$ and each $(x, u) \in X(T_1, T_2)$ which satisfies

$$I^f(T_1, T_2, x, u) \leq I^f(T_1, T_2, x_f, u_f) + M$$

the inequality

$$mes(\{t \in [T_1, T_2] : \ \rho_E(x(t), x_f(t)) > \epsilon\}) \leq L$$

is valid.

Proof In view of Theorems 8.25 and 8.26, TP implies WTP. Evidently, WTP implies (P3). Thus (TP) implies (P3).Clearly, (P3) implies (P2). In view of Theorem 8.25, in order to complete our proof it is sufficient to show that (P3) implies (P1).

Assume that (P3) holds and that $(x, u) \in X(0, \infty)$ is (f)-good. We show that

$$\lim_{t \to \infty} \rho(x(t), x_f(t)) = 0.$$

Assume the contrary. Then there exists $\epsilon \in (0, 1/4)$ and a sequence of numbers $\{t_k\}_{k=1}^{\infty}$ such that

$$t_1 \geq 8, \ t_{k+1} \geq t_k + 8 \text{ for all integers } k \geq 1, \tag{9.171}$$

$$\rho_E(x(t_k), x_f(t_k)) > \epsilon \text{ for all integers } k \geq 1. \tag{9.172}$$

Theorem 8.5 implies that there exists $M_1 > 0$ such that

$$\rho_E(x_f(t), \theta_0) \leq M_1, \ \rho_E(x(t), \theta_0) \leq M_1, \ t \in [0, \infty), \tag{9.173}$$

$$|I^f(0, t, x, u) - I^f(0, t, x_f, u_f)| < M_1, \ t \in (0, \infty). \tag{9.174}$$

Property (P3) and (9.174) imply that the following property holds:

(a) for each $\gamma > 0$ there exists $L_\gamma > 0$ such that for each $T > L_\gamma$, we have

$$\text{mes}(\{t \in [0, T] : \ \rho(x(t), x_f(t)) > \gamma\}) < L_\gamma. \tag{9.175}$$

In view of (9.76), there exists $M_2 > 0$ such that

$$I^f(S_1, S_2, x_f, u_f) \leq M_2 \text{ for every } S_1 \geq 0 \text{ and every } S_2 \in [S_1, S_1 + 1]. \tag{9.176}$$

By (9.174), for all $S > T \geq 0$,

$$|I^f(T, S, x, u) - I^f(T, S, x_f, u_f)| \leq 2M_1. \tag{9.177}$$

Equations (9.176) and (9.177) imply that

$$I^f(S_1, S_2, x, u) \leq M_2 + 2M_1 \text{ for all } S_1 \geq 0 \text{ and all } S_2 \in (S_1, S_1 + 1]. \tag{9.178}$$

Assumption (A5) implies that there exists $\delta \in (0, \epsilon)$ such that the following property holds:

(b) for each $S_1 \geq 0$, each $S_2 \in [S_1 + 2^{-1}, S_1 + 1]$, each $(x_i, u_i) \in X(S_1, S_2)$, $i = 1, 2$ satisfying

$$I^f(S_1, S_2, x_i, u_i) \leq 2M_1 + M_2, \ i = 1, 2$$

and each $\tau \in [S_1, S_2]$ satisfying

$$0 < S_2 - \tau \leq \delta, \ \rho_E(x_1(\tau), x_2(\tau)) \leq \delta$$

the inequality $\rho_E(x_1(S_2), x_2(S_2)) \leq \epsilon$ holds.

Let $k \geq 1$ be an integer. It follows from (9.171), (9.172), (9.176), (9.178) and property (b) applied with $S_1 = t_k - 1$, $S_2 = t_k$, $x_1 = x$, $x_2 = x_f$ we obtain that for each $\tau \in [t_k - \delta, t_k)$ we have

$$\rho_E(x(t), x_f(t)) > \delta.$$

Therefore

$$\cup_{k=1}^{\infty} [t_k - \delta, t_k) \subset \{t \in [0, \infty) : \ \rho_E(x(t), x_f(t)) > \delta\}.$$

This contradicts property (a). The contradiction we have reached proves that

$$\lim_{t \to \infty} \rho(x(t), x_f(t)) = 0$$

and property (P1) holds. This completes the proof of Theorem 9.10.

Assume that $\phi : E \times E \to [0, 1]$ is a continuous function satisfying $\phi(x, x) = 0$ for all $x \in E$ and such that the following properties hold:

(i) for each $\epsilon > 0$ there exists $\delta > 0$ such that if $x, y \in E$ and $\phi(x, y) \leq \delta$, then $\rho_E(x, y) \leq \epsilon$.
(ii) for each $\epsilon > 0$ there exists $\delta > 0$ such that if $x, y \in E$ and $\rho_E(x, y) \leq \delta$, then $\phi(x, y) \leq \epsilon$.

Let $f \in \mathfrak{M}_{0,reg}$, $(x_f, u_f) \in X(0, \infty)$ and $b_f > 0$ and that (8.73) and (8.74), (A1), (A4) hold and let $r \in (0, 1)$. Define

$$f_r(t, x, u) = f(t, x, u) + r\phi(x, x_t^f), \ (t, x, u) \in \mathcal{M}_0.$$

We suppose that for each $f \in \mathfrak{M}_{reg}$ and each $r \in (0, 1)$, $f_r \in \mathfrak{M}_0$ and $f_r \to f$ as $r \to 0^+$ in the strong topology.

Theorem 9.11 *Let $f \in \mathfrak{M}_{reg}$ and $r \in (0, 1)$. Then the function f_r has TP and $(x_{f_r}, u_{f_r}) = (x_f, u_f)$, $b_{f_r} = b_f$.*

Proof It is not difficult to see that (8.73), (8.74), (A2), (A4) hold. In view of Theorem 9.10, it is sufficient to show that f_r has (P3). Let $\epsilon, M > 0$. Theorem 8.5.implies that there exists $M_0 > 0$ such that for each $S_2 > S_1 \geq 0$ and each $(x, u) \in X(S_1, S_2)$,

$$I^f(S_1, S_2, x, u) + M_0 \geq I^f(S_1, S_2, x_f, u_f). \tag{9.179}$$

Property (i) implies that there exists $\delta \in (0, \epsilon)$ such that

for $(x, y) \in E \times E$ satisfying $\rho(x, y) > \epsilon$ we have $\phi(x, y) > \delta$.

Set

$$L = (r\delta)^{-1}(M_0 + M). \tag{9.180}$$

Assume that $T_1 \geq 0$, $T_2 \geq T_1 + L$, $(x, u) \in X(T_1, T_2)$ satisfies

$$I^{f_r}(T_1, T_2, x, u) \leq I^{f_r}(T_1, T_2, x_f, u_f) + M. \tag{9.181}$$

By (9.179), (9.181) and the definition of f_r,

$$I^f(T_1, T_2, x, u) + r\int_{T_1}^{T_2} \phi(x(t), x_f(t))dt$$

$$\leq I^{f_r}(T_1, T_2, x, u) \leq I^f(T_1, T_2, x_f, u_f) + M$$

$$\leq I^f(T_1, T_2, x, u) + M_0 + M.$$

Together with (9.180) and the choice of δ this implies that

$$M_0 + M \geq r\int_{T_1}^{T_2} \phi(x(t), x_f(t))dt$$

$$\geq r\delta \mathrm{mes}(\{t \in [T_1, T_2] : \ \phi_E(x(t), x_f(t)) > \epsilon\})$$

and

$$\mathrm{mes}(\{t \in [T_1, T_2] : \ \rho_E(x(t), x_f(t)) > \epsilon\}) \leq \delta^{-1}r^{-1}(M_0 + M) = L.$$

Thus f_r has (P3). Theorem 9.11 is proved.

Theorem 9.11 implies that

$$\{f_r : \ f \in \mathfrak{M}_{0,reg}, \ r \in (0, 1)\} \subset \mathfrak{M}_{reg} \subset \mathfrak{M}_{0,reg}$$

and $f_r \to f$ as $r \to 0^+$ for every $f \in \mathfrak{M}_{0,reg}$. Clearly,

$$\bar{\mathfrak{M}}_{reg} = \bar{\mathfrak{M}}_{0,reg}$$

in the strong topology. Note that Theorems 9.8 and 9.9 hold in our case.

9.11 Spaces od Smooths Integrands

We continue to use the notation, definitions, and assume the assumptions introduced in the previous section. In particular, we assume that (A5) holds. Define

$$\phi(x, y) = \min\{\rho(x, y),\ 1\},\ x, y \in E$$

Clearly, $\phi : E \times E \to [0, 1]$ is a continuous function, (i), (ii) hold, ϕ is a Lipschitz function with a Lipschitz constant 1.

We equip the space $\mathfrak{M}_\psi$ with the uniformity determined by the base

$$E_s(\epsilon) = \{(h, g) \in \mathfrak{M}_\psi \times \mathfrak{M}_\psi : |h(t, x, u) - g(t, x.u)| \le \epsilon$$

$$\text{for each } (t, x, u) \in \mathcal{M}\}$$

$$\cap \{(h, g) \in \mathfrak{M}_\psi \times \mathfrak{M}_\psi :$$

$$|(h - g)(t, x_1, u) - (h - g)(t, x_2, u)| \le \epsilon \rho_E(x_1, x_2)$$

$$\text{for each } (t, x_1, u),\ (t, x_2, u) \in \mathcal{M}\}$$

$$\cap \{(h, g) :\in \mathfrak{M}_\psi \times \mathfrak{M}_\psi :$$

$$(h - g)(t, x, u_1) = (h - g)(t, x, u_2) \text{ for each } (t, x, u_1),\ (t, x, u_2) \in \mathcal{M}\},$$

where $\epsilon > 0$. Clearly, this uniform space $\mathfrak{M}_\psi$ is Hausdorff and has a countable base. Therefore it is metrizable (by a metric d_s). It is not difficult to see that the uniform space $(\mathfrak{M}_\psi, d_s)$ is complete and that this uniformity is stronger than the uniformity determined by the base (8.10).

It is equipped with a topology induced by the uniformity which is called sometimes as a strong topology.

Denote by $\mathfrak{M}_l$ the set of all $f \in \mathfrak{M}_\psi$ such that $f(t, \cdot, \cdot)$ is a lower semicontinuous function for every $t \ge 0$ on the set $\{(x, u) \in E \times F :\ (t, x, u) \in \mathcal{M}\}$, by $\mathfrak{M}_c$ the set of all $f \in \mathfrak{M}_\psi$ such that $f(t, \cdot, \cdot)$ is a continuous function for every $t \ge 0$ on the set $\{(x, u) \in E \times F :\ (t, x, u) \in \mathcal{M}\}$, by $\mathfrak{M}_L$ the set of all $f \in \mathfrak{M}_\psi$ such that $f(t, \cdot, \cdot)$ is a Lipschitz function for every $t \ge 0$ on the set $\{(x, u) \in E \times F :\ (t, x, u) \in \mathcal{M}\}$ and by $\mathfrak{M}_{lL}$ the set of all $f \in \mathfrak{M}_\psi$ such that $f(t, \cdot, \cdot)$ is a Lipschitz function on all bounded subsets of the set $\{(x, u) \in E \times F :\ (t, x, u) \in \mathcal{M}\}$ for every $t \ge 0$. Clearly, all these sets are closed in the metric space $(\mathfrak{M}_\psi, d_s)$. Clearly, $\mathfrak{M}_0$ can be any of the spaces $\mathfrak{M}_\psi$, $\mathfrak{M}_l, \mathfrak{M}_c, \mathfrak{M}_L, \mathfrak{M}_{lL}$ and in any of them we have $f_r \to f$ as $r \to 0^+$ for each $f \in \mathfrak{M}_0$.

Now we consider spaces of smooth integrands using the notation, definitions, and assumptions of Sect. 6.15. Let $k \ge 1$ be an integer. Denote by $\tilde{\mathfrak{M}}_\psi$ the set of all borelian functions $g : [0, \infty) \times E \times F \to R^1$, which satisfy

$$g(t,x,u) \geq \psi_1(t,x,u) - a_0 \text{ for each } (t,x,u) \in [0,\infty) \times E \times F$$

(see (8.3)–(8.5)).

Assume that $(E, \|\cdot\|)$, $(F, \|\cdot\|)$ are Banach spaces, $\rho_E(x,y) = \|x-y\|$, $x, y \in E$, $\rho_F(x,y) = \|x-y\|$, $x, y \in F$, $E \times F$ is a Banach space,

$$\|(x,y)\| = \|x\| + \|y\|, \ x \in E, \ y \in F.$$

We assume that there exists a function $\lambda_* \in C^k(E, R^1)$ such that (see (6.144) with $X = E$)

$$0 \leq \lambda_*(x) \leq 1, \ x \in E,$$

$$\{x \in E : \lambda_*(x) > 0\} \subset \{x \in X : \|x\| \leq 1\},$$

$$\lambda_*(z) > 0 \text{ for some } z \in E,$$

$$\sup\{\|\lambda_*^{(j)}(z)\| : z \in E, \ j = 0, \ldots, k\} < \infty.$$

Then (see Sect. 6.15) there exists $\phi_0 \in C^k(X; R^1)$ such that

$$\phi_0(0) = 0, \ \phi_0(z) \in [0,1], \ z \in E,$$

$$\sup\{\|\phi_0^{(i)}(z)\| : z \in E, \ i = 0, \ldots, k\} < \infty$$

and that the following property holds:

for each $\epsilon > 0$ there exists $\delta > 0$ such that if $x \in E$ and $\phi(x) \leq \delta$, then $\|x\| \leq \epsilon$.

Denote by $\tilde{\mathfrak{M}}$ the set of all $f \in \mathfrak{M}_\psi$ such that for each $t \geq 0$, $f(t,\cdot,\cdot) \in C^k(E \times F)$. Denote by $\tilde{\mathfrak{M}}_l$ the set of all lower semicontinuous functions $f \in \tilde{\mathfrak{M}}$ and by $\tilde{\mathfrak{M}}_c$ the set of all continuous functions $f \in \tilde{\mathfrak{M}}$. The space $\tilde{\mathfrak{M}}$ is equipped with the uniformity determined by the base

$$E(\epsilon) = \{(h,g) :\in \tilde{\mathfrak{M}} \times \tilde{\mathfrak{M}} : \text{ for each } t \geq 0,$$

$$d^{(k)}_{E \times F}(g_1(t,\cdot,\cdot), \ g_2(t,\cdot,\cdot)) \leq \epsilon\}$$

where $\epsilon > 0$. This uniformity is metrizable (by a metric d_s) and complete and induces a topology. Clearly, $\tilde{\mathfrak{M}}_l$, $\tilde{\mathfrak{M}}_c$ are closed subsets of $\tilde{\mathfrak{M}}$. Clearly, $\mathfrak{M}_0$ can be any of the spaces $\tilde{\mathfrak{M}}$, $\tilde{\mathfrak{M}}_l$, $\tilde{\mathfrak{M}}_c$ with the metric d_s. Let $\mathfrak{M}_0$ can be any of the spaces $\tilde{\mathfrak{M}}$, $\tilde{\mathfrak{M}}_l$, $\tilde{\mathfrak{M}}_c$, $\mathcal{M}_0 = \mathcal{M}$, $\phi(x,y) = \phi_0(x-y)$, $x, y \in E$. Consider $\mathfrak{M}_{0,reg}$. It is not difficult to see that $f_r \to f$ as $r \to 0^+$ for each $f \in \mathfrak{M}_{reg}$. Thus Theorems 9.8 and 9.9 hold in our case with $\bar{\mathfrak{M}}_{reg} = \bar{\mathfrak{M}}_{0,reg}$.

9.12 Assumption (A5) for the First Class of Problems

Consider the first class of problems. (A5) follows from the next result.

Proposition 9.12 *Let $M > 0$, $\epsilon \in (0, 1)$, $\leq \tau < \tau_0 < \tau_1$. Then there exists $\delta > 0$ such that for each $T_1 \geq 0$, each $T_2 \in [T_1 + \tau_0, T_1 + \tau_1]$, each $g \in \mathfrak{M}_\psi$, each $(x_i, u_i) \in X(T_1, T_2)$, $i = 1, 2$ satisfying*

$$I^f(T_1, T_2, x_i, u_i) \leq M, \ i = 1, 2$$

and each $t_1, t_2 \in [T_1 + \tau, T_2]$ satisfying

$$0 < t_2 - t_1 \leq \delta, \ \|(x_1(t_1) - x_2(t_1)\| \leq \delta$$

the inequality $\|x_1(t_2) - x_2(t_2)\| \leq \epsilon$ holds.

Proof Proposition 8.1 implies that there exists $M_1 > M$ such that the following property holds:

(i) for each $g \in \mathfrak{M}_\psi$, each $T_1 \geq 0$, each $T_2 \in [T_1 + \tau_0, T_1 + \tau_1]$ and each $(x, u) \in X(T_1, T_2)$ satisfying

$$I^g(T_1, T_2, x, u) \leq M$$

we have

$$\|x(t)\| \leq M_1, \ t \in [T_1 + \tau, T_2].$$

Set

$$\epsilon_1 = 4^{-1}\epsilon(2\Delta_1 + 1)^{-1} \tag{9.182}$$

(see Sect. 8.2). By (8.5) and (8.20), there exists $h_0 > 0$ such that for each $(t, x, u) \in \mathcal{M}$ satisfying $\|G(t, x, u)\| - a_0\|x\| \geq h_0$ and each $g \in \mathfrak{M}_\psi$ we have

$$g(t, x, u) \geq 4\epsilon_1^{-1}(M_1 + M_2 + a_0 + 8 + a_0\tau_0)(\|G(t, x, u)\| - a_0\|x\|_+) \tag{9.183}$$

Choose $\delta \in (0, \epsilon/4)$ such that

$$\delta\Delta_1(2a_0M_0 + 2h_0 + 2) < \epsilon/4. \tag{9.184}$$

Assume that

$$g \in \mathfrak{M}_\psi, \ 0 \leq T_1, \ T_2 \in [T_1 + \tau_0, T_1 + \tau_1], \tag{9.185}$$

$$(x_i, u_i) \in X(T_1, T_2), \ i = 1, 2 \tag{9.186}$$

satisfy

$$I^f(T_1, T_2, x_i, u_i) \le M, \ i = 1, 2, \tag{9.187}$$

$$T_1 \le t_1 < t_2 \le T_2, \ t_1, t_2 \in [T_1 + \tau, T_2], \tag{9.188}$$

$$0 < t_2 - t_1 \le \delta, \ \|(x_1(t_1) - x_2(t_1)\| \le \delta. \tag{9.189}$$

In order to complete the proof it is sufficient to show that the inequality $\|(x_1(t_2) - x_2(t_2)\| \le \epsilon$.

Property (i), (9.185)–(9.187) imply that for $i = 1, 2$,

$$\|x_i(t)\| \le M_1, \ t \in [T_1 + \tau, T_2]. \tag{9.190}$$

Property (v) from Sect. 8.2 and Eq. (8.18) imply that for $i = 1, 2$,

$$x_i(t_2) = U(t_2, t_1)x_i(t_1) + \int_{t_1}^{t_2} U(t_2, s)G(s, x_i(s), u_i(s))ds. \tag{9.191}$$

It follows from (9.189) and (9.191) that

$$\begin{aligned} \|x_1(t_2) - x_2(t_2)\| &\le \|U(t_2, t_1)\| \|x_1(t_1) - x_2(t_1)\| \\ &+ \int_{t_1}^{t_2} \|U(t_2, s)\| \|G(s, x_1(s), u_1(s))\| ds \\ &+ \int_{t_1}^{t_2} \|U(t_2, s)\| \|G(s, x_2(s), u_2(s))\| ds \\ &\le \Delta_\delta \delta + \Delta_\delta \int_{t_1}^{t_2} \|G(s, x_1(s), u_1(s))\| ds \\ &+ \Delta_\delta \int_{t_1}^{t_2} \|G(s, x_2(s), u_2(s))\| ds. \end{aligned} \tag{9.192}$$

Let $i \in \{1, 2\}$. Set

$$\Omega_1 = \{t \in [t_1, t_2] : \ \|G(t, x_i(t), u_i(t))\| - a_0\|x_i(t)\| \ge h_0\}, \ \Omega_2 = [t_1, t_2] \setminus \Omega_1. \tag{9.193}$$

By (8.5), (9.187)–(9.190) and (9.193),

$$\int_{t_1}^{t_2} \|G(s, x_i(s), u_i(s))\| ds \le a_0 \int_{t_1}^{t_2} \|x_i(s)\| ds$$

$$+ \int_{t_1}^{t_2} (\|G(s, x_i(s), u_i(s))\| - a_0 \|x_i(s)\|)_+ ds$$

$$\le a_0 \delta M_1 + h_0 \delta + \int_{\Omega_1} (\|G(s, x_i(s), u_i(s))\| - a_0 \|x_i(s)\|)_+ ds$$

$$\le a_0 \delta M_1 + h_0 \delta + 4^{-1} \epsilon_1 (M_1 + M + a_0(\tau_1 + 1) + 8)^{-1} \int_{\Omega_1} g(s, x_i(s), u_i(s)) ds$$

$$\le \delta(a_0 M_1 + h_0) + (4^{-1} \epsilon_1 (M_1 + M + a_0(\tau_1 + 1) + 8)^{-1})(I^g(T_1, T_2, x_i, u_i) + a_0 \tau_1)$$

$$\le \delta(a_0 M_1 + h_0) + (4^{-1} \epsilon_1 (M_1 + M_0 + a_0(\tau_1 + 1) + 8)^{-1})(M + a_0 \tau_1)$$

$$\le \delta(a_0 M_1 + h_0) + 4^{-1} \epsilon_1. \tag{9.194}$$

In view of (9.182), (9.184), (9.192), and (9.194),

$$\|x_1(t_2) - x_2(t_2)\| \le \Delta_1 \delta + 2\Delta_1 \delta(a_0 M_1 + h_0) + 2^{-1} \epsilon_1 \Delta_1 < \epsilon.$$

Proposition 9.12 is proved.

9.13 Assumption (A5) for the Second Class of Problems

Consider the second class of problems. We assume that for this class of problem for each $(t, x, u) \in \mathcal{M}$,

$$\psi_1(t, x, u) = \max\{\psi(\|x\|),\ \psi(\|u\|^2)\}. \tag{9.195}$$

(A5) follows from the next result.

Proposition 9.13 *Let* $M_0 > 0, \epsilon \in (0, 1)$, $0 < \tau_0 < \tau_1$. *Then there exists* $\delta > 0$ *such that for each* $T_1 \ge 0$, *each* $T_2 \in [T_1 + \tau_0, T_1 + \tau_1]$, *each* $g \in \mathfrak{M}_\psi$, *each* $(x_i, u_i) \in X(T_1, T_2)$, $i = 1, 2$ *satisfying*

$$I^g(T_1, T_2, x_i, u_i) \le M,\ i = 1, 2$$

and each $t_1, t_2 \in [T_1, T_2]$ *satisfying*

$$0 < t_2 - t_1 \le \delta,\ \|x_1(t_1) - x_2(t_1)\| \le \delta \tag{9.196}$$

the inequality

$$\|x_1(t_2) - x_2(t_2)\| \le \epsilon \tag{9.197}$$

holds.

Proof Recall that $B \in \mathcal{L}(F, E_{-1})$ is an admissible operator for the semigroup $S(t) = e^{At}$, $t \ge 0$. Proposition 7.1 implies that there exist exist $M_* \ge 1$, $\omega_* \in R^1$ such that

$$\|e^{At}\| \le M_* e^{\omega_* t}, \ t \in [0, \infty) \tag{9.198}$$

and for each $\tau \ge 0$, $\Phi_\tau \in \mathcal{L}(L^2(0, \tau; F), E)$ is defined by

$$\Phi_\tau u = \int_0^\tau e^{A(\tau - s)} Bu(s)ds, \ u \in L^2(0, \tau; F) \tag{9.199}$$

(see Sect. 7.5).

Choose $\Lambda > 1$ such that

$$\Lambda^{-1/2}(\|\Phi_1\| + 1)(M + a_0\tau_1 + 1) < \epsilon/8. \tag{9.200}$$

By (9.195), there exists $c_0 > 0$ such that the following property holds:

(i) For each $(t, x, u) \in \mathcal{M}$ satisfying $\|u\| \ge c_0$ each $g \in \mathfrak{M}_\psi$ and each $t \ge 0$,

$$g(t, x, u) \ge \Lambda \|u\|^2. \tag{9.201}$$

Fix $\delta \in (0, 1)$ such that

$$M_* e^{|\omega_*|}\delta < \epsilon/8, \ \|\Phi_1\| c_0 \delta^{1/2} < \epsilon/8. \tag{9.202}$$

Assume that $g \in \mathfrak{M}_\psi$, $T_1 \ge 0$, $T_2 \in [T_1 + \tau_0, T_1 + \tau_1]$, $(x_i, u_i) \in X(T_1, T_2)$, $i = 1, 2$ satisfy

$$I^g(T_1, T_2, x_i, u_i) \le M, \ i = 1, 2, \tag{9.203}$$

$T_1 \le t_1 < t_2 \le T_2$ and that (9.196) holds. In order to complete the proof it is sufficient to show that (9.197) is true. It is clear that $(x_i, u_i) \in X(t_1, t_2)$, $i = 1, 2$.. In view of (8.24), for $i = 1, 2$,

$$x_i(t_2) = e^{A(t_2 - t_1)} x_i(t_1) + \int_{t_1}^{t_2} e^{A(t_2 - s)} Bu_i(s)ds. \tag{9.204}$$

Equations (9.129), (9.196), (9.198), (9.204) imply that

$$\|x_2(t_2) - x_1(t_2)\|$$

$$= \|e^{A(t_2-t_1)}(x_1(t_1) - x_2(t_1))\| + \|\int_{t_1}^{t_2} e^{A(t_2-s)}Bu_1(s)ds$$

$$- \int_{t_1}^{t_2} e^{A(t_2-s)}Bu_2(s)ds\|$$

$$\leq \|e^{A(t_2-t_1)}(x_1(t_1) - x_2(t_1))\| + \|\int_{t_1}^{t_2} e^{A(t_2-s)}Bu_1(s)ds\|$$

$$+ \|\int_{t_1}^{t_2} e^{A(t_2-s)}Bu_2(s)ds\|$$

$$\leq \delta M_* e^{|\omega_*|\delta} + \|\int_0^{t_2-t_1} e^{A(t_2-t_1-s)}Bu_1(t_1+s)ds\|$$

$$+ \|\int_0^{t_2-t_1} e^{A(t_2-t_1-s)}Bu_2(t_1+s)ds\|$$

$$\leq \delta M_* e^{|\omega_*|} + \|\Phi_{t_2-t_1}\|\|u_1(t_1+\cdot)\|_{L^2(0,t_2-t_1;F)}$$

$$+ \|\Phi_{t_2-t_1}\|\|u_2(t_1+\cdot)\|_{L^2(0,t_2-t_1;F)}$$

$$\leq \delta M_* e^{|\omega_*} + \|\Phi_1\|(\int_{t_1}^{t_2} \|u_1(s)\|^2 ds)^{1/2} + \|\Phi_1\|(\int_{t_1}^{t_2} \|u_2(s)\|^2 ds)^{1/2}. \tag{9.205}$$

Let $i \in \{1, 2\}$ and

$$\Omega_i = \{t \in [t_1, t_2] : \ \|u_i(t)\| \geq c_0\}. \tag{9.206}$$

Property (i) and Eqs. (9.195), (9.196), (9.203), and (9.206) imply that

$$\|\Phi_1\|(\int_{t_1}^{t_2} \|u_i(s)\|^2 ds)^{1/2} = \|\Phi_1\|(\int_{\Omega} \|u_i(s)\|^2 ds + \int_{[t_1,t_2]\setminus\Omega} \|u_i(s)\|^2 ds)^{1/2}$$

$$\leq= \|\Phi_1\|(\int_{\Omega} \|u_i(s)\|^2 ds)^{1/2} + \|\Phi_1(\int_{[t_1,t_2]\setminus\Omega} \|u_i(s)\|^2 ds)^{1/2}$$

$$\leq \|\Phi_1\|(\int_{\Omega} \Lambda^{-1}G(s, x_i(s), u_i(s))ds)^{1/2} + \|\Phi_1\|c_0\delta^{1/2}$$

$$\leq |\Phi_1\|c_0\delta^{1/2} + \|\Phi_1\|(\Lambda^{-1}I^g(T_1, T_2, x_i, u_i) + a_0\tau_1)^{1/2}$$

$$\leq \|\Phi_1\|c_0\delta^{1/2} + \|\Phi_1\|\Lambda^{-1/2}(M_1 + a_0\tau_1)^{1/2}. \tag{9.207}$$

By (9.200), (9.202), (9.205), and (9.207),

$$\|x_2(t_2) - x_1(t_2)\| \leq \delta M_* e^{|\omega_*|} + 2\|\Phi_1\|c_0\delta^{1/2}$$

$$+ 2\|\Phi_1\|\Lambda^{-1/2}(M_1 + a_0\tau_1)^{1/2} < \epsilon.$$

Proposition 9.13 is proved.

9.14 Assumption (A5) for the Third Class of Problems

Consider the third class of problems. (A5) follows from the next result.

Proposition 9.14 *Let $M > 0$, $\epsilon \in (0, 1)$, $0 < \tau_0 \leq \tau_1$. Then there exists $\delta > 0$ such that for each $T_1 \geq 0$, each $T_2 \in [T_1 + \tau_0, T_1 + \tau_1]$, each $g \in \mathfrak{M}_\psi$, each $(x_i, u_i) \in X(T_1, T_2)$, $i = 1, 2$ satisfying*

$$I^f(T_1, T_2, x_i, u_i) \leq M, \ i = 1, 2 \tag{9.208}$$

and each $t_1, t_2 \in [T_1 + \tau, T_2]$ satisfying

$$0 < |t_2 - t_1| \leq \delta$$

the inequality $\rho_E((x_1(t_2), x_2(t_2)) \leq \epsilon$ holds.

Proof Proposition 8.3 and (8.56) imply that there exists $M_1 > M$ such that the following property holds:

(i) for each $g \in \mathfrak{M}_\psi$, each $T_1 \geq 0$, each $T_2 \in [T_1 + \tau_0, T_1 + \tau_1]$ and each $(x, u) \in X(T_1, T_2)$ satisfying

$$I^g(T_1, T_2, x, u) \leq M$$

we have

$$\rho_E(x(t), \theta_0) \leq M_1, \ t \in [T_1, T_2].$$

By (8.54), there exists $h_0 > 0$ such that for each $(t, x, u) \in \mathcal{M}$ satisfying $G(t, x, u) - a_0\rho_E(x, \theta_0) \geq h_0$ and each $g \in \mathfrak{M}_\psi$ we have

$$g(t, x, u) \geq 4\epsilon_1^{-1}(M_1 + M + a_0 + 8 + a_0(\tau_0 + 1) + 8(G(t, x, u) - a_0\rho_E(x, \theta_0)_+). \tag{9.209}$$

Choose $\delta \in (0, 1)$ such that

$$\delta < \epsilon/8, \ \delta(a_0 M_1 + h_0) < \epsilon/4. \tag{9.210}$$

Assume that

$$g \in \mathfrak{M}_\psi, \ 0 \le T_1, \ T_2 \in [T_1 + \tau_0, T_1 + \tau_1],$$

$(x, u) \in X(T_1, T_2)$, (9.208) is true, and $t_1, t_2 \in [T_1 + \tau, T_2]$ satisfy

$$0 < t_2 - t_1 \le \delta. \tag{9.211}$$

Property (i), (8.55), and (9.208) imply that

$$\rho_E(x(t), \theta_0) \le M_1, \ t \in [T_1, T_2]. \tag{9.212}$$

Clearly,

$$\rho_E(x(t_2), x(t_1)) \le \int_{t_1}^{t_2} G(s, x(s), u(s))ds. \tag{9.213}$$

Set

$$\Omega = \{t \in [T_1, T_2] : \ G(t, x(s), u(s)) - a_0 \rho_E(x(s), \theta_0) \ge h_0\} \tag{9.214}$$

By (9.208)–(9.210) and (9.212)–(9.214),

$$\rho_E(x(t_2), x(t_1)) \le \int_{t_1}^{t_2} G(s, x(s), u(s))ds$$

$$\le a_0 \int_{t_1}^{t_2} \rho_E(x(s), \theta_0)ds + \int_{t_1}^{t_2} (G(s, x(s), u(s)) - a_0 \rho_E(x(s), \theta_0)_+ ds$$

$$\le a_0 \delta M_1 + h_0 \delta + \int_{\Omega} (G(s, x(s), u(s)) - a_0 \rho_E(x(s), \theta_0))_+ ds$$

$$\le \delta(a_0 M_1 + h_0) + 4^{-1}\epsilon(M_1 + M + a_0(\tau_0 + 1)) + 8^{-1} \int_{\Omega} G(s, x(s), u(s))ds$$

$$\le \delta(a_0 M_1 + h_0) + 4^{-1}\epsilon(M_1 + M + a_0(\tau_0 + 1) + 8)^{-1}(I^g(T_1, T_2, x, u) + a_0 \tau_1)$$

$$\le \delta(a_0 M_1 + h_0) + 4^{-1}\epsilon < \epsilon.$$

Proposition 9.14 is proved.

References

1. Ahmed NU (2009) Optimal control on infinite dimensional Banach spaces of neutral systems driven by vector measures. Commun Appl Nonlinear Anal 16:1–14
2. Ahmed NU (2015) Infinite dimensional uncertain dynamic systems on Banach spaces and their optimal output feedback control. Discuss Math Differ Incl Control Optim 35:65–87
3. Ahmed NU (2016) Optimal output feedback control law for a class of uncertain infinite dimensional dynamic systems. J Abstr Differ Equ Appl 7:11–29
4. Ahmed NU, Xiang X (2003) Differential inclusions on Banach spaces and their optimal control. Nonlinear Funct Anal Appl 8:461–488
5. Ambrosio L, Gigli N, Savaré G (2008) Gradient flows in metric spaces and in the space of probability measures, 2nd edn. Lectures in mathematics. ETH Zürich, Birkhäuser Verlag, Basel
6. Anderson BDO, Moore JB (1971) Linear optimal control. Prentice-Hall, Englewood Cliffs
7. Arkin VI, Evstigneev IV (1987) Stochastic models of control and economic dynamics. Academic Press, London
8. Aseev SM, Krastanov MI, Veliov VM (2017) Optimality conditions for discrete-time optimal control on infinite horizon. Pure Appl. Funct. Anal. 2:395–409
9. Aseev SM, Kryazhimskiy AV (2004) The Pontryagin maximum principle and transversality conditions for a class of optimal control problems with infinite time horizons. SIAM J Control Optim 43:1094–1119
10. Aseev SM, Veliov VM (2012) Maximum principle for infinite-horizon optimal control problems with dominating discount. Dyn. Contin Discrete Impuls Syst Series B 19:43–63
11. Atsumi H (1965) Neoclassical growth and the efficient program of capital accumulation. Rev Econ Stud 32:127–136
12. Aubin JP, Cellina A (1984) Differential inclusions. Set-valued maps and viability theory. Springer, Berlin
13. Aubin JP, Ekeland I (1984) Applied nonlinear analysis. Wiley Interscience, New York
14. Aubry S, Le Daeron PY (1983) The discrete Frenkel-Kontorova model and its extensions I. Physica D 8:381–422
15. Bachir M, Blot J (2015) Infinite dimensional infinite-horizon Pontryagin principles for discrete-time problems. Set-Valued Var Anal 23:43–54
16. Bachir M, Blot J (2017) Infinite dimensional multipliers and Pontryagin principles for discrete-time problems. Pure Appl Funct Anal 2:411–426
17. Ball J M (1977) Strong continuous semigroups, weak solutions and the variation of constants formula. Proc Am Math Soc 63:370–373

A. J. Zaslavski, *Turnpike Phenomenon in Metric Spaces*, Springer Optimization and Its Applications 201, https://doi.org/10.1007/978-3-031-27208-0

18. Ball J M (1997) Continuity properties and global attractors of generalized semiflows and the Navier-Stokes equations. J Nonlinear Sci 7:475–502
19. Barbu V (1984) Optimal control of variational inequalities. Pitman research notes in mathematics. London Boston
20. Barbu V (1993) Analysis and control of nonlinear infinite dimensional systems. Academic Press, Boston, New York
21. Barbu V, Precupanu T (2012) Convexity and optimization in Banach spaces. Springer monographs in mathematics. Springer, Dordrecht, Heidelberg, London, New York
22. Bardi M (2009) On differential games with long-time-average cost. Advances in dynamic games and their applications. Birkhauser, Basel, pp 3–18
23. Baumeister J, Leitao A, Silva GN (2007) On the value function for nonautonomous optimal control problem with infinite horizon. Syst Control Lett 56:188–196
24. Blot J (2009) Infinite-horizon Pontryagin principles without invertibility. J Nonlinear Convex Anal 10:177–189
25. Blot J, Cartigny P (2000) Optimality in infinite-horizon variational problems under sign conditions. J Optim Theory Appl 106:411–419
26. Blot J, Hayek N (2000) Sufficient conditions for infinite-horizon calculus of variations problems. ESAIM Control Optim Calc Var 5:279–292
27. Blot J, Hayek N (2014) Infinite-horizon optimal control in the discrete-time framework. SpringerBriefs in optimization. New York
28. Bright I (2012) A reduction of topological infinite-horizon optimization to periodic optimization in a class of compact 2-manifolds. J Math Anal Appl 394:84–101
29. Brock WA (1970) On existence of weakly maximal programmes in a multi-sector economy. Rev Econ Stud 37:275–280
30. Carlson DA (1990) The existence of catching-up optimal solutions for a class of infinite horizon optimal control problems with time delay. SIAM J Control Optim 28:402–422
31. Carlson DA, Haurie A, Leizarowitz A (1991) Infinite horizon optimal control. Springer, Berlin
32. Carlson DA, Jabrane A, Haurie A (1987) Existence of overtaking solutions to infinite dimensional control problems on unbounded time intervals. SIAM J Control Optim 25:1517–1541
33. Cartigny P, Michel P (2003) On a sufficient transversality condition for infinite horizon optimal control problems. Autom. J. IFAC 39:1007–1010
34. Cellina A, Colombo G (1990) On a classical problem of the calculus of variations without convexity assumptions. Ann Inst H Poincare Anal Non Lineaire 7:97–106
35. Cesari L (1983) Optimization—theory and applications. Springer, Berlin-Heidelberg-New York
36. Coleman BD, Marcus M, Mizel VJ (1992) On the thermodynamics of periodic phases. Arch Ration Mech Anal 117:321–347
37. Coron JM (2007) Control and nonlinearity. AMS, Providence, Rhode Island
38. Damm T, Grune L, Stieler M, Worthmann K (2014) An exponential turnpike theorem for dissipative discrete time optimal control problems. SIAM J Control Optim 52:1935–1957
39. De Oliveira VA, Silva GN (2009) Optimality conditions for infinite horizon control problems with state constraints. Nonlinear Anal 71:1788–1795
40. Deville R, Godefroy R, Zizler V (1993) Smoothness and renorming in Banach spaces. Longman Scientific and Technical, Harlow
41. Eells J (1966) A setting for global analysis. Bull Am Math Soc 72:751–807
42. Evstigneev IV, Flam SD (1998) Rapid growth paths in multivalued dynamical systems generated by homogeneous convex stochastic operators. Set-Valued Anal 6:61–81
43. Gaitsgory V, Grune L, Thatcher N (2015) Stabilization with discounted optimal control. Syst Control Lett 82:91–98
44. Gaitsgory V, Mammadov M, Manic L (2017) On stability under perturbations of long-run average optimal control problems. Pure Appl Funct Anal 2:461–476

45. Gaitsgory V, Parkinson, A, Shvartsman, I (2017) Linear programming formulations of deterministic infinite horizon optimal control problems in discrete time. Discrete Contin Dyn Syst Ser B 22:3821–3838
46. Gaitsgory V, Rossomakhine S, Thatcher N (2012) Approximate solution of the HJB inequality related to the infinite horizon optimal control problem with discounting. Dyn Contin Discrete Impuls Syst Series B 19:65–92
47. Gale D (1967) On optimal development in a multi-sector economy. Rev Econ Stud 34:1–18
48. Ghosh MK, Mallikarjuna Rao KS (2005) Differential games with ergodic payoff. SIAM J Control Optim 43:2020–2035
49. Glizer VY, Kelis O (2017) Upper value of a singular infinite horizon zero-sum linear-quadratic differential game. Pure Appl Funct Anal 2:511–534
50. Grune L, Guglielmi R (2018) Turnpike properties and strict dissipativity for discrete time linear quadratic optimal control problems. SIAM J Control Optim 56:1282–1302
51. Gugat M (2019) A turnpike result for convex hyperbolic optimal boundary control problems. Pure Appl Funct Anal 4:849–866
52. Gugat M, Hante FM (2019) On the turnpike phenomenon for optimal boundary control problems with hyperbolic systems. SIAM J Control Optim 57:264–289
53. Gugat M, Trelat E, Zuazua, E (2016) Optimal Neumann control for the 1D wave equation: finite horizon, infinite horizon, boundary tracking terms and the turnpike property. Syst Control Lett 90:61–70
54. Guo X, Hernandez-Lerma O (2005) Zero-sum continuous-time Markov games with unbounded transition and discounted payoff rates. Bernoulli 11:1009–1029
55. Hayek N (2011) Infinite horizon multiobjective optimal control problems in the discrete time case. Optimization 60:509–529
56. Hernandez-Lerma O, Lasserre JB (2001) Zero-sum stochastic games in Borel spaces: average payoff criteria. SIAM J Control Optim 39:1520–1539
57. Hille E, Phillips RS (1957) Functional analysis and semigroups. AMS, Providence
58. Hritonenko N, Yatsenko Yu (1997) Turnpike theorems in an integral dynamic model of economic restoration. Cybernet Syst Anal 33:259–273
59. Hritonenko N, Yatsenko Yu (2005) Turnpike and optimal trajectories in integral dynamic models with endogenous delay. J Optim Theory Appl 127:109–127
60. Jasso-Fuentes H, Hernandez-Lerma O (2008) Characterizations of overtaking optimality for controlled diffusion processes. Appl Math Optim 57:349–369
61. Jiang Y-R, Huang, N-J, Yao, J-C (2017) Solvability and optimal control of semilinear nonlocal fractional evolution inclusion with Clarke subdifferential. Appl Anal 96:2349–2366
62. Khan MA, Zaslavski AJ (2010) On two classical turnpike results for the Robinson-Solow-Srinivasan (RSS) model. Adv Math Econ 13:47–97
63. Khlopin DV (2013) Necessity of vanishing shadow price in infinite horizon control problems. J Dyn Control Syst 19:519–552
64. Khlopin DV (2017) On Lipschitz continuity of value functions for infinite horizon problem. Pure Appl Funct Anal 2:535–552
65. Kien BT, Thi TN, Wong MM, Yao J-C (2012) Lower semicontinuity of the solution mapping to a parametric optimal control problem. SIAM J Control Optim 50:2889–2906
66. Kolokoltsov V, Yang W (2012) The turnpike theorems for Markov games. Dyn Games Appl 2:294–312
67. Lasiecka I, Triggiani R (2000) Control theory for partial differential equations: continuous and approximation theories; vol 1: abstract parabolic systems. Encyclopedia of mathematics and its applications series. Cambridge University Press, Cambridge
68. Lasiecka I, Triggiani R (2000) Control theory for partial differential equations: continuous and approximation theories; vol 2: abstract hyperbolic-like systems over a finite time horizon. Encyclopedia of mathematics and its applications series. Cambridge University Press, Cambridge
69. Leizarowitz A (1985) Infinite horizon autonomous systems with unbounded cost. Appl Math Opt 13:19–43

70. Leizarowitz A (1986) Tracking nonperiodic trajectories with the overtaking criterion. Appl Math Opt 14:155–171
71. Leizarowitz A, Mizel VJ (1989) One dimensional infinite horizon variational problems arising in continuum mechanics. Arch Ration Mech Anal 106:161–194
72. Leizarowitz A, Zaslavski AJ (2005) On a class of infinite horizon optimal control problems with periodic cost functions. J Nonlinear Convex Anal 6:71–91
73. Li X, Yong J (1995) Optimal control theory for infinite dimensional systems. Birkhauser, Boston Basel Berlin
74. Lykina V, Pickenhain S, Wagner M (2008) Different interpretations of the improper integral objective in an infinite horizon control problem. J Math Anal Appl 340:498–510
75. Makarov VL, Rubinov AM (1977) Mathematical theory of economic dynamics and equilibria. Springer, New York
76. Malinowska AB, Martins N, Torres DFM (2011) Transversality conditions for infinite horizon variational problems on time scales. Optim Lett 5:41–53
77. Mammadov M (2014) Turnpike theorem for an infinite horizon optimal control problem with time delay. SIAM J Control Optim 52:420–438
78. Marcus M, Zaslavski AJ (1999) On a class of second order variational problems with constraints. Israel J Math 111:1–28
79. Marcus M, Zaslavski AJ (1999) The structure of extremals of a class of second order variational problems. Ann Inst H Poincaré Anal Non linéaire 16:593–629
80. Marcus M, Zaslavski AJ (2002) The structure and limiting behavior of locally optimal minimizers. Ann Inst H Poincaré Anal Non linéaire 19:343–370
81. McKenzie LW (1976) Turnpike theory. Econometrica 44:841–866
82. Melnik, VS, Valero, J (1998) On attractors of multivalued semi-flows and differential inclusions, Set-Valued Anal 6 1998:83–111
83. Mordukhovich BS (1988) Approximation methods in optimization and control. Nauka, Moscow
84. Mordukhovich BS (1990) Minimax design for a class of distributed parameter systems. Autom Remote Control 50:1333–1340
85. Mordukhovich BS (1999) Existence theorems in nonconvex optimal control. Calculus of variations and optimal control. CRC Press, Boca Raton, pp 175–197
86. Mordukhovich BS (2011) Optimal control and feedback design of state-constrained parabolic systems in uncertainly conditions. Appl Anal 90:1075–1109
87. Mordukhovich BS, Shvartsman I (2004) Optimization and feedback control of constrained parabolic systems under uncertain perturbations, optimal control, stabilization and nonsmooth analysis. Lecture Notes Control Inform. Sci. Springer, Berlin, pp 121–132
88. Moser J (1986) Minimal solutions of variational problems on a torus. Ann Inst H Poincaré Anal Non linéaire 3:229–272
89. Nitecki Z (1971) Differentiable dynamics. An introduction to the orbit structure of diffeomorphisms. The MIT Press, Cambridge
90. Ocana Anaya E, Cartigny P, Loisel P (2009) Singular infinite horizon calculus of variations. Applications to fisheries management. J Nonlinear Convex Anal 10:157–176
91. Pickenhain S, Lykina V, Wagner M (2008) On the lower semicontinuity of functionals involving Lebesgue or improper Riemann integrals in infinite horizon optimal control problems. Control Cybernet 37:451–468
92. Porretta A, Zuazua E (2013) Long time versus steady state optimal control. SIAM J Control Optim 51:4242–4273
93. Prieto-Rumeau T, Hernandez-Lerma O (2005) Bias and overtaking equilibria for zero-sum continuous-time Markov games. Math Methods Oper Res 61:437–454
94. Reich S, Zaslavski AJ (2014) Genericity in nonlinear analysis. Springer, New York
95. Rockafellar RT (1970) Convex analysis. Princeton University Press, Princeton
96. Rubinov AM (1980) Turnpike sets in discrete disperse dynamical systems. Sib Math J 21:136–146
97. Rubinov AM (1984) Economic dynamics. J Sov Math 26:1975–2012

98. Sagara N (2018) Recursive variational problems in nonreflexive Banach spaces with an infinite horizon: an existence result. Discrete Contin Dyn Syst Ser S 11:1219–1232
99. Samuelson PA (1965) A catenary turnpike theorem involving consumption and the golden rule. Am Econ Rev 55:486–496
100. Tonelli L (1921) Fondamenti di Calcolo delle Variazioni. Zanicelli, Bolonia
101. Trelat E, Zhang C, Zuazua E (2018) Optimal shape design for 2D heat equations in large time. Pure Appl Funct Anal 3:255–269
102. Trelat E, Zhang C, Zuazua E (2018) Steady-state and periodic exponential turnpike property for optimal control problems in Hilbert spaces. SIAM J Control Optim 56:1222–1252
103. Troltzsch F (2010) Optimal control of partial differential equations. Theory, methods and applications. American Mathematical Society, Providence
104. Tucsnak M, Weiss G (2009) Observation and control for operator semigroups. Birkhauser, Basel
105. von Weizsacker CC (1965) Existence of optimal programs of accumulation for an infinite horizon. Rev Econ Stud 32:85–104
106. Yosida K (1980) Functional analysis, 6th edn. Springer, Berlin
107. Zaslavski AJ (1987) Ground states in Frenkel-Kontorova model. Math. USSR Izvestiya 29:323–354
108. Zaslavski AJ (1995) Optimal programs on infinite horizon 1. SIAM J Control Optim 33:1643–1660
109. Zaslavski AJ (1995) Optimal programs on infinite horizon 2. SIAM J Control Optim 33:1661–1686
110. Zaslavski AJ (1996) Dynamic properties of optimal solutions of variational problems. Nonlinear Anal 27:895–931
111. Zaslavski AJ (1998) Turnpike theorem for convex infinite dimensional discrete-time control systems. J. Convex Anal 5:237–248
112. Zaslavski AJ (2000) The turnpike property for extremals of nonautonomous variational problems with vector-valued functions. Nonlinear Anal 42:1465–1498
113. Zaslavski AJ (2000) Existence and structure of optimal solutions of infinite dimensional control systems. Appl Math Optim 42:291–313
114. Zaslavski AJ (2000) Turnpike theorem for nonautonomous infinite dimensional discrete-time control systems. Optimization 48:69–92
115. Zaslavski AJ (2004) The structure of approximate solutions of variational problems without convexity. J Math Anal Appl 296:578–593
116. Zaslavski AJ (2005) The turnpike property of discrete-time control problems arising in economic dynamics. Discrete Contin Dyn Syst B 5:861–880
117. Zaslavski AJ (2006) Turnpike properties in the calculus of variations and optimal control. Springer, New York
118. Zaslavski AJ (2006) The turnpike result for approximate solutions of nonautonomous variational problems. J Aust Math Soc 80:105–130
119. Zaslavski AJ (2007) Turnpike results for a discrete-time optimal control systems arising in economic dynamics. Nonlinear Anal 67:2024–2049
120. Zaslavski AJ (2009) Two turnpike results for a discrete-time optimal control systems. Nonlinear Anal 71:902–909
121. Zaslavski AJ (2010) Stability of a turnpike phenomenon for a discrete-time optimal control systems. J Optim Theory Appl 145:597–612
122. Zaslavski AJ (2010) Optimization on metric and normed spaces. Springer, New York
123. Zaslavski AJ (2011) Turnpike properties of approximate solutions for discrete-time control systems. Commun Math Anal 11:36–45
124. Zaslavski AJ (2011) Structure of approximate solutions for a class of optimal control systems. J Math Appl 34:1–14
125. Zaslavski AJ (2011) Stability of a turnpike phenomenon for a class of optimal control systems in metric spaces. Numer Algebra Control Optim 1:245–260

126. Zaslavski AJ (2011) The existence and structure of approximate solutions of dynamic discrete time zero-sum games. J Nonlinear Convex Anal 12:49–68
127. Zaslavski AJ (2012) A generic turnpike result for a class of discrete-time optimal control systems. Dyn Contin Discrete Impuls Syst Series B 19:225–265
128. Zaslavski AJ (2012) Existence and turnpike properties of solutions of dynamic discrete time zero-sum games. Commun Appl Anal 16:261–276
129. Zaslavski AJ (2013) Nonconvex optimal control and variational problems. Springer optimization and its applications. New York
130. Zaslavski AJ (2013) Structure of approximate solutions of optimal control problems. SpringerBriefs in optimization. New York
131. Zaslavski AJ (2013) Necessary and sufficient conditions for turnpike properties of solutions of optimal control systems arising in economic dynamics. Dyn Contin Discrete Impulsive Syst Ser B Appl Algorithms 20:391–420
132. Zaslavski AJ (2014) Turnpike properties of approximate solutions of nonconcave discrete-time optimal control problems. J Convex Anal 21:681–701
133. Zaslavski AJ (2014) Turnpike phenomenon and infinite horizon optimal control. Springer optimization and its applications. New York
134. Zaslavski AJ (2014) Turnpike properties for nonconcave problems. Adv Math Econ 18:101–134
135. Zaslavski AJ (2014) Structure of solutions of discrete time optimal control problems in the regions close to the endpoints. Set-Valued Var Anal 22:809–842
136. Zaslavski AJ (2014) Turnpike theory for dynamic zero-sum games. In: Proceedings of the workshop "Variational and optimal control problems on unbounded domains". Haifa. 2012. Contemporary mathematics, vol 619, pp 225–247
137. Zaslavski AJ (2014) Turnpike properties of approximate solutions of dynamic discrete time zero-sum games. J Dyn Games 2014:299–330
138. Zaslavski AJ (2014) Stability of the turnpike phenomenon in discrete-time optimal control problems. SpringerBriefs in optimization. New York
139. Zaslavski AJ (2015) Convergence of solutions of optimal control problems with discounting on large intervals in the regions close to the endpoints. Set-Valued Var Anal 23:191–204
140. Zaslavski AJ (2015) Structure of approximate solutions of discrete time optimal control Bolza problems on large intervals. Nonlinear Anal 123–124:23–55
141. Zaslavski AJ (2015) Discrete time optimal control problems on large intervals. Adv Math Econ 19:91–135
142. Zaslavski AJ (2015) Turnpike theory of continuous-time linear optimal control problems. Springer optimization and its applications. Cham, Heidelberg-New York-Dordrecht-London
143. Zaslavski AJ (2016) Convergence of solutions of concave discrete optimal control problems in the regions close to the endpoints. Commun Appl Nonlinear Anal 23:1–10
144. Zaslavski AJ (2016) Structure of solutions of optimal control problems on large intervals: a survey of recent results. Pure Appl Funct Anal 1:123–158
145. Zaslavski AJ (2016) Linear control systems with nonconvex integrands on large intervals. Pure Appl Funct Anal 1:441–474
146. Zaslavski AJ (2017) Bolza optimal control problems with linear equations and nonconvex integrands on large intervals. Pure Appl Funct Anal 2:153–182
147. Zaslavski AJ (2017) Discrete-time optimal control and games on large intervals. Springer optimization and its applications. Springer, Berlin
148. Zaslavski AJ (2019) Turnpike conditions in infinite dimensional optimal control. Springer optimization and its applications. Springer, Berlin
149. Zaslavski AJ (2021) Optimal control problems related to the Robinson-Solow-Srinivasan model. Monographs in mathematical economics. Springer, Singapore
150. Zaslavski AJ (2022) Turnpike properties of solutions of a differential inclusion with a Lyapunov function, I. Pure Appl Funct Anal 7:1085–11021
151. Zaslavski AJ (2022) Turnpike properties of solutions of a differential inclusion with a Lyapunov function, II. Pure Appl Funct Anal 7 (in press)

Index

A. J. Zaslavski, *Turnpike Phenomenon in Metric Spaces*, Springer Optimization and Its Applications 201, https://doi.org/10.1007/978-3-031-27208-0

The manufacturer's authorised representative in the EU is Springer Nature Customer Service Centre GmbH, Europaplatz 3, 69115 Heidelberg, Germany. If you have any concerns regarding our products, please contact ProductSafety@springernature.com

Printed and bound by CPI Group (UK) Ltd, Croydon, CR0 4YY
01/10/2026
02225865-0013